AF718598

ENICHEM AMERICAS, INC.
LIBRARY
2000 PRINCETONPARK CORPORATE CENTER
MONMOUTH JUNCTION, NJ 08852

# Polymer Blends and Alloys

GUIDEBOOK TO COMMERCIAL PRODUCTS

TECHNOMIC PUBLISHING CO., INC.
LANCASTER · BASEL

*Published in the Western Hemisphere by*
Technomic Publishing Company, Inc.
851 New Holland Avenue
Box 3535
Lancaster, Pennsylvania 17604 U.S.A.

*Distributed in the Rest of the World by*
Technomic Publishing AG

©1988 by Technomic Publishing Company, Inc.
All rights reserved

No part of this publication may be reproduced, stored in a retrieval system, or transmitted, in any form or by any means, electronic, mechanical, photocopying, recording, or otherwise, without the prior written permission of the publisher.

Printed in the United States of America
10 9 8 7 6 5 4 3 2 1

Main entry under title:
Polymer Blends and Alloys—Guidebook to Commercial Products

A Technomic Publishing Company book
Bibliography: p.
Includes index p. 557

Library of Congress Card No. 88-50188
ISBN No. 87762-566-2

# Foreword

One of the rapidly emerging areas in polymer technology is that of polymer blends and alloys. While physical mixtures of polymers are not new ("Polyblends" have been around for a good many years), many advances have been made recently in the combining of two or more dissimilar polymers, and the use of these techniques is growing.

Simple blends generally show average properties of the components; "alloys" may show some synergism of properties as well as a single second order transition. Any number of combinations—and combinations of these—may be tailored to meet specific end-use demands by judicious choice of components and proportions. Questions of compatibility, partial miscibility, and immiscibility are best dealt with on a case-by-case basis and in consideration of end-uses.

While the number of combinations of "conventional" polymers has by no means been fully explored, the ongoing addition to polymer technology of new, highly stable polymers with more and more unique characteristics guarantees that this is a field of increasing interest, growth and commercial development.

For this reason, Technomic Publishing Company has compiled this first Guidebook designed to provide practitioners with a somewhat broader overview, as this empirical technology continues to evolve from a new branch of polymer science to a wider realm of industrial applications.

The editors believe this book will be useful to processors seeking technical information about polymer blends and alloys. It is the desire of the editors that companies having commercial blend and alloy products will send information on these products to the Editor, *Polymer Blends and Alloys—Guidebook to Commercial Products,* for future editions. And finally, we invite your comments and suggestions so that future editions will be ever more useful to the plastics industry.

# Acknowledgment

Technomic Publishing Company, Inc. is indebted to Mr. F. Melvin Sweeney, who served as the Editor for this first edition of *Polymer Blends and Alloys—Guidebook to Commercial Products.*

Mr. Sweeney's industrial experience includes 20 years of varied assignments with Armstrong World Industries' Research and Development Center; earlier he was product manager for Krauss-Maffei, GmbH. Currently a consultant specializing in polyurethane technology and Reaction Injection Molding, in recent years he has consulted with the Soundesign Corporation in Jersey City, New Jersey, and the Dynasty Chemical Company in Lu Kang, Taiwan. Formerly a professor at the State University of New York/Morrisville College, where he taught in the plastics technology program, Mr. Sweeney has also conducted special programs for the New York State Private Industry Council and the New York State Education Foundation.

# Introduction

## POLYMER BLENDS AND ALLOYS: EXPANDING OPPORTUNITIES FOR PLASTICS

Sarah Y. Kienzle
CHEM SYSTEMS, INC., 303 South Broadway, Tarrytown, NY 10591-5487

Plastics have enjoyed strong growth in demand, from less than 20 billion pounds in 1970 to nearly 50 billion pounds consumed in the United States in 1986, largely as a result of their substitution for traditional materials. Plastics have replaced metals, glass, ceramics, wood, paper, and natural fibers in a wide variety of industries including packaging, consumer products, automobiles, building and construction, electronics, electric equipment, appliances, furniture, piping and heavy industrial equipment. The success of various polymers in such a broad spectrum of applications is attributable to the wide range of properties available with plastic materials. These include light weight, breakage resistance, the ability to consolidate functional parts, inexpensive, automated production of complex components and, often, lower finished part cost by comparison to traditional materials.

Because of the reproducible properties and cost effectiveness of polymeric materials, much of the easy substitution for traditional materials has already been accomplished. With slow economic and population growth in the developed regions of the world, coupled with fewer substitution opportunities, polymer growth rates have begun to slow. Additionally, polymer downgrading has become a significant trend. As engineers and designers have gained familiarity with the use of plastics, over-engineered plastic components are increasingly being redesigned with less costly, lower propertied polymer compounds. Thus, intercompetition among polymers has sharply increased and many suppliers suggest that the development risks of synthesizing a truly new polymer often outweigh the potential commercial rewards.

The combination of two or more commercially available polymers through alloying or blending, however, represents an inexpensive route to product differentiation for suppliers. Existing equipment may be utilized and the properties and chemical behavior of the constituents are generally well understood. For the processor and end user, alloying and blending technology permits tailoring of a polymer compound to their specific application requirements, often at a lower cost than the current material and over a short developmental period. Alloy and blend development, therefore, is typically market driven and requires an ongoing dialogue between supplier and customer to ensure commercial success.

### ALLOY AND BLEND TECHNOLOGY

The terms alloy and blend are often used synonymously, yet the two material types differ in the level of thermodynamic compatibility between their components and the resulting effects on physical and mechanical properties. Some level of thermodynamic compatibility is necessary between components in all polymer combinations to prevent phase separation in processing and use. Beyond this level of compatibility, greater attractive forces between constituents serve to enhance the resultant property profile.

In general, two component polymer mixtures may be described by the following equation:

$$P = P_1C_1 + P_2C_2 + IP_1P_2$$

where $P$ is a property value of the alloy or blend, $P_1$ and $P_2$ are the property values of the isolated components, and $C_1$ and $C_2$ are the concentrations of the two constituents. $I$ is an interaction coefficient that describes the level of synergism, or thermodynamic compatibility, of the components in the mixture.

In mixtures where the interaction coefficient, $I$, has a positive value ($I>0$), the resulting polymer combination exhibits better property values than the weighted arithmetic average of the components' properties and is termed synergistic. If $I$ equals zero, the properties of the combination are equal to the weighted arithmetic average of the constituents' properties. In this case, an additive blend results. When $I$ takes a negative value ($I<0$), with resultant blend properties below those predicted by the components' weighted arithmetic property averages, a nonsynergistic blend results. The effects of various values of the interaction coefficient on polymer combinations are graphically depicted in Figure 1.

Alloys are synergistic polymer combinations with real property advantages derived from a high level of thermodynamic compatibility between components. Alloys exhibit strong intermolecular forces and form single-phase systems with unique glass transition temperatures. The most significant commercial alloys are the polystyrene-modified polyphenylene oxide and ether products developed by General Electric (NORYL) and Borg-Warner/Mitsubishi Gas Chemicals (PREVEX), respectively. Here the polymers are mutually soluble and the poly-

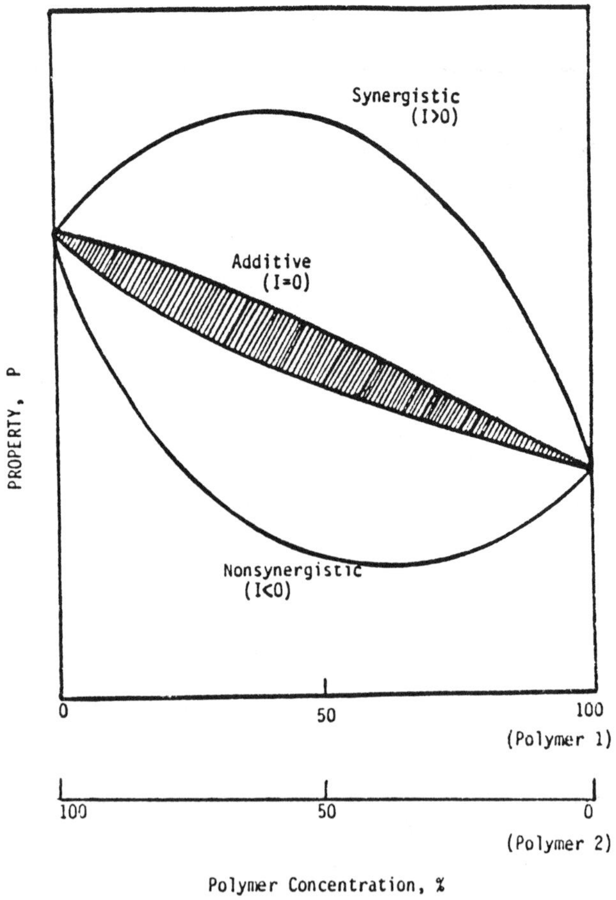


**Figure 1.** Property relationship in alloys and blends as a function of concentration.

styrene reduces the PPO or PPE melt viscosity sufficiently to allow ease of processing over a broad range of conditions. The phenylene-based constituent contributes strength, impact resistance, flame retardancy, and superior hydrolytic stability, which are minimally reduced by the presence of polystyrene.

Polymer blends, by comparison, have less intense thermodynamic compatibility than alloys. The blends exhibit discrete polymer phases and multiple glass transition temperatures. In general, the properties of a blend reflect the weighted arithmetic averages of the properties of its constituents ($I = 0$), as shown in Table 1. Here, commercially available blends of polycarbonate and ABS are compared with the theoretical "additive" property profile of a 50/50 PC/ABS blend. PC/ABS blends contain 50 percent or more polycarbonate, while ABS/PC blends contain less than 50 percent polycarbonate. Blends may also be nonsynergistic ($I<0$), athough this represents the least desirable outcome for the formulator.

## Preparation of Alloys and Blends

Alloys and blends may be manufactured by a variety of means, including melt blending, solution blending, and latex or dispersion mixing. Melt blending, in intensive mixing extrusion equipment, is the predominant commercial method of alloy and blend preparation.

In melt blending, two or more polymers plus any desired fillers, reinforcements, and additives are metered, by weight, into a shear intensive extruder. The constituents are mixed at elevated temperatures (i.e., above the melt points of the polymer constituents) by the extruder screw, which exerts mechanical shearing forces and ensures even distribution and thorough blending of the alloy and blend elements. Polymers may be blended in powder or pellet form, and the resulting alloy or blend is typically stranded, cooled, and chopped into pellets.

A compatibilizing agent may be incorporated in the alloy or blend to enhance the synergism of the mixture. By incorporating different functional groups in one chemical compound that can mate readily with the dissimilar polymers to be blended, the compatibilizing agent acts as a fastener, preventing phase separation and often enhancing the physical and mechanical properties of the resulting alloy or blend. Typical alloying and blending aids include block copolymers, graft copolymers, and polymeric plasticizers.

Melt blending has several advantages over solution blending and latex mixing. A liquid blending or dispersion agent is eleminiated, thus negating costs associated with solvent removal, recovery, and losses. Additionally, by combining only those elements desired in the final mixture, melt blending reduces the likelihood of contamination in the alloy or blend. Also, polymer producers can use in-place equipment. For example, sharing extruders across product lines is possible and may reduce the capital investment necessary to introduce a new product line. The use of elevated temperatures in melt blending, however, poses the possibility of undesired crosslinking or grafting of the polymers. Additionally, polymer degradation, caused by chain scission, can result in color shifts and poor mechanical properties, if processing temperatures are not carefully controlled.

Solution blending reduces the likelihood of thermally or mechanically induced polymer degradation, since alloy or blend mixing occurs in a solvent, at relatively low temperature and with low shear forces. Contamination of the alloy or blend with residual solvent and the potential for phase separation or selective precipitation are hazards of this preparation method.

Latex or dispersion mixing uses coagulation to give an intimate alloy or blend mixture. Acrylonitrile butadiene styrene (ABS) polymers may be prepared by this method, which is also applicable for alloys and blends, in some cases. Dispersion mixing is followed with melt mixing (or compounding) to produce polymer strands for pelletizing.

## Selecting a Polymer Combination

Polymers to be combined in alloys and blends are generally selected to complement each other in one or more of the following property categories: cost, processability, mechanical properties, warpage resistance, chemical resistance, and thermal performance. Blends are typically viewed as cost saving devices, whereby an expensive polymer may be combined with a less costly plastic to provide adequate performance at a significantly reduced price to the consumer. Although alloys may also represent cost reductions to prospective consumers, the driv-

**Table 1. Properties of ABS and Polycarbonate Blends vs. Properties of Their Components**

| | Components | | Theoretical Blend | Commercial Blends | |
|---|---|---|---|---|---|
| | ABS* | PC | 50/50 ABS/PC | PC/ABS | ABS/PC |
| Specific gravity | 1.07 | 1.2 | 1.14 | 1.12 | 1.10 |
| Water absorption, 24 hr, % | 0.32 | 0.15 | 0.24 | 0.21 | — |
| Coeff. of thermal exp., $10^{-5}/°F$ | 6.0 | 6.8 | 6.4 | 3.5 | 5.1 |
| HDT at 264 psi, °F | 230** | 270 | 250 | 240 | 190 |
| Tensile strength, psi | 6,200 | 9,500 | 7,850 | 8,500 | 6,300 |
| Flexural strength, psi | 11,000 | 13,500 | 12,250 | 13,000 | 10,500 |
| Flexural modulus, $10^3$ psi | 350 | 340 | 345 | 350 | 330 |
| Notched Izod impact, ft-lb/in. | 4.3 | 16.0 | 10.2 | 10.5 | 8.5 |
| Rockwell hardness | R112 | M70 | — | R117 | R105 |

*Heat resistant.
**Annealed.

**Table 2. Classification of Thermoplastic Polymers**

| Commodity | Transitional | Engineering | Performance |
|---|---|---|---|
| PET (unfilled)<br>Polyethylene<br>Polypropylene<br>Polystyrene<br>Polyvinylchloride | ABS/SAN<br>Acrylics<br>SMA copolymer | Modified PPO<br>Nylon<br>PBT<br>PET<br>Polyacetal<br>Polycarbonate<br>Alloys blends<br>SMA terpolymer | Fluoropolymers<br>Polyamide-imide<br>Polyarylate<br>Polyetheretherketone<br>Polyetherimide<br>Polyethersulfone<br>Polyimide<br>Polyphenylenesulfone<br>Polysulfone |

← VOLUME

PRICE →

**Table 3. Crystalline and Amorphous Polymers**

| Crystalline | Amorphous |
|---|---|
| Properties<br>• Excellent chemical resistance<br>• Sharp melting point<br>• Low viscosity melt<br>• Significant tensile flexural, and heat distortion improvement with reinforcement | Properties<br>• Dimensional stability and warpage resistance<br>• Maintenance of properties at elevated temperatures<br>• Clarity<br>• Generally good impact strength<br>• Melting range |
| Polymers<br>• Polypropylene<br>• Polyacetal<br>• Nylon 6/6<br>• Polybutylene terephthalate (PBT) | Polymers<br>• ABS<br>• Polycarbonate<br>• Polysulfone<br>• Polyarylate |

ing forces for their development and use are more typically the synergistic property profiles achievable.

Polymers can be segmented into four families, as described in Table 2. By combining polymers from two different families (e.g., commodity with transitional, transitional with engineering, engineering with performance), the property profile of the more costly polymer may be partially exploited, with the less costly polymer reducing overall compound cost. Many commercial blends, such as ABS/polycarbonate, ABS/polysulfone, and polyethersulfone/polyetheretherketone, have been developed for this reason.

A second common method of selecting polymers for combination into alloys or blends is the choice of one crystalline and one amorphous polymer. As described in Table 3, crystalline polymers have excellent chemical resistance, shown large improvements in mechanical properties with the addition of reinforcing agents (e.g., glass or carbon fiber), and exhibit low viscosity above their melting points, which aids in ease of melt processing. Amorphous polymers, such as polycarbonate, polyarylate, and polysulfone, provide dimensional stability, freedom from warpage, and excellent impact strength. When combining a crystalline and an amorphous polymer, the resulting blend or alloy provides good dimensional stability, ease of processing, and chemical resistance, as well as mechanical properties that can be tailored for a specific application.

Much of the current work in developing engineering thermoplastic blends and alloys for automotive body panels relies on crystalline/amorphous polymer combinations. Blends and alloys, such as nylon/PPO, PBT/polyarylate and PBT/polycarbonate, provide the combination of warpage resistance necessary for large part molding, chemical resistance to automotive and road chemicals, impact resistance, energy absorption characteristics, and stiffness necessary for body panel applications. This combination of properties is currently unobtainable with a single commercial polymer.

The combination of polymers with elastomers represents yet another group of alloys and blends. Typically, the polymer provides strength, stiffness, and thermal resistance, while an elastomer, such as polyurethane, the styrene block copolymers, copolyester elastomers, or EPDM, provides impact modification. The blending of polypropylene with EPDM for bumper covers and automotive fascia is a significant commercial example.

## Alloy and Blend Supply

Several driving forces have prompted polymer suppliers to intensify their research and development efforts in alloy and blend technology. Increased interpolymer competition, with resultant losses in market share and margins, has caused polymer suppliers to focus their attention on providing polymer systems that offer a unique competitive advantage to their customers. The advantage may be solely a cost reduction from the previously specified material or the alloy or blend may provide a specific property advantage, such as improved impact strength or flame retardancy.

The development costs and risks for commercialization of a new base polymer are becoming nearly prohibitive for many suppliers. Chem Systems estimates that successful commercialization of a new polymer will require 15 to 20 years of development effort. By contrast, provision of an alloy or blend based on a supplier's commercial base polymers may require less than half that time for development and commercialization. Additionally, raw materials and manufacturing equipment for alloys and blends are readily available to suppliers from other product lines (e.g., an extruder is often used as the reactor), thus reducing procurement and capital investment risks.

Alloys and blends represent a route to portfolio diversification as well. For commodity plastic suppliers, engineering or transitional polymer alloys and blends offer a route to penetrate higher margin applications. For engineering plastic suppliers, alloys and blends may serve as product line extensions, to allow competition in new markets or represent a buffer against interpolymer competition losses in traditional product lines.

In general, it is important for the alloy and blend supplier to be basic in at least one of the alloy or blend components (preferably, the predominant one). As illustrated in Table 4, a supplier producing a blend based on market priced materials is at a significant cost disadvantage by comparison to a basic supplier with favorable material transfer costs. Blends are relatively easy to duplicate and patents are routinely circumnavigated by development of products with similar properties but compositions outside patent limits. Thus, a producer's cost position represents one of the few effective methods of deterring competition.

A second method of maintaining competitive dominance requires development of highly sophisticated blends and alloys through the use of proprietary technology. General Electric, DuPont and Monsanto, for example, have developed compatibilizing agents to achieve synergistic properties in a variety of polymer alloys. These additives, as well as the alloys containing them, may be patented and the property synergism derived is typically not simple for competitors to duplicate at comparable cost.

Custom compounders have established business based on production of production of proprietary polymer compositions. The compounder offers the advantage of being able to combine nearly any commercially available polymers with a plethora of fillers and additives. Proprietary technology and specialized, flexible manufacturing equipment allow the compounder to rapidly tailor formulations to meet specific application requirements. With smaller equipment, the compounder may be able to manufacture an alloy or blend more efficiently than larger suppliers, particularly for the small run sizes that characterize the alloy and blend business.

To assist their customers, successful alloy and blend suppliers provide a high level of technical service, in addition to tailoring the properties of the alloy or blend for specific applications. Market development activities of the suppliers include assisting the end user and fabricator with concept and design analysis, computer aided engineering and stress analysis, prototyping, tooling design, and manufacturing scale-up. Since many applications re-

**Table 4. PC/ABS Cost of Production Economics, 1986**
**(10 Million Pounds per Year, U.S. Gulf Coast)**

| Capital summary | $ million | |
|---|---|---|
| Inside battery limits | 2.9 | |
| Outside battery limits | 0.7 | |
| Total fixed investment | 3.6 | |
| Working capital | 1.3 | |
| Total investment | 4.9 | |
| | **Market*** | **Transfer**** |
| Cost of production | (¢/lb) | (¢/lb) |
| Raw materials | | |
| ABS (@ 89¢/lb) | 22.3 | 13.0 |
| Polycarbonate (@ 145¢/lb) | 108.7 | 65.3 |
| Others | 1.7 | 1.7 |
| Total raw materials | 132.7 | 80.0 |
| Utilities | 0.6 | 0.6 |
| Operating costs | 5.0 | 5.0 |
| Overhead expenses | 5.2 | 5.2 |
| By-product credit | 0 | 0 |
| Cash cost of production | 143.5 | 90.8 |
| Depreciation | 6.5 | 6.5 |
| Net cost of production | 150.0 | 97.3 |
| Return @ 20% on investment | 9.7 | 9.7 |
| Cost plus 20% return | 159.7 | 107.0 |
| Market price | 135.0 | 135.0 |
| List price | 146.0 | 146.0 |

*ABS at market price of 89¢/lb. PC at market price of 145¢/lb.
**ABS at transfer price of 52¢/lb. PC at transfer price of 87¢/lb.

**Table 5. Alloy and Blend Programs of Several U.S. Suppliers**

| **Amoco** | **Celanese** | **DuPont** | **ICI** |
|---|---|---|---|
| PSO/ABS<br>PSO/Polyester<br>Nylon/Polyarylate | PBT/PET<br>Nylon/Elastomer<br>PBT/Elastomer<br>Polyester/Polyarylate<br>Acetal/Elastomer | Nylon/Elastomer<br>PET/Elastomer<br>Acetal/Elastomer<br>PBT/PET<br>Nylon/Polyolefin<br>Polyarylate/Polyester | PEEK/PES |
| **ARCO**<br>SMA/PC | **Dow**<br>PC/ABS | **General Electric**<br>PPO/PS<br>PC/ASA<br>PPO/Nylon<br>PPO/Polyester<br>PC/Polyester<br>PBT/PET<br>PVC/ASA | **Mobay**<br>PC/ABS<br>PBT/PET<br>PC/Polyester<br>PC/Polyester<br>PC/TPU |
| **Borg-Warner**<br>PPE/PS<br>ABS/PC<br>ABS/Nylon<br>PPE/Nylon | | | **Monsanto**<br>ABS/Nylon |

quires material approval not only by the end user, but also by government, private, and trade regulatory bodies, suppliers also often assist in satisfying these requirements through development of polymer, and sometimes end product, performance data.

Since application requirements may not be well represented by standard material test data, it is incumbent on the blend supplier to simulate end-use tests and environments to ensure the end user of the viability of the polymeric materal. Pin insertion in a connector is not well described by tensile strength or flexural modulus, just as dart impact strength is not representative of a polymeric bat hitting a baseball, for example. Thus, the material supplier may develop specific tests to simulate conditions encountered in product use and work with the end user to refine alloy or blend performance to meet these requirements.

Despite successful alloy and blend commercialization, suppliers are continually developing new alloy and blend products. Development of alloys and blends is a dynamic arena in which interpolymer substitution is a continued threat. Longevity of alloys and blends in specific applications appears to be somewhat shorter than with traditional polymers, largely, we believe, as a result of intense competition for the same target markets. Successful suppliers are following a pattern of product development, commercialization, and immediate new product development to provide consumers with state-of-the-art products and to forestall replacement with competitors' alloys and blends.

### Commercial Significance of Alloys and Blends

Alloying and blending of polymers is not new. In 1986, engineering polymer alloys and blends, for example, represented nearly 300 million pounds of commercial sales in the United States. Chem Systems estimates that by 1995, U.S. consumption of engineering alloys and blends will reach approximately 700 million pounds. Alloy and blend demand growth will outstrip that for engineering base polymers by nearly two to one and will average at least 9 percent annually. Modified polyphenylene oxide and ether alloys enjoy the largest demand, but have begun to lose market share to other polymers and blends (e.g., ABS/PC) in a number of cost reduction programs.

Alloys and blends represent inexpensive routes to satisfy both end-user material requirements and suppliers' desires for competitive product differentiation. The high level of alloy and blend development activity is apparent from the list of products already commercially available, as described in Table 5. Polymer alloys and blends are well positioned for significant demand growth through the end of the century. As suppliers and consumers more fully understand the dynamics of alloy and blend development, including the necessity for rapid and continued grade modifications and detailed knowledge of end product performance requirements, product development times may be shortened and manufacturing competitiveness enhanced through the use of polymer blends and alloys.

# Amoco Chemicals Corp.

## Contact Address:

Amoco Chemicals Corp.
Engineering Resins
200 East Randolph Drive
Chicago, IL 60601

Phone: 1 (800) 621-4557

# TORLON 4203L

## General Information:

TORLON engineering polymers are composites of polyamide-imide polymers alloyed with fluorocarbon polymers and reinforced with fillers to produce a family of high performance, high temperature molding resins. Products molded using TORLON have reliable performance at high temperature, and under high stress.

## Product Information:

TORLON 4203L is formulated with ½% fluorocarbon and 3% $TiO_2$ as reinforcement. This molding resin has the best impact resistance and the most elongation for the TORLON family of resins. The resin's natural lubricity provides excellent release from the mold.

TORLON 4203 L has excellent electrical properties and molded products include parts for switches, relays, and other electrical products. The natural lubricity and excellent wear characteristics of TORLON 4203L moldings are important in such products as thrust washers, mechanical linkages, bushings, bearings, rollers, cams, and valve components. The high temperature performance is essential for thermal insulators and other applications in high temperature service.

## Product Data:

**TORLON 4203L**

| Property | ASTM Test Method | English Units | |
|---|---|---|---|
| **Mechanical** | | | |
| Tensile strength | D1708 | $10^3$ psi | |
| −321°F | | | 31.5 |
| 73°F | | | 27.8 |
| 275°F | | | 16.9 |
| 450°F | | | 9.5 |
| Tensile elongation | D1708 | % | |
| −321°F | | | 6 |
| 73°F | | | 15 |
| 275°F | | | 21 |
| 450°F | | | 22 |

*(continued)*

NOTE – Information provided in this book has been condensed. The publisher recommends that the product manufacturer be contacted, at the address or phone number provided, for more complete information on product safety, processing parameters, and applications not provided in this information sheet.

## TORLON 4203L *(continued)*

| Property | ASTM Test Method | English Units | |
|---|---|---|---|
| **Mechanical** | | | |
| Tensile modulus<br>73°F | D1708 | $10^5$ psi | 7.0 |
| Flexural strength<br>−321°F<br>73°F<br>275°F<br>450°F | D790 | $10^3$ psi | 41.0<br>34.9<br>24.8<br>17.1 |
| Flexural modulus<br>−321°F<br>73°F<br>275°F<br>450°F | D790 | $10^5$ psi | 11.4<br>7.3<br>5.6<br>5.2 |
| Compressive strength | D695 | $10^3$ psi | 32.1 |
| Compressive modulus | D695 | $10^5$ psi | |
| Shear strength<br>73°F | D732 | $10^3$ psi | 18.5 |
| Izod impact strength (1/8 in)<br>notched<br>unnotched | D256 | ft·lb/in | 2.7<br>20.0 |
| Poisson's ratio | | | 0.45 |
| **Thermal** | | | |
| Deflection temperature<br>264 psi | D648 | °F | 532 |
| Coefficient of linear thermal expansion | D696 | $10^{-6}$in/in·°F | 17 |
| Thermal conductivity | C177 | Btu·in/hr·ft$^2$°F | 1.8 |
| Flammability<br>Underwriters Laboratories | | | 94 V-O |
| Limiting oxygen index | D2863 | % | 45 |
| **Electrical** | | | |
| Dielectric constant<br>$10^3$Hz<br>$10^5$Hz | D150 | | 4.2<br>3.9 |
| Dissipation factor<br>$10^3$Hz<br>$10^5$Hz | D150 | | 0.026<br>0.031 |
| Volume resistivity | D257 | ohm·in | $8 \times 10^{16}$ |
| Surface resistivity | D257 | ohm | $5 \times 10^{15}$ |
| Dielectric strength (0.40 in) | D149 | V/mil | 580 |

*(continued)*

TORLON 4203L *(continued)*

| Property | ASTM Test Method | English Units | |
|---|---|---|---|
| **General** | | | |
| Density | D792 | lb/in$^3$ | 0.050 |
| Hardness, Rockwell E | D785 | | 86 |
| Water absorption | D570 | % | 0.33 |

TORLON 4203L

| Property | ASTM Test Method | Metric Units | |
|---|---|---|---|
| **Mechanical** | | | |
| Tensile strength | D1708 | N/mm$^2$ | |
| −196°C | | | 218 |
| 23°C | | | 192 |
| 135°C | | | 117 |
| 232°C | | | 66 |
| Tensile elongation | D1708 | % | |
| −196°C | | | 6 |
| 23°C | | | 15 |
| 135°C | | | 21 |
| 232°C | | | 22 |
| Tensile modulus | D1708 | $10^3$ N/mm$^2$ | |
| 23°C | | | 4.9 |
| Flexural strength | D790 | N/mm$^2$ | |
| −196°C | | | 287 |
| 23°C | | | 244 |
| 135°C | | | 174 |
| 232°C | | | 120 |
| Flexural modulus | D790 | $10^3$ N/mm$^2$ | |
| −196°C | | | 7.9 |
| 23°C | | | 5.0 |
| 135°C | | | 3.9 |
| 232°C | | | 3.6 |
| Compressive strength | D695 | N/mm$^2$ | 220 |
| Compressive modulus | D695 | $10^3$ N/mm$^2$ | 4.0 |
| Shear strength | D732 | N/mm$^2$ | |
| 23°C | | | 128 |
| Izod impact strength (3.2 mm) | D256 | J/m | |
| notched | | | 142 |
| unnotched | | | 1062 |
| Poisson's ratio | | | 0.45 |

*(continued)*

TORLON 4203L *(continued)*

| Property | ASTM Test Method | Metric Units | |
|---|---|---|---|
| **Thermal** | | | |
| Deflection temperature<br>1.82 N/mm$^2$ | D648 | °C | 278 |
| Coefficient of linear thermal expansion | D696 | $\mu$·m/m·K | 30.6 |
| Thermal conductivity | C177 | W/m·K | 0.26 |
| Flammability<br>Underwriters Laboratories | | | 94 V-O |
| Limiting oxygen index | D2863 | % | 45 |
| **Electrical** | | | |
| Dielectric constant<br>$10^3$Hz<br>$10^5$Hz | D150 | | <br>4.2<br>3.9 |
| Dissipation factor<br>$10^3$Hz<br>$10^5$Hz | D150 | | <br>0.026<br>0.031 |
| Volume resistivity | D257 | ohm·m | $2 \times 10^{16}$ |
| Surface resistivity | D257 | ohm | $5 \times 10^{15}$ |
| Dielectric strength (1.02 mm) | D149 | kV/mm | 23.6 |
| **General** | | | |
| Density | D792 | g/cm$^3$ | 1.38 |
| Hardness, Rockwell E | D785 | | 86 |
| Water absorption | D570 | % | 0.33 |

# Amoco Chemicals Corp.

## Contact Address:

Amoco Chemicals Corp.
Engineering Resins
200 East Randolph Drive
Chicago, IL 60601

Phone: 1 (800) 621-4557

## TORLON 4275

### General Information:

TORLON engineering polymers are composites of polyamide-imide polymers alloyed with fluorocarbon polymers and reinforced with fillers to produce a family of high performance, high temperature molding resins. Products molded using TORLON have reliable performance at high temperature, and under high stress.

### Product Information:

TORLON 4275 has a polyamide-imide binder modified with 3% fluorocarbon and 20% graphite binder. Similar to TORLON 4301, the molded resin has better wear resistance at higher speeds. Applications include bearings, thrust washers, wear pads, guide strips, piston rings, seals, valve seats, etc.

### Product Data:

**TORLON 4275**

| Property | ASTM Test Method | English Units | |
|---|---|---|---|
| **Mechanical** | | | |
| Tensile strength | D1708 | $10^3$ psi | |
| −321°F | | | 18.8 |
| 73°F | | | 22.0 |
| 275°F | | | 16.3 |
| 450°F | | | 8.1 |
| Tensile elongation | D1708 | % | |
| −321°F | | | 3 |
| 73°F | | | 7 |
| 275°F | | | 15 |
| 450°F | | | 17 |
| Tensile modulus | D1708 | $10^5$ psi | |
| 73°F | | | 11.3 |

*(continued)*

NOTE—Information provided in this book has been condensed. The publisher recommends that the product manufacturer be contacted, at the address or phone number provided, for more complete information on product safety, processing parameters, and applications not provided in this information sheet.

### TORLON 4275 *(continued)*

| Property | ASTM Test Method | English Units | |
|---|---|---|---|
| **Mechanical** | | | |
| Flexural strength | D790 | $10^3$ psi | |
| −321°F | | | 29.0 |
| 73°F | | | 30.2 |
| 275°F | | | 22.4 |
| 450°F | | | 15.8 |
| Flexural modulus | D790 | $10^5$ psi | |
| −321°F | | | 13.9 |
| 73°F | | | 10.6 |
| 275°F | | | 8.1 |
| 450°F | | | 7.4 |
| Compressive strength | D695 | $10^3$ psi | 17.8 |
| Compressive modulus | D695 | $10^5$ psi | 5.8 |
| Shear strength | D732 | $10^3$ psi | |
| 73°F | | | 11.1 |
| Izod impact strength (1/8 in) | D256 | ft·lb/in | |
| notched | | | 1.6 |
| unnotched | | | 4.7 |
| Poisson's ratio | | | 0.39 |
| **Thermal** | | | |
| Deflection temperature | D648 | °F | |
| 264 psi | | | 536 |
| Coefficient of linear thermal expansion | D696 | $10^{-6}$in/in·°F | 14 |
| Thermal conductivity | C177 | Btu·in/hr·ft²°F | |
| Flammability Underwriters Laboratories | | | 94 V-O |
| Limiting oxygen index | D2863 | % | 45 |
| **Electrical** | | | |
| Dielectric constant | D150 | | |
| $10^3$Hz | | | 7.3 |
| $10^5$Hz | | | 6.6 |
| Dissipation factor | D150 | | |
| $10^3$Hz | | | 0.059 |
| $10^5$Hz | | | 0.063 |
| Volume resistivity | D257 | ohm·in | $3 \times 10^{15}$ |
| Surface resistivity | D257 | ohm | $4 \times 10^{17}$ |
| Dielectric strength (0.40 in) | D149 | V/mil | |

*(continued)*

TORLON 4275 *(continued)*

| Property | ASTM Test Method | English Units | |
|---|---|---|---|
| **General** | | | |
| Density | D792 | lb/in$^3$ | 0.052 |
| Hardness, Rockwell E | D785 | | 70 |
| Water absorption | D570 | % | 0.33 |

TORLON 4275

| Property | ASTM Test Method | Metric Units | |
|---|---|---|---|
| **Mechanical** | | | |
| Tensile strength | D1708 | N/mm$^2$ | |
| −196°C | | | 130 |
| 23°C | | | 152 |
| 135°C | | | 113 |
| 232°C | | | 56 |
| Tensile elongation | D1708 | % | |
| −196°C | | | 3 |
| 23°C | | | 7 |
| 135°C | | | 15 |
| 232°C | | | 17 |
| Tensile modulus | D1708 | $10^3$ N/mm$^2$ | |
| 23°C | | | 7.8 |
| Flexural strength | D790 | N/mm$^2$ | |
| −196°C | | | 203 |
| 23°C | | | 212 |
| 135°C | | | 157 |
| 232°C | | | 111 |
| Flexural modulus | D790 | $10^3$ N/mm$^2$ | |
| −196°C | | | 9.6 |
| 23°C | | | 7.3 |
| 135°C | | | 5.6 |
| 232°C | | | 5.1 |
| Compressive strength | D695 | N/mm$^2$ | 120 |
| Compressive modulus | D695 | $10^3$ N/mm$^2$ | |
| Shear strength | D732 | N/mm$^2$ | |
| 23°C | | | 77 |
| Izod impact strength (3.2 mm) | D256 | J/m | |
| notched | | | 84 |
| unnotched | | | 250 |
| Poisson's ratio | | | 0.39 |

*(continued)*

**TORLON 4275 *(continued)***

| Property | ASTM Test Method | Metric Units | |
|---|---|---|---|
| **Thermal** | | | |
| Deflection temperature<br>1.82 N/mm² | D648 | °C | 280 |
| Coefficient of linear thermal expansion | D696 | μ·m/m·K | 25.2 |
| Thermal conductivity | C177 | W/m·K | |
| Flammability<br>Underwriters Laboratories | | | 94 V-O |
| Limiting oxygen index | D2863 | % | 45 |
| **Electrical** | | | |
| Dielectric constant<br>$10^3$Hz<br>$10^5$Hz | D150 | | <br>7.3<br>6.6 |
| Dissipation factor<br>$10^3$Hz<br>$10^5$Hz | D150 | | <br>0.059<br>0.063 |
| Volume resistivity | D257 | ohm·m | $8 \times 10^{13}$ |
| Surface resistivity | D257 | ohm | $4 \times 10^{17}$ |
| Dielectric strength (1.02 mm) | D149 | kV/mm | |
| **General** | | | |
| Density | D792 | g/cm³ | 1.44 |
| Hardness, Rockwell E | D785 | | 70 |
| Water absorption | D570 | % | 0.33 |

# Amoco Chemicals Corp.

## Contact Address:

Amoco Chemicals Corp.
Engineering Resins
200 East Randolph Drive
Chicago, IL 60601

Phone: 1 (800) 621-4557

## TORLON 4301

### General Information:

TORLON engineering polymers are composites of polyamide-imide polymers alloyed with fluorocarbon polymers and reinforced with fillers to produce a family of high performance, high temperature molding resins. Products molded using TORLON have reliable performance at high temperature, and under high stress.

### Product Information:

TORLON 4301 has a polyamide-imide binder modified with 3% fluorocarbon and reinforced with 12% graphite powder. Similar to TORLON 4347, this resin is designed for bearing use where good wear resistance, low coefficient of friction, and high compressive strength are important. Products suggested are bearings, thrust washers, wear pads, guide strips, seals, vanes, and valve parts.

### Product Data:

**TORLON 4301**

| **Property** | **ASTM Test Method** | **English Units** | |
|---|---|---|---|
| **Mechanical** | | | |
| Tensile strength | D1708 | $10^3$ psi | |
| −321°F | | | |
| 73°F | | | 23.7 |
| 275°F | | | 16.3 |
| 450°F | | | 10.6 |
| Tensile elongation | D1708 | % | |
| −321°F | | | |
| 73°F | | | 7 |
| 275°F | | | 20 |
| 450°F | | | 17 |
| Tensile modulus | D1708 | $10^5$ psi | |
| 73°F | | | 9.5 |

*(continued)*

NOTE–Information provided in this book has been condensed. The publisher recommends that the product manufacturer be contacted, at the address or phone number provided, for more complete information on product safety, processing parameters, and applications not provided in this information sheet.

**TORLON 4301 *(continued)***

| Property | ASTM Test Method | English Units | |
|---|---|---|---|
| **Mechanical** | | | |
| Flexural strength | D790 | $10^3$ psi | |
| −321°F | | | |
| 73°F | | | 31.2 |
| 275°F | | | 23.5 |
| 450°F | | | 16.2 |
| Flexural modulus | D790 | $10^5$ psi | |
| −321°F | | | |
| 73°F | | | 10.0 |
| 275°F | | | 7.9 |
| 450°F | | | 7.2 |
| Compressive strength | D695 | $10^3$ psi | 24.1 |
| Compressive modulus | D695 | $10^5$ psi | 7.7 |
| Shear strength | D732 | $10^3$ psi | |
| 73°F | | | 16.1 |
| Izod impact strength (1/8 in) | D256 | ft·lb/in | |
| notched | | | 1.2 |
| unnotched | | | 7.6 |
| Poisson's ratio | | | 0.39 |
| **Thermal** | | | |
| Deflection temperature | D648 | °F | |
| 264 psi | | | 534 |
| Coefficient of linear thermal expansion | D696 | $10^{-6}$in/in·°F | 14 |
| Thermal conductivity | C177 | Btu·in/hr·ft²°F | 3.7 |
| Flammability Underwriters Laboratories | | | 94 V-O |
| Limiting oxygen index | D2863 | % | 44 |
| **Electrical** | | | |
| Dielectric constant | D150 | | |
| $10^3$Hz | | | 6.0 |
| $10^6$Hz | | | 5.4 |
| Dissipation factor | D150 | | |
| $10^3$Hz | | | 0.037 |
| $10^6$Hz | | | 0.042 |
| Volume resistivity | D257 | ohm·in | $3 \times 10^{15}$ |
| Surface resistivity | D257 | ohm | $8 \times 10^{17}$ |
| Dielectric strength (0.40 in) | D149 | V/mil | |

*(continued)*

TORLON 4301 *(continued)*

| Property | ASTM Test Method | English Units | |
|---|---|---|---|
| **General** | | | |
| Density | D792 | lb/in$^3$ | 0.052 |
| Hardness, Rockwell E | D785 | | 72 |
| Water absorption | D570 | % | 0.28 |

TORLON 4301

| Property | ASTM Test Method | Metric Units | |
|---|---|---|---|
| **Mechanical** | | | |
| Tensile strength | D1708 | N/mm$^2$ | |
| −196°C | | | |
| 23°C | | | 164 |
| 135°C | | | 113 |
| 232°C | | | 73 |
| Tensile elongation | D1708 | % | |
| −196°C | | | |
| 23°C | | | 7 |
| 135°C | | | 20 |
| 232°C | | | 17 |
| Tensile modulus | D1708 | $10^3$ N/mm$^2$ | |
| 23°C | | | 6.6 |
| Flexural strength | D790 | N/mm$^2$ | |
| −196°C | | | |
| 23°C | | | 219 |
| 135°C | | | 165 |
| 232°C | | | 113 |
| Flexural modulus | D790 | $10^3$ N/mm$^2$ | |
| −196°C | | | |
| 23°C | | | 6.9 |
| 135°C | | | 5.5 |
| 232°C | | | 4.5 |
| Compressive strength | D695 | N/mm$^2$ | 170 |
| Compressive modulus | D695 | $10^3$ N/mm$^2$ | 5.3 |
| Shear strength | D732 | N/mm$^2$ | |
| 23°C | | | 112 |
| Izod impact strength (3.2 mm) | D256 | J/m | |
| notched | | | 63 |
| unnotched | | | 404 |
| Poisson's ratio | | | 0.39 |

*(continued)*

**TORLON 4301 *(continued)***

| Property | ASTM Test Method | Metric Units | |
|---|---|---|---|
| **Thermal** | | | |
| Deflection temperature<br>1.82 $N/mm^2$ | D648 | °C | 279 |
| Coefficient of linear thermal expansion | D696 | $\mu$·m/m·K | 25.2 |
| Thermal conductivity | C177 | W/m·K | 0.54 |
| Flammability<br>Underwriters Laboratories | | | 94 V-O |
| Limiting oxygen index | D2863 | % | 44 |
| **Electrical** | | | |
| Dielectric constant<br>$10^3$Hz<br>$10^5$Hz | D150 | | <br>6.0<br>5.4 |
| Dissipation factor<br>$10^3$Hz<br>$10^5$Hz | D150 | | <br>0.037<br>0.042 |
| Volume resistivity | D257 | ohm·m | $8 \times 10^{13}$ |
| Surface resistivity | D257 | ohm | $8 \times 10^{17}$ |
| Dielectric strength (1.02 mm) | D149 | kV/mm | |
| **General** | | | |
| Density | D792 | $g/cm^3$ | 1.43 |
| Hardness, Rockwell E | D785 | | 72 |
| Water absorption | D570 | % | 0.28 |

# Amoco Chemicals Corp.

## Contact Address:

Amoco Chemicals Corp.
Engineering Resins
200 East Randolph Drive
Chicago, IL 60601

Phone: 1 (800) 621-4557

## TORLON 4347

### General Information:

TORLON engineering polymers are composites of polyamide-imide polymers alloyed with fluorocarbon polymers and reinforced with fillers to produce a family of high performance, high temperature molding resins. Products molded using TORLON have reliable performance at high temperature, and under high stress.

### Product Information:

TORLON 4347 has a polyamide-imide binder modified with 8% fluorocarbon and reinforced with 12% graphite powder. Formulated to meet specifications for wear resistant products, such as bearings, thrust washers, wear pads, guide strips, piston rings, and seals, the product combines the best wear resistance for reciprocating motion or bearing use subject to high load at low speed.

### Product Data:

**TORLON 4347**

| Property | ASTM Test Method | English Units | |
|---|---|---|---|
| **Mechanical** | | | |
| Tensile strength | D1708 | $10^3$ psi | |
| −321°F | | | |
| 73°F | | | 17.8 |
| 275°F | | | 15.1 |
| 450°F | | | 7.8 |
| Tensile elongation | D1708 | % | |
| −321°F | | | |
| 73°F | | | 9 |
| 275°F | | | 21 |
| 450°F | | | 15 |
| Tensile modulus | D1708 | $10^5$ psi | |
| 73°F | | | 8.7 |

*(continued)*

NOTE – Information provided in this book has been condensed. The publisher recommends that the product manufacturer be contacted, at the address or phone number provided, for more complete information on product safety, processing parameters, and applications not provided in this information sheet.

**TORLON 4347 *(continued)***

| Property | ASTM Test Method | English Units | |
|---|---|---|---|
| **Mechanical** | | | |
| Flexural strength | D790 | $10^3$ psi | |
| −321°F | | | |
| 73°F | | | 27.0 |
| 275°F | | | 20.5 |
| 450°F | | | 14.3 |
| Flexural modulus | D790 | $10^5$ psi | |
| −321°F | | | |
| 73°F | | | 9.1 |
| 275°F | | | 6.4 |
| 450°F | | | 6.2 |
| Compressive strength | D695 | $10^3$ psi | 18.3 |
| Compressive modulus | D695 | $10^5$ psi | |
| Shear strength | D732 | $10^3$ psi | |
| 73°F | | | 11.5 |
| Izod impact strength (1/8 in) | D256 | ft·lb/in | |
| notched | | | 1.3 |
| unnotched | | | |
| Poisson's ratio | | | |
| **Thermal** | | | |
| Deflection temperature | D648 | °F | |
| 264 psi | | | 532 |
| Coefficient of linear thermal expansion | D696 | $10^{-6}$in/in·°F | 15 |
| Thermal conductivity | C177 | Btu·in/hr·ft²°F | |
| Flammability Underwriters Laboratories | | | 94 V-O |
| Limiting oxygen index | D2863 | % | 46 |
| **Electrical** | | | |
| Dielectric constant | D150 | | |
| $10^3$Hz | | | 6.8 |
| $10^5$Hz | | | 6.0 |
| Dissipation factor | D150 | | |
| $10^3$Hz | | | 0.037 |
| $10^5$Hz | | | 0.071 |
| Volume resistivity | D257 | ohm·in | $3 \times 10^{15}$ |
| Surface resistivity | D257 | ohm | $1 \times 10^{18}$ |
| Dielectric strength (0.40 in) | D149 | V/mil | |

*(continued)*

TORLON 4347 *(continued)*

| Property | ASTM Test Method | English Units | |
|---|---|---|---|
| **General** | | | |
| Density | D792 | lb/in$^3$ | 0.052 |
| Hardness, Rockwell E | D785 | | 66 |
| Water absorption | D570 | % | 0.17 |

TORLON 4347

| Property | ASTM Test Method | Metric Units | |
|---|---|---|---|
| **Mechanical** | | | |
| Tensile strength | D1708 | N/mm$^2$ | |
| −196°C | | | |
| 23°C | | | 123 |
| 135°C | | | 104 |
| 232°C | | | 54 |
| Tensile elongation | D1708 | % | |
| −196°C | | | |
| 23°C | | | 9 |
| 135°C | | | 21 |
| 232°C | | | 15 |
| Tensile modulus | D1708 | 10$^3$ N/mm$^2$ | |
| 23°C | | | 6.0 |
| Flexural strength | D790 | N/mm$^2$ | |
| −196°C | | | |
| 23°C | | | 189 |
| 135°C | | | 144 |
| 232°C | | | 100 |
| Flexural modulus | D790 | 10$^3$ N/mm$^2$ | |
| −196°C | | | |
| 23°C | | | 6.3 |
| 135°C | | | 4.4 |
| 232°C | | | 4.3 |
| Compressive strength | D695 | N/mm$^2$ | 130 |
| Compressive modulus | D695 | 10$^3$ N/mm$^2$ | |
| Shear strength | D732 | N/mm$^2$ | |
| 23°C | | | 80 |
| Izod impact strength (3.2 mm) | D256 | J/m | |
| notched | | | 69 |
| unnotched | | | |
| Poisson's ratio | | | |

*(continued)*

TORLON 4347 *(continued)*

| Property | ASTM Test Method | Metric Units | |
|---|---|---|---|
| **Thermal** | | | |
| Deflection temperature<br>1.82 N/mm$^2$ | D648 | °C | 278 |
| Coefficient of linear thermal expansion | D696 | $\mu$·m/m·K | 27.0 |
| Thermal conductivity | C177 | W/m·K | |
| Flammability<br>Underwriters Laboratories | | | 94 V-O |
| Limiting oxygen index | D2863 | % | 46 |
| **Electrical** | | | |
| Dielectric constant<br>$10^3$Hz<br>$10^5$Hz | D150 | | 6.8<br>6.0 |
| Dissipation factor<br>$10^3$Hz<br>$10^5$Hz | D150 | | 0.037<br>0.071 |
| Volume resistivity | D257 | ohm·m | $8 \times 10^{13}$ |
| Surface resistivity | D257 | ohm | $1 \times 10^{18}$ |
| Dielectric strength (1.02 mm) | D149 | kV/mm | |
| **General** | | | |
| Density | D792 | g/cm$^3$ | 1.44 |
| Hardness, Rockwell E | D785 | | 66 |
| Water absorption | D570 | % | 0.17 |

# Amoco Chemicals Corp.

## Contact Address:

Amoco Chemicals Corp.
Engineering Resins
200 East Randolph Drive
Chicago, IL 60601

Phone: 1 (800) 621-4557

## TORLON 5030

### General Information:

TORLON engineering polymers are composites of polyamide-imide polymers alloyed with fluorocarbon polymers and reinforced with fillers to produce a family of high performance, high temperature molding resins. Products molded using TORLON have reliable performance at high temperature, and under high stress.

### Product Information:

TORLON 5030 is a polyamide-imide polymer modified with 1% fluorocarbon and reinforced with 30% glass fiber to produce a high strength molding resin with high strength, high stiffness, and very low creep, even at elevated temperatures.

These properties lend themselves to such applications as burn in sockets, gears, valve plates, fairings, tube clamps, impellers, rotor housings, terminal strips, insulators, and brackets.

### Product Data:

**TORLON 5030**

| Property | ASTM Test Method | English Units | |
|---|---|---|---|
| **Mechanical** | | | |
| Tensile strength | D1708 | $10^3$ psi | |
| −321°F | | | 29.5 |
| 73°F | | | 29.7 |
| 275°F | | | 23.1 |
| 450°F | | | 16.3 |
| Tensile elongation | D1708 | % | |
| −321°F | | | 4 |
| 73°F | | | 7 |
| 275°F | | | 15 |
| 450°F | | | 12 |
| Tensile modulus | D1708 | $10^5$ psi | |
| 73°F | | | 15.6 |

*(continued)*

NOTE—Information provided in this book has been condensed. The publisher recommends that the product manufacturer be contacted, at the address or phone number provided, for more complete information on product safety, processing parameters, and applications not provided in this information sheet.

TORLON 5030 *(continued)*

| Property | ASTM Test Method | English Units | |
|---|---|---|---|
| **Mechanical** | | | |
| Flexural strength | D790 | $10^3$ psi | |
| −321°F | | | 54.4 |
| 73°F | | | 48.3 |
| 275°F | | | 35.9 |
| 450°F | | | 26.2 |
| Flexural modulus | D790 | $10^5$ psi | |
| −321°F | | | 20.4 |
| 73°F | | | 17.0 |
| 275°F | | | 15.5 |
| 450°F | | | 14.3 |
| Compressive strength | D695 | $10^3$ psi | 38.3 |
| Compressive modulus | D695 | $10^5$ psi | 11.5 |
| Shear strength | D732 | $10^3$ psi | |
| 73°F | | | 20.1 |
| Izod impact strength (1/8 in) | D256 | ft·lb/in | |
| notched | | | 1.5 |
| unnotched | | | 9.5 |
| Poisson's ratio | | | 0.43 |
| **Thermal** | | | |
| Deflection temperature | D648 | °F | |
| 264 psi | | | 539 |
| Coefficient of linear thermal expansion | D696 | $10^{-6}$in/in·°F | 9 |
| Thermal conductivity | C177 | Btu·in/hr·ft²°F | 2.5 |
| Flammability Underwriters Laboratories | | | 94 V-O |
| Limiting oxygen index | D2863 | % | 51 |
| **Electrical** | | | |
| Dielectric constant | D150 | | |
| $10^3$Hz | | | 4.4 |
| $10^5$Hz | | | 6.5 |
| Dissipation factor | D150 | | |
| $10^3$Hz | | | 0.022 |
| $10^5$Hz | | | 0.023 |
| Volume resistivity | D257 | ohm·in | $6 \times 10^{16}$ |
| Surface resistivity | D257 | ohm | $1 \times 10^{18}$ |
| Dielectric strength (0.40 in) | D149 | V/mil | 840 |

*(continued)*

TORLON 5030 *(continued)*

| Property | ASTM Test Method | English Units | |
|---|---|---|---|
| **General** | | | |
| Density | D792 | lb/in$^3$ | 0.056 |
| Hardness, Rockwell E | D785 | | 94 |
| Water absorption | D570 | % | 0.24 |

TORLON 5030

| Property | ASTM Test Method | Metric Units | |
|---|---|---|---|
| **Mechanical** | | | |
| Tensile strength<br>−196°C<br>23°C<br>135°C<br>232°C | D1708 | N/mm$^2$ | <br>204<br>205<br>160<br>113 |
| Tensile elongation<br>−196°C<br>23°C<br>135°C<br>232°C | D1708 | % | <br>4<br>7<br>15<br>12 |
| Tensile modulus<br>23°C | D1708 | $10^3$ N/mm$^2$ | <br>10.8 |
| Flexural strength<br>−196°C<br>23°C<br>135°C<br>232°C | D790 | N/mm$^2$ | <br>381<br>338<br>251<br>184 |
| Flexural modulus<br>−196°C<br>23°C<br>135°C<br>232°C | D790 | $10^3$ N/mm$^2$ | <br>14.1<br>11.7<br>10.7<br>9.9 |
| Compressive strength | D695 | N/mm$^2$ | 260 |
| Compressive modulus | D695 | $10^3$ N/mm$^2$ | 7.9 |
| Shear strength<br>23°C | D732 | N/mm$^2$ | <br>140 |
| Izod impact strength (3.2 mm)<br>notched<br>unnotched | D256 | J/m | <br>79<br>504 |
| Poisson's ratio | | | 0.43 |

*(continued)*

TORLON 5030 *(continued)*

| Property | ASTM Test Method | Metric Units | |
|---|---|---|---|
| **Thermal** | | | |
| Deflection temperature<br>1.82 $N/mm^2$ | D648 | °C | 282 |
| Coefficient of linear thermal expansion | D696 | $\mu$·m/m·K | 16.2 |
| Thermal conductivity | C177 | W/m·K | 0.37 |
| Flammability<br>Underwriters Laboratories | | | 94 V-O |
| Limiting oxygen index | D2863 | % | 51 |
| **Electrical** | | | |
| Dielectric constant<br>$10^3$Hz<br>$10^5$Hz | D150 | | <br>4.4<br>6.5 |
| Dissipation factor<br>$10^3$Hz<br>$10^5$Hz | D150 | | <br>0.022<br>0.023 |
| Volume resistivity | D257 | ohm·m | $2 \times 10^{15}$ |
| Surface resistivity | D257 | ohm | $1 \times 10^{18}$ |
| Dielectric strength (1.02 mm) | D149 | kV/mm | 32.6 |
| **General** | | | |
| Density | D792 | $g/cm^3$ | 1.56 |
| Hardness, Rockwell E | D785 | | 94 |
| Water absorption | D570 | % | 0.24 |

# Amoco Chemicals Corp.

## Contact Address:

Amoco Chemicals Corp.
Engineering Resins
200 East Randolph Drive
Chicago, IL 60601

Phone: 1 (800) 621-4557

# TORLON 5030M

## General Information:

TORLON engineering polymers are composites of polyamide-imide polymers alloyed with fluorocarbon polymers and reinforced with fillers to produce a family of high performance, high temperature molding resins. Products molded using TORLON have reliable performance at high temperature, and under high stress.

## Product Information:

TORLON 5030M, like TORLON 5040, this resin consists of polyamide-imide polymer with 1% fluorocarbon and 30% glass fiber reinforcement. Proprietary modifications of the basic composite make this resin a better choice when the part to be molded is thick.

Applications are similar to those listed for TORLON 5030.

## Product Data:

**TORLON 5030M**

| Property | ASTM Test Method | English Units | |
|---|---|---|---|
| **Mechanical** | | | |
| Tensile strength | D1708 | $10^3$ psi | |
| −321 °F | | | |
| 73 °F | | | 27.0 |
| 275 °F | | | 15.7 |
| 450 °F | | | 9.3 |
| Tensile elongation | D1708 | % | |
| −321 °F | | | |
| 73 °F | | | 6 |
| 275 °F | | | 4 |
| 450 °F | | | 5 |
| Tensile modulus | D1708 | $10^5$ psi | |
| 73 °F | | | 15.2 |

*(continued)*

NOTE – Information provided in this book has been condensed. The publisher recommends that the product manufacturer be contacted, at the address or phone number provided, for more complete information on product safety, processing parameters, and applications not provided in this information sheet.

## TORLON 5030M *(continued)*

| Property | ASTM Test Method | English Units | |
|---|---|---|---|
| **Mechanical** | | | |
| Flexural strength | D790 | $10^3$ psi | |
| −321°F | | | |
| 73°F | | | 41.5 |
| 275°F | | | 26.3 |
| 450°F | | | 20.4 |
| Flexural modulus | D790 | $10^5$ psi | |
| −321°F | | | |
| 73°F | | | 16.6 |
| 275°F | | | 12.7 |
| 450°F | | | 12.9 |
| Compressive strength | D695 | $10^3$ psi | 30.3 |
| Compressive modulus | D695 | $10^5$ psi | |
| Shear strength | D732 | $10^3$ psi | |
| 73°F | | | 14.6 |
| Izod impact strength (1/8 in) | D256 | ft·lb/in | |
| notched | | | 1.3 |
| unnotched | | | |
| Poisson's ratio | | | |
| **Thermal** | | | |
| Deflection temperature | D648 | °F | |
| 264 psi | | | 524 |
| Coefficient of linear thermal expansion | D696 | $10^{-6}$in/in·°F | 10 |
| Thermal conductivity | C177 | Btu·in/hr·ft²°F | |
| Flammability Underwriters Laboratories | | | 94 V-O |
| Limiting oxygen index | D2863 | % | 47 |
| **Electrical** | | | |
| Dielectric constant | D150 | | |
| $10^3$Hz | | | 4.0 |
| $10^5$Hz | | | 4.3 |
| Dissipation factor | D150 | | |
| $10^3$Hz | | | 0.040 |
| $10^5$Hz | | | 0.030 |
| Volume resistivity | D257 | ohm·in | $2 \times 10^{16}$ |
| Surface resistivity | D257 | ohm | $8 \times 10^{16}$ |
| Dielectric strength (0.40 in) | D149 | V/mil | 730 |

*(continued)*

### TORLON 5030M *(continued)*

| Property | ASTM Test Method | English Units | |
|---|---|---|---|
| **General** | | | |
| Density | D792 | lb/in$^3$ | 0.057 |
| Hardness, Rockwell E | D785 | | 87 |
| Water absorption | D570 | % | 0.21 |

### TORLON 5030M

| Property | ASTM Test Method | Metric Units | |
|---|---|---|---|
| **Mechanical** | | | |
| Tensile strength | D1708 | N/mm$^2$ | |
| −196°C | | | |
| 23°C | | | 187 |
| 135°C | | | 108 |
| 232°C | | | 64 |
| Tensile elongation | D1708 | % | |
| −196°C | | | |
| 23°C | | | 6 |
| 135°C | | | 4 |
| 232°C | | | 5 |
| Tensile modulus | D1708 | 10$^3$ N/mm$^2$ | |
| 23°C | | | 10.5 |
| Flexural strength | D790 | N/mm$^2$ | |
| −196°C | | | |
| 23°C | | | 291 |
| 135°C | | | 184 |
| 232°C | | | 143 |
| Flexural modulus | D790 | 10$^3$ N/mm$^2$ | |
| −196°C | | | |
| 23°C | | | 11.5 |
| 135°C | | | 8.8 |
| 232°C | | | 8.9 |
| Compressive strength | D695 | N/mm$^2$ | 210 |
| Compressive modulus | D695 | 10$^3$ N/mm$^2$ | |
| Shear strength | D732 | N/mm$^2$ | |
| 23°C | | | 101 |
| Izod impact strength (3.2 mm) | D256 | J/m | |
| notched | | | 69 |
| unnotched | | | |
| Poisson's ratio | | | |

*(continued)*

**TORLON 5030M *(continued)***

| Property | ASTM Test Method | Metric Units | |
|---|---|---|---|
| **Thermal** | | | |
| Deflection temperature<br>1.82 $N/mm^2$ | D648 | °C | 273 |
| Coefficient of linear thermal expansion | D696 | $\mu$·m/m·K | 18.0 |
| Thermal conductivity | C177 | W/m·K | |
| Flammability<br>Underwriters Laboratories | | | 94 V-O |
| Limiting oxygen index | D2863 | % | 47 |
| **Electrical** | | | |
| Dielectric constant<br>$10^3$Hz<br>$10^5$Hz | D150 | | <br>4.1<br>4.3 |
| Dissipation factor<br>$10^3$Hz<br>$10^5$Hz | D150 | | <br>0.040<br>0.030 |
| Volume resistivity | D257 | ohm·m | $5 \times 10^{14}$ |
| Surface resistivity | D257 | ohm | $8 \times 10^{16}$ |
| Dielectric strength (1.02 mm) | D149 | kV/mm | 28.5 |
| **General** | | | |
| Density | D792 | $g/cm^3$ | 1.59 |
| Hardness, Rockwell E | D785 | | 87 |
| Water absorption | D570 | % | 0.21 |

# Amoco Chemicals Corp.

## Contact Address:

Amoco Chemicals Corp.
Engineering Resins
200 East Randolph Drive
Chicago, IL 60601

Phone: 1 (800) 621-4557

## TORLON 5430

### General Information:

TORLON engineering polymers are composites of polyamide-imide polymers alloyed with fluorocarbon polymers and reinforced with fillers to produce a family of high performance, high temperature molding resins. Products molded using TORLON have reliable performance at high temperature, and under high stress.

### Product Information:

TORLON 5430 has a polyamide-imide binder modified with 1% fluorocarbon; 30% glass fiber provides reinforcement; 4% $TiO_2$ improves the processibility of the molding composite.

Like TORLON 5030, TORLON 5430 provides high strength, high stiffness, and low creep. These properties are retained at elevated temperatures. The resin finds application in structural components, electrical devices, valve parts, and metal replacement products.

### Product Data:

**TORLON 5430**

| Property | ASTM Test Method | English Units | |
|---|---|---|---|
| **Mechanical** | | | |
| Tensile strength | D1708 | $10^3$ psi | |
| −321°F | | | |
| 73°F | | | 27.7 |
| 275°F | | | 19.0 |
| 450°F | | | 12.9 |
| Tensile elongation | D1708 | % | |
| −321°F | | | |
| 73°F | | | 12 |
| 275°F | | | 10 |
| 450°F | | | 8 |
| Tensile modulus | D1708 | $10^5$ psi | |
| 73°F | | | 15.8 |

*(continued)*

NOTE—Information provided in this book has been condensed. The publisher recommends that the product manufacturer be contacted, at the address or phone number provided, for more complete information on product safety, processing parameters, and applications not provided in this information sheet.

TORLON 5430 *(continued)*

| Property | ASTM Test Method | English Units | |
|---|---|---|---|
| **Mechanical** | | | |
| Flexural strength | D790 | $10^3$ psi | |
| −321 °F | | | |
| 73 °F | | | 39.7 |
| 275 °F | | | 29.7 |
| 450 °F | | | 20.7 |
| Flexural modulus | D790 | $10^5$ psi | |
| −321 °F | | | |
| 73 °F | | | 16.7 |
| 275 °F | | | 14.2 |
| 450 °F | | | 13.2 |
| Compressive strength | D695 | $10^3$ psi | 34.1 |
| Compressive modulus | D695 | $10^5$ psi | |
| Shear strength | D732 | $10^3$ psi | |
| 73 °F | | | 19.1 |
| Izod impact strength (1/8 in) | D256 | ft·lb/in | |
| notched | | | 1.3 |
| unnotched | | | 8.8 |
| Poisson's ratio | | | |
| **Thermal** | | | |
| Deflection temperature | D648 | °F | |
| 264 psi | | | 534 |
| Coefficient of linear thermal expansion | D696 | $10^{-6}$in/in·°F | 10 |
| Thermal conductivity | C177 | Btu·in/hr·ft² °F | |
| Flammability Underwriters Laboratories | | | 94 V-O |
| Limiting oxygen index | D2863 | % | 48 |
| **Electrical** | | | |
| Dielectric constant | D150 | | |
| $10^3$Hz | | | 4.2 |
| $10^5$Hz | | | 4.0 |
| Dissipation factor | D150 | | |
| $10^3$Hz | | | 0.020 |
| $10^5$Hz | | | 0.036 |
| Volume resistivity | D257 | ohm·in | $2 \times 10^{16}$ |
| Surface resistivity | D257 | ohm | $3 \times 10^{17}$ |
| Dielectric strength (0.40 in) | D149 | V/mil | 650 |

*(continued)*

TORLON 5430 *(continued)*

| Property | ASTM Test Method | English Units | |
|---|---|---|---|
| **General** | | | |
| Density | D792 | lb/in$^3$ | 0.059 |
| Hardness, Rockwell E | D785 | | 93 |
| Water absorption | D570 | % | 0.31 |

TORLON 5430

| Property | ASTM Test Method | Metric Units | |
|---|---|---|---|
| **Mechanical** | | | |
| Tensile strength | D1708 | N/mm$^2$ | |
| −196°C | | | |
| 23°C | | | 191 |
| 135°C | | | 131 |
| 232°C | | | 89 |
| Tensile elongation | D1708 | % | |
| −196°C | | | |
| 23°C | | | 12 |
| 135°C | | | 10 |
| 232°C | | | 8 |
| Tensile modulus | D1708 | $10^3$ N/mm$^2$ | |
| 23°C | | | 11.0 |
| Flexural strength | D790 | N/mm$^2$ | |
| −196°C | | | |
| 23°C | | | 278 |
| 135°C | | | 208 |
| 232°C | | | 145 |
| Flexural modulus | D790 | $10^3$ N/mm$^2$ | |
| −196°C | | | |
| 23°C | | | 11.5 |
| 135°C | | | 9.8 |
| 232°C | | | 9.1 |
| Compressive strength | D695 | N/mm$^2$ | 240 |
| Compressive modulus | D695 | $10^3$ N/mm$^2$ | |
| Shear strength | D732 | N/mm$^2$ | |
| 23°C | | | 133 |
| Izod impact strength (3.2 mm) | D256 | J/m | |
| notched | | | 69 |
| unnotched | | | 467 |
| Poisson's ratio | | | |

*(continued)*

TORLON 5430 *(continued)*

| Property | ASTM Test Method | Metric Units | |
|---|---|---|---|
| **Thermal** | | | |
| Deflection temperature<br>1.82 N/mm$^2$ | D648 | °C | 279 |
| Coefficient of linear thermal expansion | D696 | $\mu$·m/m·K | 18.0 |
| Thermal conductivity | C177 | W/m·K | |
| Flammability<br>Underwriters Laboratories | | | 94 V-O |
| Limiting oxygen index | D2863 | % | 48 |
| **Electrical** | | | |
| Dielectric constant<br>$10^3$Hz<br>$10^5$Hz | D150 | | 4.2<br>4.0 |
| Dissipation factor<br>$10^3$Hz<br>$10^5$Hz | D150 | | 0.020<br>0.036 |
| Volume resistivity | D257 | ohm·m | $5 \times 10^{14}$ |
| Surface resistivity | D257 | ohm | $3 \times 10^{17}$ |
| Dielectric strength (1.02 mm) | D149 | kV/mm | 25.4 |
| **General** | | | |
| Density | D792 | g/cm$^3$ | 1.62 |
| Hardness, Rockwell E | D785 | | 93 |
| Water absorption | D570 | % | 0.31 |

# Amoco Chemicals Corp.

## Contact Address:

Amoco Chemicals Corp.
Engineering Resins
200 East Randolph Drive
Chicago, IL 60601

Phone: 1 (800) 621-4557

## TORLON 7130

### General Information:

TORLON engineering polymers are composites of polyamide-imide polymers alloyed with fluorocarbon polymers and reinforced with fillers to produce a family of high performance, high temperature molding resins. Products molded using TORLON have reliable performance at high temperature, and under high stress.

### Product Information:

TORLON 7130 has a polyamide-imide binder system modified with 1% fluorocarbon, and reinforced with 30% graphite fiber.

This resin provides higher stiffness than TORLON 5030, and has the best high temperature stiffness properties and fatigue resistance in the TORLON family of resins. The product is electrically conductive, so is not a candidate for products where its electrical insulation properties are a factor.

TORLON 7130 finds application as a high strength replacement for metal in housings, linkages, gears, cargo rollers, valve parts, seals, and other applications where self lubrication and high wear are important. The resin has potential for EMI shielding.

### Product Data:

**TORLON 7130**

| Property | ASTM Test Method | English Units | |
|---|---|---|---|
| **Mechanical** | | | |
| Tensile strength | D1708 | $10^3$ psi | |
| −321°F | | | 22.8 |
| 73°F | | | 29.4 |
| 275°F | | | 22.8 |
| 450°F | | | 15.7 |
| Tensile elongation | D1708 | % | |
| −321°F | | | 3 |
| 73°F | | | 6 |
| 275°F | | | 14 |
| 450°F | | | 11 |

*(continued)*

NOTE—Information provided in this book has been condensed. The publisher recommends that the product manufacturer be contacted, at the address or phone number provided, for more complete information on product safety, processing parameters, and applications not provided in this information sheet.

TORLON 7130 *(continued)*

| Property | ASTM Test Method | English Units | |
|---|---|---|---|
| **Mechanical** | | | |
| Tensile modulus<br>73F | D1708 | $10^5$ psi | <br>32.2 |
| Flexural strength<br>−321°F<br>73°F<br>275°F<br>450°F | D790 | $10^3$ psi | <br>45.0<br>50.7<br>37.6<br>25.2 |
| Flexural modulus<br>−321°F<br>73°F<br>275°F<br>450°F | D790 | $10^5$ psi | <br>35.7<br>28.8<br>27.2<br>22.8 |
| Compressive strength | D695 | $10^3$ psi | 36.9 |
| Compressive modulus | D695 | $10^5$ psi | 14.3 |
| Shear strength<br>73°F | D732 | $10^3$ psi | <br>17.3 |
| Izod impact strength (1/8 in)<br>notched<br>unnotched | D256 | ft·lb/in | <br>0.9<br>6.4 |
| Poisson's ratio | | | 0.39 |
| **Thermal** | | | |
| Deflection temperature<br>264 psi | D648 | °F | <br>540 |
| Coefficient of linear thermal expansion | D696 | $10^{-6}$in/in·°F | 5 |
| Thermal conductivity | C177 | Btu·in/hr·ft²°F | 3.6 |
| Flammability<br>Underwriters Laboratories | | | 94 V-O |
| Limiting oxygen index | D2863 | % | 52 |
| **Electrical** | | | |
| Dielectric constant<br>$10^3$Hz<br>$10^5$Hz | D150 | | |
| Dissipation factor<br>$10^3$Hz<br>$10^5$Hz | D150 | | |
| Volume resistivity | D257 | ohm·in | |
| Surface resistivity | D257 | ohm | |
| Dielectric strength (0.40 in) | D149 | V/mil | |

*(continued)*

TORLON 7130 *(continued)*

| Property | ASTM Test Method | English Units | |
|---|---|---|---|
| **General** | | | |
| Density | D792 | $lb/in^3$ | 0.051 |
| Hardness, Rockwell E | D785 | | 94 |
| Water absorption | D570 | % | 0.26 |

TORLON 7130

| Property | ASTM Test Method | Metric Units | |
|---|---|---|---|
| **Mechanical** | | | |
| Tensile strength | D1708 | $N/mm^2$ | |
| −196°C | | | 158 |
| 23°C | | | 203 |
| 135°C | | | 158 |
| 232°C | | | 108 |
| Tensile elongation | D1708 | % | |
| −196°C | | | 3 |
| 23°C | | | 6 |
| 135°C | | | 14 |
| 232°C | | | 11 |
| Tensile modulus | D1708 | $10^3$ $N/mm^2$ | |
| 23°C | | | 22.3 |
| Flexural strength | D790 | $N/mm^2$ | |
| −196°C | | | 315 |
| 23°C | | | 355 |
| 135°C | | | 263 |
| 232°C | | | 177 |
| Flexural modulus | D790 | $10^3$ $N/mm^2$ | |
| −196°C | | | 24.6 |
| 23°C | | | 19.9 |
| 135°C | | | 18.8 |
| 232°C | | | 15.7 |
| Compressive strength | D695 | $N/mm^2$ | 250 |
| Compressive modulus | D695 | $10^3$ $N/mm^2$ | 9.9 |
| Shear strength | D732 | $N/mm^2$ | |
| 23°C | | | 120 |
| Izod impact strength (3.2 mm) | D256 | J/m | |
| notched | | | 47 |
| unnotched | | | 340 |
| Poisson's ratio | | | 0.39 |

*(continued)*

## TORLON 7130 *(continued)*

| Property | ASTM Test Method | Metric Units | |
|---|---|---|---|
| **Thermal** | | | |
| Deflection temperature<br>1.82 N/mm² | D648 | °C | 282 |
| Coefficient of linear thermal expansion | D696 | $\mu$·m/m·K | 9.0 |
| Thermal conductivity | C177 | W/m·K | 0.53 |
| Flammability<br>Underwriters Laboratories | | | 94 V-O |
| Limiting oxygen index | D2863 | % | 52 |
| **Electrical** | | | |
| Dielectric constant<br>$10^3$Hz<br>$10^5$Hz | D150 | | |
| Dissipation factor<br>$10^3$Hz<br>$10^5$Hz | D150 | | |
| Volume resistivity | D257 | ohm·m | |
| Surface resistivity | D257 | ohm | |
| Dielectric strength (1.02 mm) | D149 | kV/mm | |
| **General** | | | |
| Density | D792 | g/cm³ | 1.42 |
| Hardness, Rockwell E | D785 | | 94 |
| Water absorption | D570 | % | 0.26 |

# Amoco Chemicals Corp.

## Contact Address:

Amoco Chemicals Corp.
Engineering Resins
200 East Randolph Drive
Chicago, IL 60601

Phone: 1 (800) 621-4557

## TORLON 7233

### General Information:

TORLON engineering polymers are composites of polyamide-imide polymers alloyed with fluorocarbon polymers and reinforced with fillers to produce a family of high performance, high temperature molding resins. Products molded using TORLON have reliable performance at high temperature, and under high stress.

### Product Information:

TORLON 7233 has polyamide-imide binder modified with 1% fluorocarbon and 33% graphite fiber reinforcement. This modification of TORLON 7130 is a better candidate for moldings having thick sections, but the added reinfocement reduces mechanical properties somewhat. Like TORLON 7130, the product is electrically conductive.

TORLON 7233 is selected when parts having thick sections cause molding problems, and optimum physical properties are of secondary importance.

### Product Data:

**TORLON 7233**

| Property | ASTM Test Method | English Units | |
|---|---|---|---|
| **Mechanical** | | | |
| Tensile strength | D1708 | $10^3$ psi | |
| −321°F | | | |
| 73°F | | | 19.8 |
| 275°F | | | |
| 450°F | | | 10.3 |
| Tensile elongation | D1708 | % | |
| −321°F | | | |
| 73°F | | | 22 |
| 275°F | | | |
| 450°F | | | |

*(continued)*

NOTE—Information provided in this book has been condensed. The publisher recommends that the product manufacturer be contacted, at the address or phone number provided, for more complete information on product safety, processing parameters, and applications not provided in this information sheet.

## TORLON 7233 *(continued)*

| Property | ASTM Test Method | English Units | |
|---|---|---|---|
| **Mechanical** | | | |
| Tensile modulus | D1708 | $10^5$ psi | |
| 73F | | | 25.8 |
| Flexural strength | D790 | $10^3$ psi | |
| −321°F | | | |
| 73°F | | | |
| 275°F | | | |
| 450°F | | | |
| Flexural modulus | D790 | $10^5$ psi | |
| −321°F | | | |
| 73°F | | | 23.5 |
| 275°F | | | |
| 450°F | | | 18.7 |
| Compressive strength | D695 | $10^3$ psi | |
| Compressive modulus | D695 | $10^5$ psi | |
| Shear strength | D732 | $10^3$ psi | |
| 73°F | | | 12.6 |
| Izod impact strength (1/8 in) | D256 | ft·lb/in | |
| notched | | | 1.1 |
| unnotched | | | |
| Poisson's ratio | | | |
| **Thermal** | | | |
| Deflection temperature | D648 | °F | |
| 264 psi | | | 530 |
| Coefficient of linear thermal expansion | D696 | $10^{-6}$in/in·°F | 6 |
| Thermal conductivity | C177 | Btu·in/hr·ft$^2$°F | 1.2 |
| Flammability Underwriters Laboratories | | | 94 V-O |
| Limiting oxygen index | D2863 | % | |
| **Electrical** | | | |
| Dielectric constant | D150 | | |
| $10^3$Hz | | | |
| $10^5$Hz | | | |
| Dissipation factor | D150 | | |
| $10^3$Hz | | | |
| $10^5$Hz | | | |
| Volume resistivity | D257 | ohm·in | |
| Surface resistivity | D257 | ohm | |
| Dielectric strength (0.40 in) | D149 | V/mil | |

*(continued)*

TORLON 7233 *(continued)*

| Property | ASTM Test Method | English Units | |
|---|---|---|---|
| **General** | | | |
| Density | D792 | lb/in³ | |
| Hardness, Rockwell E | D785 | | 92 |
| Water absorption | D570 | % | |

TORLON 7233

| Property | ASTM Test Method | Metric Units | |
|---|---|---|---|
| **Mechanical** | | | |
| Tensile strength | D1708 | N/mm² | |
| −196°C | | | |
| 23°C | | | 137 |
| 135°C | | | |
| 232°C | | | 71 |
| Tensile elongation | D1708 | % | |
| −196°C | | | |
| 23°C | | | 22 |
| 135°C | | | |
| 232°C | | | |
| Tensile modulus | D1708 | 10³ N/mm² | |
| 23°C | | | 17.9 |
| Flexural strength | D790 | N/mm² | |
| −196°C | | | |
| 23°C | | | |
| 135°C | | | |
| 232°C | | | |
| Flexural modulus | D790 | 10³ N/mm² | |
| −196°C | | | |
| 23°C | | | 16.2 |
| 135°C | | | |
| 232°C | | | 12.9 |
| Compressive strength | D695 | N/mm² | |
| Compressive modulus | D695 | 10³ N/mm² | |
| Shear strength | D732 | N/mm² | |
| 23°C | | | 87 |
| Izod impact strength (3.2 mm) | D256 | J/m | |
| notched | | | 58 |
| unnotched | | | |
| Poisson's ratio | | | |

*(continued)*

## TORLON 7233 *(continued)*

| Property | ASTM Test Method | Metric Units | |
|---|---|---|---|
| **Thermal** | | | |
| Deflection temperature<br>1.82 $N/mm^2$ | D648 | °C | 277 |
| Coefficient of linear thermal expansion | D696 | $\mu$·m/m·K | 11.0 |
| Thermal conductivity | C177 | W/m·K | 0.17 |
| Flammability<br>Underwriters Laboratories | | | 94 V-O |
| Limiting oxygen index | D2863 | % | |
| **Electrical** | | | |
| Dielectric constant<br>$10^3$Hz<br>$10^5$Hz | D150 | | |
| Dissipation factor<br>$10^3$Hz<br>$10^5$Hz | D150 | | |
| Volume resistivity | D257 | ohm·m | |
| Surface resistivity | D257 | ohm | |
| Dielectric strength (1.02 mm) | D149 | kV/mm | |
| **General** | | | |
| Density | D792 | $g/cm^3$ | |
| Hardness, Rockwell E | D785 | | 92 |
| Water absorption | D570 | % | |

# Amoco Chemicals Corp.

## Contact Address:

Amoco Chemicals Corp.
Engineering Resins
200 East Randolph Drive
Chicago, IL 60601

Phone: 1 (800) 621-4557

## TORLON 7330

### General Information:

TORLON engineering polymers are composites of polyamide-imide polymers alloyed with fluorocarbon polymers and reinforced with fillers to produce a family of high performance, high temperature molding resins. Products molded using TORLON have reliable performance at high temperature, and under high stress.

### Product Information:

TORLON 7330 is a proprietary blend of polyamide-imide polymer, fluorocarbon, and carbon fiber reinforcement. The product provides high stiffness and good lubricity. It finds applications that combine stiffness, wear, and lubricity, such as sliding vanes and wear surfaces. The product has potential for EMI shielding.

### Product Data:

**TORLON 7330**

| Property | ASTM Test Method | English Units | |
|---|---|---|---|
| **Mechanical** | | | |
| Tensile strength | D1708 | $10^3$ psi | |
| −321°F | | | |
| 73°F | | | 26.0 |
| 275°F | | | |
| 450°F | | | |
| Tensile elongation | D1708 | % | |
| −321°F | | | |
| 73°F | | | 6 |
| 275°F | | | |
| 450°F | | | |
| Tensile modulus | D1708 | $10^5$ psi | |
| 73°F | | | |

*(continued)*

NOTE—Information provided in this book has been condensed. The publisher recommends that the product manufacturer be contacted, at the address or phone number provided, for more complete information on product safety, processing parameters, and applications not provided in this information sheet.

TORLON 7330 *(continued)*

| Property | ASTM Test Method | English Units | |
|---|---|---|---|
| **Mechanical** | | | |
| Flexural strength | D790 | $10^3$ psi | |
| −321°F | | | |
| 73°F | | | 40.1 |
| 275°F | | | 31.6 |
| 450°F | | | 23.0 |
| Flexural modulus | D790 | $10^5$ psi | |
| −321°F | | | |
| 73°F | | | 24.4 |
| 275°F | | | 20.0 |
| 450°F | | | 18.7 |
| Compressive strength | D695 | $10^3$ psi | |
| Compressive modulus | D695 | $10^5$ psi | |
| Shear strength | D732 | $10^3$ psi | |
| 73°F | | | |
| Izod impact strength (1/8 in) | D256 | ft·lb/in | |
| notched | | | 1.0 |
| unnotched | | | |
| Poisson's ratio | | | |
| **Thermal** | | | |
| Deflection temperature | D648 | °F | |
| 264 psi | | | 534 |
| Coefficient of linear thermal expansion | D696 | $10^{-6}$in/in·°F | 7 |
| Thermal conductivity | C177 | Btu·in/hr·ft²°F | |
| Flammability Underwriters Laboratories | | | 94 V-O |
| Limiting oxygen index | D2863 | % | |
| **Electrical** | | | |
| Dielectric constant | D150 | | |
| $10^3$Hz | | | |
| $10^5$Hz | | | |
| Dissipation factor | D150 | | |
| $10^3$Hz | | | |
| $10^5$Hz | | | |
| Volume resistivity | D257 | ohm·in | |
| Surface resistivity | D257 | ohm | |
| Dielectric strength (0.40 in) | D149 | V/mil | |

*(continued)*

**TORLON 7330 *(continued)***

| Property | ASTM Test Method | English Units | |
|---|---|---|---|
| **General** | | | |
| Density | D792 | lb/in³ | 0.054 |
| Hardness, Rockwell E | D785 | | |
| Water absorption | D570 | % | |

**TORLON 7330**

| Property | ASTM Test Method | Metric Units | |
|---|---|---|---|
| **Mechanical** | | | |
| Tensile strength | D1708 | N/mm² | |
| −196°C | | | |
| 23°C | | | 180 |
| 135°C | | | |
| 232°C | | | |
| Tensile elongation | D1708 | % | |
| −196°C | | | |
| 23°C | | | 6 |
| 135°C | | | |
| 232°C | | | |
| Tensile modulus | D1708 | 10³ N/mm² | |
| 23°C | | | |
| Flexural strength | D790 | N/mm² | |
| −196°C | | | |
| 23°C | | | 281 |
| 135°C | | | 221 |
| 232°C | | | 161 |
| Flexural modulus | D790 | 10³ N/mm² | |
| −196°C | | | |
| 23°C | | | 16.8 |
| 135°C | | | 13.8 |
| 232°C | | | 12.9 |
| Compressive strength | D695 | N/mm² | |
| Compressive modulus | D695 | 10³ N/mm² | |
| Shear strength | D732 | N/mm² | |
| 23°C | | | |
| Izod impact strength (3.2 mm) | D256 | J/m | |
| notched | | | 53 |
| unnotched | | | |
| Poisson's ratio | | | |

*(continued)*

**TORLON 7330 *(continued)***

| Property | ASTM Test Method | Metric Units | |
|---|---|---|---|
| **Thermal** | | | |
| Deflection temperature<br>1.82 $N/mm^2$ | D648 | °C | 279 |
| Coefficient of linear thermal expansion | D696 | $\mu$·m/m·K | 12.6 |
| Thermal conductivity | C177 | W/m·K | |
| Flammability<br>Underwriters Laboratories | | | 94 V-O |
| Limiting oxygen index | D2863 | % | |
| **Electrical** | | | |
| Dielectric constant<br>$10^3$Hz<br>$10^5$Hz | D150 | | |
| Dissipation factor<br>$10^3$Hz<br>$10^5$Hz | D150 | | |
| Volume resistivity | D257 | ohm·m | |
| Surface resistivity | D257 | ohm | |
| Dielectric strength (1.02 mm) | D149 | kV/mm | |
| **General** | | | |
| Density | D792 | $g/cm^3$ | 1.50 |
| Hardness, Rockwell E | D785 | | |
| Water absorption | D570 | % | |

# Amoco Chemicals Corp.

## Contact Address:

Amoco Chemicals Corp.
Engineering Resins
200 East Randolph Drive
Chicago, IL 60601

Phone: 1 (800) 621-4557

# TORLON 9040

## General Information:

TORLON engineering polymers are composites of polyamide-imide polymers alloyed with fluorocarbon polymers and reinforced with fillers to produce a family of high performance, high temperature molding resins. Products molded using TORLON have reliable performance at high temperature, and under high stress.

## Product Information:

TORLON 9040 has a polyamide-imide binder reinforced with 40% glass fiber. This composite is designed for applications where economics is important, and where optimum physical properties are not essential. Similar to TORLON 5030, this resin finds application in switch parts, relay components, terminal strips, wear bands, back up rings, component housings, impellers, brackets, and thermal insulators.

## Product Data:

**TORLON 9040**

| Property | ASTM Test Method | English Units | |
|---|---|---|---|
| **Mechanical** | | | |
| Tensile strength | D1708 | $10^3$ psi | |
| −321°F | | | |
| 73°F | | | 31.8 |
| 275°F | | | 25.0 |
| 450°F | | | 19.9 |
| Tensile elongation | D1708 | % | |
| −321°F | | | |
| 73°F | | | 7 |
| 275°F | | | 6 |
| 450°F | | | 8 |
| Tensile modulus | D1708 | $10^5$ psi | |
| 73°F | | | 20.2 |

*(continued)*

NOTE—Information provided in this book has been condensed. The publisher recommends that the product manufacturer be contacted, at the address or phone number provided, for more complete information on product safety, processing parameters, and applications not provided in this information sheet.

## TORLON 9040 *(continued)*

| Property | ASTM Test Method | English Units | |
|---|---|---|---|
| **Mechanical** | | | |
| Flexural strength | D790 | $10^3$ psi | |
| −321°F | | | |
| 73°F | | | 52.0 |
| 275°F | | | 43.6 |
| 450°F | | | 38.9 |
| Flexural modulus | D790 | $10^5$ psi | |
| −321°F | | | |
| 73°F | | | 21.0 |
| 275°F | | | 20.0 |
| 450°F | | | 20.0 |
| Compressive strength | D695 | $10^3$ psi | |
| Compressive modulus | D695 | $10^5$ psi | |
| Shear strength | D732 | $10^3$ psi | |
| 73°F | | | 22.7 |
| Izod impact strength (1/8 in) | D256 | ft·lb/in | |
| notched | | | 1.5 |
| unnotched | | | 1.8 |
| Poisson's ratio | | | |
| **Thermal** | | | |
| Deflection temperature | D648 | °F | |
| 264 psi | | | 536 |
| Coefficient of linear thermal expansion | D696 | $10^{-6}$in/in·°F | 7 |
| Thermal conductivity | C177 | Btu·in/hr·ft²°F | |
| Flammability Underwriters Laboratories | | | 94 V-O |
| Limiting oxygen index | D2863 | % | 47 |
| **Electrical** | | | |
| Dielectric constant | D150 | | |
| $10^3$Hz | | | 4.3 |
| $10^6$Hz | | | 4.6 |
| Dissipation factor | D150 | | |
| $10^3$Hz | | | 0.040 |
| $10^6$Hz | | | 0.028 |
| Volume resistivity | D257 | ohm·in | $2 \times 10^{16}$ |
| Surface resistivity | D257 | ohm | $9 \times 10^{17}$ |
| Dielectric strength (0.40 in) | D149 | V/mil | 490 |

*(continued)*

TORLON 9040 *(continued)*

| Property | ASTM Test Method | English Units | |
|---|---|---|---|
| **General** | | | |
| Density | D792 | lb/in$^3$ | 0.060 |
| Hardness, Rockwell E | D785 | | 107 |
| Water absorption | D570 | % | 0.21 |

TORLON 9040

| Property | ASTM Test Method | Metric Units | |
|---|---|---|---|
| **Mechanical** | | | |
| Tensile strength | D1708 | N/mm$^2$ | |
| −196°C | | | |
| 23°C | | | 220 |
| 135°C | | | 172 |
| 232°C | | | 137 |
| Tensile elongation | D1708 | % | |
| −196°C | | | |
| 23°C | | | 7 |
| 135°C | | | 6 |
| 232°C | | | 8 |
| Tensile modulus | D1708 | 10$^3$ N/mm$^2$ | |
| 23°C | | | 13.9 |
| Flexural strength | D790 | N/mm$^2$ | |
| −196°C | | | |
| 23°C | | | 364 |
| 135°C | | | 298 |
| 232°C | | | 279 |
| Flexural modulus | D790 | 10$^3$ N/mm$^2$ | |
| −196°C | | | |
| 23°C | | | 14.5 |
| 135°C | | | 13.8 |
| 232°C | | | 13.8 |
| Compressive strength | D695 | N/mm$^2$ | |
| Compressive modulus | D695 | 10$^3$ N/mm$^2$ | |
| Shear strength | D732 | N/mm$^2$ | |
| 23°C | | | 150 |
| Izod impact strength (3.2 mm) | D256 | J/m | |
| notched | | | 79 |
| unnotched | | | 97 |
| Poisson's ratio | | | |

*(continued)*

TORLON 9040 *(continued)*

| Property | ASTM Test Method | Metric Units | |
|---|---|---|---|
| **Thermal** | | | |
| Deflection temperature<br>1.82 N/mm² | D648 | °C | 280 |
| Coefficient of linear thermal expansion | D696 | μ·m/m·K | 12.6 |
| Thermal conductivity | C177 | W/m·K | |
| Flammability<br>Underwriters Laboratories | | | 94 V-O |
| Limiting oxygen index | D2863 | % | 47 |
| **Electrical** | | | |
| Dielectric constant<br>$10^3$Hz<br>$10^5$Hz | D150 | | <br>4.3<br>4.6 |
| Dissipation factor<br>$10^3$Hz<br>$10^5$Hz | D150 | | <br>0.040<br>0.028 |
| Volume resistivity | D257 | ohm·mn | $5 \times 10^{14}$ |
| Surface resistivity | D257 | ohm | $9 \times 10^{17}$ |
| Dielectric strength (1.02 mm) | D149 | kV/mm | 19.5 |
| **General** | | | |
| Density | D792 | g/cm³ | 1.66 |
| Hardness, Rockwell E | D785 | | 107 |
| Water absorption | D570 | % | 0.21 |

# ARCO Chemical Company

Contact Address:

ARCO Chemical Company
3801 West Chester Pike
Newtown Square, PA 19073

Phone: 1 (800) 354-1500

## ARLOY® 1000 GRADE

Product Information:

ARLOY® Engineering Resins are blends of DYLARK® Engineering Resins and polycarbonate. The three resin compositions—1000, 1100 and 1200 have an excellent balance of physical properties, superior toughness, and high heat resistance, yet are easy to process.

Products which have been molded using ARLOY resins include the following: one-piece automobile instrument panels, skuff plates, seat belt retractor housings, speaker grills, tool housings, camera components, hospital feeding trays and surgical appliances.

ARLOY resins retain their excellent physical properties after extended or repeated high temperature exposure, such as sterilizing autoclaves. They also retain cold temperature impact strength.

ARLOY grades can be injection molded, extruded, and thermoformed. Molded products can be decorated by hot stamping, painting, and vacuum metalizing. Molded component parts can be joined by adhesives, selected solvents, and sonic welding processes.

Product Data:

**ARLOY® 1000 GRADE**

| Property | ASTM Method | |
|---|---|---|
| **Mechanical** | | |
| Gardner Impact | | |
| N-m @ 23°C/73°F | | >54 |
| @ −29°C/−20°F | | 53 |
| In-lbs @ 23°C/73°C | | >480 |
| @ −29°C/−20°F | | 470 |
| Notched Izod Impact | | |
| J/m @ 23°C/73°F | D256 | 640 |
| @ −29°C/−20°F | | 320 |
| Ft-lbs/in @ 23°C/73°F | | 12 |
| @ −29°C/−20°F | | 6 |
| Tensile Strength, MPa | D638 | 45 |
| psi | | 6,600 |

*(continued)*

NOTE—Information provided in this book has been condensed. The publisher recommends that the product manufacturer be contacted, at the address or phone number provided, for more complete information on product safety, processing parameters, and applications not provided in this information sheet.

ARLOY® 1000 GRADE *(continued)*

| Property | ASTM Method | |
|---|---|---|
| **Mechanical** | | |
| Elongation, % | D638 | 80 |
| Flexural Strength, MPa<br>psi | D790 | 83<br>12,300 |
| Flexural Modulus, MPa<br>psi | D790 | 2,200<br>334,000 |
| **Thermal** | | |
| Deflection Temperature Under Load<br>°C @ 18.6 kg/mc[2]<br>3.2 mm bar, unconditioned<br>3.2 mm bar, conditioned[1]<br>12.7 mm bar, unconditioned<br>12.7 mm bar, conditioned[1]<br>°F @ 264 psi<br>1/8 inch bar, unconditioned<br>1/8 inch bar, conditioned[1]<br>1/2 inch bar, unconditioned<br>1/2 inch bar, conditioned[1] | D648 | <br><br>108<br>114<br>115<br>123<br><br>228<br>248<br>240<br>254 |
| Vicat Softening Point, °C<br>°F | D1525 | 138<br>280 |
| **General** | | |
| Melt Flow, g/10 min. Condition "L" | D1238 | 1.0 |
| Specific Gravity, 23/23°C | D792 | 1.12 |
| Mold Shrinkage, in/in | D955 | 0.007 |
| Water Absorption, % in 24 hrs. | D570 | 0.20 |
| Visible Light Transmission, % | D1003 | Opaque |

[1]Bars are conditioned for 48 hours @ 90°C (194°F).

# ARCO Chemical Company

Contact Address:

ARCO Chemical Company
3801 West Chester Pike
Newtown Square, PA 19073

Phone: 1 (800) 354-1500

## ARLOY® 1100 GRADE

Description:

ARLOY® Engineering Resins are blends of DYLARK® Engineering Resins copolymers and polycarbonate. The three resin compositions—1000, 1100 and 1200 have an excellent balance of physical properties, superior toughness, and high heat resistance, yet are easy to process.

Products which have been molded using ARLOY resins include the following: one-piece automobile instrument panels, skuff plates, seat belt retractor housings, speaker grills, tool housings, camera components, hospital feeding trays and surgical appliances.

ARLOY resins retain their excellent physical properties after extended or repeated high temperature exposure, such as sterilizing autoclaves. They also retain cold temperature impact strength.

ARLOY grades can be injection molded, extruded, and thermoformed. Molded products can be decorated by hot stamping, painting, and vacuum metalizing. Molded component parts can be joined by adhesives, selected solvents, and sonic welding processes.

Product Data:

**ARLOY® 1100 GRADE**

| | ASTM Method | |
|---|---|---|
| **Mechanical** | | |
| Gardner Impact | | |
| N-m @ 23°C/73°F | | >54 |
| @ −29°C/−20°F | | 52 |
| In-lbs @ 23°C/73°C | | >480 |
| @ −29°C/−20°F | | 460 |
| Notched Izod Impact | | |
| J/m @ 23°C/73°F | D256 | 640 |
| @ −29°C/−20°F | | 530 |
| Ft-lbs/in @ 23°C/73°F | | 12 |
| @ −29°C/−20°F | | 10 |
| Tensile Strength, MPa | D638 | 48 |
| psi | | 7,000 |

*(continued)*

NOTE–Information provided in this book has been condensed. The publisher recommends that the product manufacturer be contacted, at the address or phone number provided, for more complete information on product safety, processing parameters, and applications not provided in this information sheet.

**ARLOY® 1100 GRADE *(continued)***

| Property | ASTM Method | |
|---|---|---|
| **Mechanical** | | |
| Elongation, % | D638 | 110 |
| Flexural Strength, MPa<br>psi | D790 | 90<br>13,000 |
| Flexural Modulus, MPa<br>psi | D790 | 2,300<br>344,000 |
| **Thermal** | | |
| Deflection Temperature Under Load | D648 | |
| °C @ 18.6 kg/mc² | | |
| 3.2 mm bar, unconditioned | | 113 |
| 3.2 mm bar, conditioned[1] | | 120 |
| 12.7 mm bar, unconditioned | | 122 |
| 12.7 mm bar, conditioned[1] | | 129 |
| °F @ 264 psi | | |
| 1/8 inch bar, unconditioned | | 235 |
| 1/8 inch bar, conditioned[1] | | 257 |
| 1/2 inch bar, unconditioned | | 252 |
| 1/2 inch bar, conditioned[1] | | 264 |
| Vicat Softening Point, °C<br>°F | D1525 | 143<br>290 |
| **General** | | |
| Melt Flow, g/10 min. Condition "L" | D1238 | 1.0 |
| Specific Gravity, 23/23°C | D792 | 1.13 |
| Mold Shrinkage, in/in | D955 | 0.007 |
| Water Absorption, % in 24 hrs. | D570 | 0.20 |
| Visible Light Transmission, % | D1003 | Opaque |

[1]Bars are conditioned for 48 hours @ 90°C (194°F).

# ARCO Chemical Company

## Contact Address:

ARCO Chemical Company
3801 West Chester Pike
Newtown Square, PA 19073

Phone: 1 (800) 354-1500

## ARLOY 1200

### Description:

ARLOY® Engineering Resins are blends of DYLARK® Engineering Resins and polycarbonate. The three resin compositions—1000, 1100 and 1200 have an excellent balance of physical properties, superior toughness, and high heat resistance, yet are easy to process.

Products which have been molded using ARLOY resins include the following: one-piece automobile instrument panels, skuff plates, seat belt retractor housings, speaker grills, tool housings, camera components, hospital feeding trays and surgical appliances.

ARLOY resins retain their excellent physical properties after extended or repeated high temperature exposure, such as sterilizing autoclaves. They also retain cold temperature impact strength.

ARLOY grades can be injection molded, extruded, and thermoformed. Molded products can be decorated by hot stamping, painting, and vacuum metalizing. Molded component parts can be joined by adhesives, selected solvents, and sonic welding processes.

### Product Data:

**ARLOY® 1200 GRADE**

| | ASTM Method | |
|---|---|---|
| **Mechanical** | | |
| Gardner Impact | | |
| N-m @ 23°C/73°F | | 54 |
| @ −29°C/−20°F | | 25 |
| In-lbs @ 23°C/73°C | | >480 |
| @ −29°C/−20°F | | 220 |
| Notched Izod Impact | | |
| J/m @ 23°C/73°F | D256 | 530 |
| @ −29°C/−20°F | | 160 |
| Ft-lbs/in @ 23°C/73°F | | 13 |
| @ −29°C/−20°F | | 3 |
| Tensile Strength, MPa | D638 | 55 |
| psi | | 7,700 |

*(continued)*

NOTE – Information provided in this book has been condensed. The publisher recommends that the product manufacturer be contacted, at the address or phone number provided, for more complete information on product safety, processing parameters, and applications not provided in this information sheet.

## ARLOY® 1200 GRADE *(continued)*

| Property | ASTM Method | |
|---|---|---|
| **Mechanical** | | |
| Elongation, % | D638 | 100 |
| Flexural Strength, MPa<br>psi | D790 | 103<br>15,000 |
| Flexural Modulus, MPa<br>psi | D790 | 2,700<br>390,000 |
| **Thermal** | | |
| Deflection Temperature Under Load | D648 | |
| °C @ 18.6 kg/mc² | | |
| 3.2 mm bar, unconditioned | | 117 |
| 3.2 mm bar, conditioned[1] | | 123 |
| 12.7 mm bar, unconditioned | | 126 |
| 12.7 mm bar, conditioned[1] | | 131 |
| °F @ 264 psi | | |
| 1/8 inch bar, unconditioned | | 241 |
| 1/8 inch bar, conditioned[1] | | 261 |
| 1/2 inch bar, unconditioned | | 259 |
| 1/2 inch bar, conditioned[1] | | 267 |
| Vicat Softening Point, °C<br>°F | D1525 | 143<br>290 |
| **General** | | |
| Melt Flow, g/10 min. Condition "L" | D1238 | 1.1 |
| Specific Gravity, 23/23°C | D792 | 1.15 |
| Mold Shrinkage, in/in | D955 | 0.007 |
| Water Absorption, % in 24 hrs. | D570 | 0.20 |
| Visible Light Transmission, % | D1003 | Opaque |

[1]Bars are conditioned for 48 hours @ 90°C (194°F).

# Azdel, Inc.

## Contact Address:

Azdel, Inc.
General Electric Co.
One Plastics Ave.
Pittsfield, MA 01201

Phone: 1 (413) 965-4973

# AZDEL

## Description:

AZDEL is a unique product. The company manufacturing AZDEL is a joint venture for the General Electric Company and PPG Industries. The product is sold in sheet form.

The product, a glass reinforced thermoplastic polyester alloy composite, is sold in sheet form, or die-cut preforms for compression molding of large parts, primarily for automotive applications. Azdel, described in company literature as a "technopolymer" is compression molded. Preforms of the composite sheet are cut as required by the cavity configuration, pre-heated as a flat sheet, and compression molded to shape or form a contained panel, such as a car body panel. The part, depending on part geometry and size, can be compression molded in 20 to 50 seconds. There is no scrap loss. No post cure is required.

The parts are 30% lighter than equivalent SMC parts. Regrind from die-cut frame scrap is 100% reusable. It can be painted directly without priming, and can be made resistant to ultra-violet light without painting. Having high strength, it is not subject to damage in shipment. Adhesive bonding and mechanical fastening of components presents no problem.

## Product Data:

**Thermoplastic Polyester Composite Sheet**

| Property | ASTM Method | English Units | SI Units |
|---|---|---|---|
| **Mechanical** | | | |
| Tensile Strength | D638 | 15,000 psi | 103 N/mm² |
| Tensile Elongation @ Break | D638 | 2.1% | 2.1% |
| Flexural Strength | D790 | 32,000 psi | 221 N/mm² |
| Flexural Modulus | D790 | 1,200,000 psi | 8274 N/mm² |
| Izod Impact Strength, Notched @ 73°F (23°C) 1/8″ (3.2mm) | D256 | 13.0 ft-lb/in | 699 J/m |
| Falling Dart Impact Strength (1½″ Dia Dart) | | 130 ft-lb/in | 6990 J/m |

*(continued)*

NOTE—Information provided in this book has been condensed. The publisher recommends that the product manufacturer be contacted, at the address or phone number provided, for more complete information on product safety, processing parameters, and applications not provided in this information sheet.

**Thermoplastic Polyester Composite Sheet *(continued)***

| Property | ASTM Method | English Units | SI Units |
|---|---|---|---|
| **Thermal** | | | |
| Heat Deflection Temperature @ 264 psi (1.82 N/mm²) | D648 | 425°F | 218°C |
| Coefficient of Thermal Expansion, −40 to 400°F (−40 to 204°C) | D696 | $2.0 \times 10^{-5}$ in/in-°F | $3.6 \times 10^{-5}$ m/m-°C |
| **Physical** | | | |
| Specific Gravity | D792 | 1.59 | 1.59 |
| Glass Content (% by weight) | | 35% | 35% |

**AZDEL® P-100 (Natural)**
**AZDEL® P-110 (Grey Black)**
**AZDEL® P-112 (Black)**

| Property | ASTM Method | English Units | SI Units |
|---|---|---|---|
| **Mechanical** | | | |
| Tensile Strength | D638 | 14,000 psi | 97 N/mm² |
| Tensile Elongation @ Break | D638 | 2.1% | 2.1% |
| Flexural Strength | D790 | 24,000 psi | 165 N/mm² |
| Flexural Modulus | D790 | 800,000 psi | 5520 N/mm² |
| Izod Impact Strength, Notched @ 73°F (23°C) 1/8″ (3.2mm) | D256 | 14.0 ft-lb/in | 752 J/m |
| Falling Dart Impact Strength (1½″ Dia Dart) | | 165 ft-lb/in | 8870 J/m |
| **Thermal** | | | |
| Heat Deflection Temperature @ 264 psi (1.82 N/mm²) | D648 | 310°F | 154°C |
| Coefficient of Thermal Expansion, −40 to 200°F (−40 to 293°C) | D696 | $1.5 \times 10^{-5}$ in/in-°F | $2.7 \times 10^{-5}$ m/m-°C |
| **Physical** | | | |
| Specific Gravity | D792 | 1.19 | 1.19 |
| Glass Fiber Content (% by weight) | | 40% | 40% |

# BASF Corporation

## Contact Address:

BASF Corporation
P.O. Box 405
Bridgeport, NJ 08014

Phone: 1 (609) 467-2456

## TERBLEND S KR 2861/1

### Product Information:

TERBLEND S KR 2861/1 (formerly LURAN® S KR 2861) resin is a blend of thermoplastics, viz. LURAN S ASA and a polycarbonate, and is equally suitable for injection molding and extrusion.

Benefits that it offers are very high resistance to heat deformation, outstanding toughness and good flow. It also retains another advantage of LURAN S ASA, viz. its resistance to yellowing and outdoor exposure. TERBLEND S KR 2861/1 moldings display good rigidity and hardness and are dimensionally stable and antistatic.

TERBLEND S KR 2861/1 is tougher and has even greater resistance to heat deformation than a high-heat ASA molding compound; and it has decidedly better resistance to outdoor exposure and yellowing than a high-heat ABS polymer.

### Processing:

TERBLEND S KR 2861/1 can be processed on all the injection molding machines and extruders normally encountered on the market.

TERBLEND S KR 2861/1 sheet can be readily thermoformed. Beforehand, precautions should be taken to prevent the absorption of moisture, similar to those adopted for ABS and ASA.

### Product Data:

**TERBLEND™ S KR 2861/1 ASA/PC Blend**

| Property | Test Method | Unit |
|---|---|---|
| **Mechanical** | | |
| Tensile strength<br>at yield<br>at break | D638 | <br>7500 psi<br>7500 psi |
| Elongation<br>at yield<br>at break | D638 | <br>5 %<br>100 % |

*(continued)*

NOTE – Information provided in this book has been condensed. The publisher recommends that the product manufacturer be contacted, at the address or phone number provided, for more complete information on product safety, processing parameters, and applications not provided in this information sheet.

**TERBLEND™ S KR 2861/1 ASA/PC Blend *(continued)***

| Property | Test Method | Unit |
|---|---|---|
| **Mechanical** | | |
| Flexural modulus | D790 | 320,000 psi |
| Izod notched impact strength (specimen thickness 1/8 in.) | D256 | 11.2 ft. lb/in. |
| Rockwell Hardness | D785 | R109 |
| **Thermal** | | |
| Vicat softening temperature, VST/A | D1525 | 262°F |
| Deflection temperature | | |
| 264 psi Method A | D648 | 226°F |
| 66 psi Method B | D648 | 260°F |
| Maximum permissible service temperature | | 230°F |
| Thermal coefficient of linear expansion | D696 | $7.5–9.0 \cdot 10^{-5}$ 1/K |
| **Dielectric** | | |
| Dielectric constant (1 MHz) | D150 | 3.2 |
| Dissipation factor (1 MHz) | D150 | 0.019 |
| Volume resistivity | D257 | $10^{13}$ ohm Ω-cm |
| Surface resistivity $R_{OA}$ | DIN 53482 | $10^{13}$ ohmΩ |
| Dielectric strength | DIN 53481 | 95 kV/mm |
| **Miscellaneous** | | |
| Mass density | D792 | 1.15 g/cm$^3$ |
| Melt index | D1238 | |
| MFI 260/5 | | 15 g/10 min |
| MFI 220/10 | | 4 g/10 min |
| Mold shrinkage | | 0.3–0.6 %<br>0.003–0.006 in/in |
| Water absorption Method A (24 hrs.) | D750 | 0.35 % |
| Flammability | UL Standard 94 (1/32 in. specimen) | 94 HB Class |

# Borg-Warner Chemicals Inc.

## Contact Address:

Borg-Warner Chemicals Inc.
International Center
Parkersburg, WV 26102

Phone: 1 (304) 424-5411

## PREVEX P2A

### Introduction:

Prevex polymers, offered by Borg-Warner Chemicals, Inc., are engineering thermoplastics that balance high impact resistance with excellent performance at elevated temperatures. These resins provide a unique combination of characteristics including Underwriters' Laboratory (UL) flammability classification, Relative Thermal Index (RTI) up to 110°C, electrical properties for direct support, and excellent dimensional stability. Easy flow, minimum drying and excellent thermal stability during molding combine for good processability leading to consistent part production.

To meet the range of design and processing demands presented by today's technology, Prevex polymers are available in numerous grades for injection and structural-foam molding as well as extrusion processing. New grades are being developed to expand the range of options already available from Borg-Warner Chemicals.

### Grade Characteristics:

Prevex P2A is a 20% glass-reinforced grade that features high flexural modulus with low mold shrinkage. Prevex P2A is an attractive replacement for die cast metal parts in such applications as pumps, fluid meters, and other plumbing applications.

### Product Data:

**PREVEX P2A Glass Reinforced Grade**

| Properties | Test Methods | Test Conditions | English Units |
|---|---|---|---|
| **Mechanical** | | | |
| Impact Resistance<br>Izod Impact<br>0.125″ bar, notched<br>reverse notched | ASTM D 256<br>Method A | 73°F | 1.8 ft-lb/in<br>5.5 ft-lb/in |
| Gardner Impact<br>0.125″ specimen | ASTM D 3029<br>Method G | 73°F | |
| Dynatup, Total Energy | | 73°F | |

*(continued)*

NOTE – Information provided in this book has been condensed. The publisher recommends that the product manufacturer be contacted, at the address or phone number provided, for more complete information on product safety, processing parameters, and applications not provided in this information sheet.

**PREVEX P2A Glass Reinforced Grade *(continued)***

| Properties | Test Methods | Test Conditions | English Units |
|---|---|---|---|
| **Thermal** | | | |
| Deflection Temperature under load 0.5 × 0.5 × 5.0″ bar | ASTM D 648 | 264 psi<br>64 psi | 290°F |
| Relative Thermal Index<br>Without Impact<br>With Impact | UL Standard 746 B | | |
| Vicat Softening Temperature | ASTM D 1525 | | |
| Coefficient of Linear Thermal Expansion | ASTM D 696 | −22°F to 86°F | $10^{-5}$ in/in/°F |
| Thermal Conductivity | ASTM C 177 | | BTU in/hr ft$^2$°F |
| **Flame-Retardant** | | | |
| Flame Class Rating | UL Standard 94 | 0.062″ thickness | 94HB[1] |
| **Electrical** | | | |
| Hot Wire Ignition | UL Standard 746 A | 0.062″<br>0.125″ | seconds to ignite |
| High Amperage Arc Ignition | UL Standard 746 A | 0.062″<br>0.125″ | arcs |
| High Voltage Tracking Rate | UL Standard 746 A | 0.062″<br>0.125″ | in/min |
| Arc Resistance | ASTM D 495 | 0.062″<br>0.125″ | seconds |
| Comparative Tracking Index | UL Standard 746 A | 0.062″<br>0.125″ | volts |
| Dielectric Strength | ASTM D 149 | 73°F | volts/mil |
| Dissipation Factor | ASTM D 150 | 73°F<br>100 cycles | |
| Dielectric Constant | ASTM D 150 | 73°F<br>100 cycles | |
| Volume Resistivity | ASTM D 257 | 73°F | $10^9$ ohm/cm |
| **Other Physical Properties** | | | |
| Specific Gravity[2] | ASTM D 792 Method A | 73°F | 1.20 |
| Mold Shrinkage | ASTM D 955 | | 0.002–0.004 in/in |
| Water Absorption | ASTM D 570 | 24 hrs at 73°F | % weight |

[1]BWC Test based on UL Standard 94 test methods at 0.062 inch; UL classification pending.
[2]Compression molded. All other specimens injection molded.

### PREVEX P2A Glass Reinforced Grade *(continued)*

| Properties | Test Methods | Test Conditions | English Units |
|---|---|---|---|
| **Electrical** | | | |
| Dissipation Factor | ASTM D 150 | 73°F<br>100 cycles | |
| Dielectric Constant | ASTM D 150 | 73°F<br>100 cycles | |
| Volume Resistivity | ASTM D 257 | 73°F | $10^9$ ohm/cm |
| **Other Physical Properties** | | | |
| Specific Gravity[2] | ASTM D 792<br>Method A | 73°F | 1.20 |
| Mold Shrinkage | ASTM D 955 | | 0.002–0.004 in/in |
| Water Absorption | ASTM D 570 | 24 hrs at<br>73°F | % weight |

[1]BWC Test based on UL Standard 94 test methods at 0.062 inch; UL classification pending.
[2]Compression molded. All other specimens injection molded.

### PREVEX P2A Glass Reinforced Grade

| Properties | Test Methods | Test Conditions | Metric Units |
|---|---|---|---|
| **Mechanical** | | | |
| Impact Resistance<br>Izod Impact<br>3.2mm bar, notched<br>reverse notched | ASTM D 256<br>Method A | 23°C | 96 J/m<br>294 J/m |
| Gardner Impact<br>3.2mm specimen | ASTM D 3029<br>Method G | 23°C | |
| Dynatup, Total Energy | | 23°C | |
| Tensile Strength<br>3.2mm thickness | ASTM D 638 | 23°C | 93 MPa |
| Tensile Modulus<br>3.2mm thickness | ASTM D 638 | 23°C | 4.8 GPa |
| Flexural Yield Strength<br>3.2 × 13 × 130mm bar | ASTM D 790 | 23°C | 138 MPa |
| Flexural Modulus<br>3.2 × 13 × 130mm bar | ASTM D 638 | 23°C | 5.2 GPa |
| Rockwell Hardness* | ASTM D 785 | | |

*Compression molded. All other specimens injection molded.

*(continued)*

PREVEX P2A Glass Reinforced Grade *(continued)*

| Properties | Test Methods | Test Conditions | Metric Units |
|---|---|---|---|
| **Thermal** | | | |
| Deflection Temperature under load 13 × 13 × 130mm bar | ASTM D 648 | 1820 kPa<br>440 kPa | 143°C |
| Relative Thermal Index<br>Without Impact<br>With Impact | UL Standard 746 B | | |
| Vicat Softening Temperature | ASTM D 1525 | | |
| Coefficient of Linear Thermal Expansion | ASTM D 696 | −30°C to +30°C | $10^{-5}$ cm/cm/°C |
| Thermal Conductivity | ASTM C 177 | | W/km |
| **Flame-Retardant** | | | |
| Flame Class Rating | UL Standard 94 | 1.6mm thickness | 94HB[1] |
| **Electrical** | | | |
| Hot Wire Ignition | UL Standard 746 A | 1.6mm<br>3.2mm | seconds to ignite |
| High Amperage Arc Ignition | UL Standard 746 A | 1.6mm<br>3.2mm | arcs |
| High Voltage Tracking Rate | UL Standard 746 A | 1.6mm<br>3.2mm | in/min |
| Arc Resistance | ASTM D 495 | 1.6mm<br>3.2mm | seconds |
| Comparative Tracking Index | UL Standard 746 A | 1.6mm<br>3.2mm | volts |
| Dielectric Strength | ASTM D 149 | 23°C | kV/mm |
| Dissipation Factor | ASTM D 150 | 23°C<br>100 cycles | |
| Dielectric Constant | ASTM D 150 | 23°C<br>100 cycles | |
| Volume Resistivity | ASTM D 257 | 23°C | $10^9$ ohm/cm |
| **Other Physical Properties** | | | |
| Specific Gravity[2] | ASTM D 792 Method A | 23°C | 1.20 |
| Mold Shrinkage | ASTM D 955 | | 0.002–0.004 cm/cm |
| Water Absorption | ASTM D 570 | 24 hrs at 23°C | % weight |

[1]BWC Test based on UL Standard 94 test methods at 1.6 mm; UL classification pending.
[2]Compression molded. All other specimens injection molded.

*(continued)*

# Borg-Warner Chemicals Inc.

## Contact Address:

Borg-Warner Chemicals Inc.
International Center
Parkersburg, WV 26102

Phone: 1 (304) 424-5411

## PREVEX P3A

### Introduction:

Prevex polymers, offered by Borg-Warner Chemicals, Inc., are engineering thermoplastics that balance high impact resistance with excellent performance at elevated temperatures. These resins provide a unique combination of characteristics including Underwriters' Laboratory (UL) flammability classification, Relative Thermal Index (RTI) up to 110°C, electrical properties for direct support, and excellent dimensional stability. Easy flow, minimum drying and excellent thermal stability during molding combine for good processability leading to consistent part production.

To meet the range of design and processing demands presented by today's technology, Prevex polymers are available in numerous grades for injection and structural-foam molding as well as extrusion processing. New grades are being developed to expand the range of options already available from Borg-Warner Chemicals.

### Grade Characteristics:

Prevex P3A is a 30% glass-reinforced grade with an even higher flexural modulus ($1.1 \times 10^6$ psi). Prevex P3A is especially suited as a die cast metal replacement in more demanding pump, fluid meter, and other plumbing applications.

### Product Data:

**PREVEX P3A Glass Reinforced Grade**

| Properties | Test Methods | Test Conditions | English Units |
|---|---|---|---|
| **Mechanical** | | | |
| Impact Resistance<br>Izod Impact<br>0.125″ bar, notched<br>reverse notched | ASTM D 256<br>Method A | 73°F | 1.8 ft-lb/in<br>5.5 ft-lb/in |
| Gardner Impact<br>0.125″ specimen | ASTM D 3029<br>Method G | 73°F | |
| Dynatup, Total Energy | | 73°F | |

*(continued)*

NOTE—Information provided in this book has been condensed. The publisher recommends that the product manufacturer be contacted, at the address or phone number provided, for more complete information on product safety, processing parameters, and applications not provided in this information sheet.

**PREVEX P3A Glass Reinforced Grade *(continued)***

| Properties | Test Methods | Test Conditions | English Units |
|---|---|---|---|
| **Mechanical** | | | |
| Tensile Strength<br>0.125″ thickness | ASTM D 638 | 73°F | 14,500 psi |
| Tensile Modulus<br>0.125″ thickness | ASTM D 638 | 73°F | 9.0 $10^5$ psi |
| Flexural Yield Strength<br>0.125 × 0.5 × 5.0″ bar | ASTM D 790 | 73°F | 21,000 psi |
| Flexural Modulus<br>0.125 × 0.5 × 5.0″ bar | ASTM D 638 | 73°F | 11.0 $10^5$ psi |
| Rockwell Hardness | ASTM D 785 | | |
| **Thermal** | | | |
| Deflection Temperature under load<br>0.5 × 0.5 × 5.0″ bar | ASTM D 648 | 264 psi<br>64 psi | 300°F |
| Relative Thermal Index<br>Without Impact<br>With Impact | UL Standard 746 B | | |
| Vicat Softening Temperature | ASTM D 1525 | | |
| Coefficient of Linear Thermal Expansion | ASTM D 696 | −22°F to 86°F | $10^{-5}$ in/in/°F |
| Thermal Conductivity | ASTM C 177 | | BTU in/hr ft²°F |
| **Flame-Retardant** | | | |
| Flame Class Rating | UL Standard 94 | 0.062″ thickness | 94HB[1] |
| **Electrical** | | | |
| Hot Wire Ignition | UL Standard 746 A | 0.062″<br>0.125″ | seconds to ignite |
| High Amperage Arc Ignition | UL Standard 746 A | 0.062″<br>0.125″ | arcs |
| High Voltage Tracking Rate | UL Standard 746 A | 0.062″<br>0.125″ | in/min |
| Arc Resistance | ASTM D 495 | 0.062″<br>0.125″ | seconds |
| Comparative Tracking Index | UL Standard 746 A | 0.062″<br>0.125″ | volts |
| Dielectric Strength | ASTM D 149 | 73°F | volts/mil |

[1]BWC Test based on UL Standard 94 test methods at 0.062 inch; UL classification pending.

*(continued)*

### PREVEX P3A Glass Reinforced Grade *(continued)*

| Properties | Test Methods | Test Conditions | English Units |
|---|---|---|---|
| **Electrical** | | | |
| Dissipation Factor | ASTM D 150 | 73°F<br>100 cycles | |
| Dielectric Constant | ASTM D 150 | 73°F<br>100 cycles | |
| Volume Resistivity | ASTM D 257 | 73°F | $10^9$ ohm/cm |
| **Other Physical Properties** | | | |
| Specific Gravity[2] | ASTM D 792<br>Method A | 73°F | 1.27 |
| Mold Shrinkage | ASTM D 955 | | 0.001–0.003 in/in |
| Water Absorption | ASTM D 570 | 24 hrs at<br>73°F | % weight |

[2]Compression molded. All other specimens injection molded.

### PREVEX P3A Glass Reinforced Grade

| Properties | Test Methods | Test Conditions | Metric Units |
|---|---|---|---|
| **Mechanical** | | | |
| Impact Resistance<br>Izod Impact<br>3.2mm bar, notched<br>reverse notched | ASTM D 256<br>Method A | 23°C | 96 J/m<br>294 J/m |
| Gardner Impact<br>3.2mm specimen | ASTM D 3029<br>Method G | 23°C | |
| Dynatup, Total Energy | | 23°C | |
| Tensile Strength<br>3.2mm thickness | ASTM D 638 | 23°C | 145 MPa |
| Tensile Modulus<br>3.2mm thickness | ASTM D 638 | 23°C | 6.2 GPa |
| Flexural Yield Strength<br>3.2 × 13 × 130mm bar | ASTM D 790 | 23°C | 145 MPa |
| Flexural Modulus<br>3.2 × 13 × 130mm bar | ASTM D 638 | 23°C | 7.6 GPa |
| Rockwell Hardness* | ASTM D 785 | | |

*Compression molded. All other specimens injection molded.

*(continued)*

## PREVEX P3A Glass Reinforced Grade *(continued)*

| Properties | Test Methods | Test Conditions | Metric Units |
|---|---|---|---|
| **Thermal** | | | |
| Deflection Temperature under load 13 × 13 × 130mm bar | ASTM D 648 | 1820 kPa<br>440 kPa | 149°C |
| Relative Thermal Index<br>Without Impact<br>With Impact | UL Standard 746 B | | |
| Vicat Softening Temperature | ASTM D 1525 | | |
| Coefficient of Linear Thermal Expansion | ASTM D 696 | −30°C to +30°C | $10^{-5}$ cm/cm/°C |
| Thermal Conductivity | ASTM C 177 | | W/km |
| **Flame-Retardant** | | | |
| Flame Class Rating | UL Standard 94 | 1.6mm thickness | 94HB[1] |
| **Electrical** | | | |
| Hot Wire Ignition | UL Standard 746 A | 1.6mm<br>3.2mm | seconds to ignite |
| High Amperage Arc Ignition | UL Standard 746 A | 1.6mm<br>3.2mm | arcs |
| High Voltage Tracking Rate | UL Standard 746 A | 1.6mm<br>3.2mm | in/min |
| Arc Resistance | ASTM D 495 | 1.6mm<br>3.2mm | seconds |
| Comparative Tracking Index | UL Standard 746 A | 1.6mm<br>3.2mm | volts |
| Dielectric Strength | ASTM D 149 | 23°C | kV/mm |
| Dissipation Factor | ASTM D 150 | 23°C<br>100 cycles | |
| Dielectric Constant | ASTM D 150 | 23°C<br>100 cycles | |
| Volume Resistivity | ASTM D 257 | 23°C | $10^9$ ohm/cm |
| **Other Physical Properties** | | | |
| Specific Gravity[2] | ASTM D 792 Method A | 23°C | 1.27 |
| Mold Shrinkage | ASTM D 955 | | 0.001–0.003 cm/cm |
| Water Absorption | ASTM D 570 | 24 hrs at 23°C | % weight |

[1]BWC Test based on UL Standard 94 test methods at 1.6 mm; UL classification pending.
[2]Compression molded. All other specimens injection molded.

# Borg-Warner Chemicals Inc.

## Contact Address:

Borg-Warner Chemicals Inc.
International Center
Parkersburg, WV 26102

Phone: 1 (304) 424-5411

# PREVEX PMA

## Introduction:

Prevex polymers, offered by Borg-Warner Chemicals, Inc., are engineering thermoplastics that balance high impact resistance with excellent performance at elevated temperatures. These resins provide a unique combination of characteristics including Underwriters' Laboratory (UL) flammability classification, Relative Thermal Index (RTI) up to 110°C, electrical properties for direct support, and excellent dimensional stability. Easy flow, minimum drying and excellent thermal stability during molding combine for good processability leading to consistent part production.

To meet the range of design and processing demands presented by today's technology, Prevex polymers are available in numerous grades for injection and structural-foam molding as well as extrusion processing. New grades are being developed to expand the range of options already available from Borg-Warner Chemicals.

## Grade Characteristics:

Prevex PMA, a general purpose grade, provides good heat resistance (DTUL of 245°F) and dimensional stability. This property balance can be utilized in small appliances, lawn care tools, power tools, and many industrial devices.

## Product Data:

**PREVEX PMA Molding Grade**

| Properties | Test Methods | Test Conditions | English Units |
|---|---|---|---|
| **Mechanical** | | | |
| Impact Resistance<br>Izod Impact<br>0.125″ bar, notched<br>reverse notched | ASTM D 256<br>Method A | 73°F | 5.0 ft-lb/in |
| Gardner Impact<br>0.125″ specimen | ASTM D 3029<br>Method G | 73°F | 20 ft-lb |
| Dynatup, Total Energy | | 73°F | |

*(continued)*

NOTE—Information provided in this book has been condensed. The publisher recommends that the product manufacturer be contacted, at the address or phone number provided, for more complete information on product safety, processing parameters, and applications not provided in this information sheet.

**PREVEX PMA Molding Grade *(continued)***

| Properties | Test Methods | Test Conditions | English Units |
|---|---|---|---|
| **Mechanical** | | | |
| Tensile Strength<br>0.125″ thickness | ASTM D 638 | 73°F | 7,300 psi |
| Tensile Modulus<br>0.125″ thickness | ASTM D 638 | 73°F | 3.5 $10^5$ psi |
| Flexural Yield Strength<br>0.125 × 0.5 × 5.0″ bar | ASTM D 790 | 73°F | 12,000 psi |
| Flexural Modulus<br>0.125 × 0.5 × 5.0″ bar | ASTM D 638 | 73°F | 3.6 $10^5$ psi |
| Rockwell Hardness | ASTM D 785 | | 114 R |
| **Thermal** | | | |
| Deflection Temperature under load<br>0.5 × 0.5 × 5.0″ bar | ASTM D 648 | 264 psi<br>64 psi | 245°F |
| Relative Thermal Index<br>Without Impact<br>With Impact | UL Standard 746 B | | |
| Vicat Softening Temperature | ASTM D 1525 | | 277°F |
| Coefficient of Linear Thermal Expansion | ASTM D 696 | −22°F to 86°F | 4.3 $10^{-5}$ in/in/°F |
| Thermal Conductivity | ASTM C 177 | | BTU in/hr ft$^2$°F |
| **Flame-Retardant** | | | |
| Flame Class Rating | UL Standard 94 | 0.062″ thickness | 94HB |
| **Electrical** | | | |
| Hot Wire Ignition | UL Standard 746 A | 0.062″<br>0.125″ | seconds to ignite |
| High Amperage Arc Ignition | UL Standard 746 A | 0.062″<br>0.125″ | arcs<br>arcs |
| High Voltage Tracking Rate | UL Standard 746 A | 0.062″<br>0.125″ | in/min<br>in/min |
| Arc Resistance | ASTM D 495 | 0.062″<br>0.125″ | seconds |
| Comparative Tracking Index | UL Standard 746 A | 0.062″<br>0.125″ | volts |
| Dielectric Strength | ASTM D 149 | 73°F | volts/mil |

*(continued)*

**PREVEX PMA Molding Grade *(continued)***

| Properties | Test Methods | Test Conditions | English Units |
|---|---|---|---|
| **Electrical** | | | |
| Dissipation Factor | ASTM D 150 | 73°F<br>100 cycles | |
| Dielectric Constant | ASTM D 150 | 73°F<br>100 cycles | |
| Volume Resistivity | ASTM D 257 | 73°F | $10^9$ ohm/cm |
| **Other Physical Properties** | | | |
| Specific Gravity* | ASTM D 792<br>Method A | 73°F | 1.06 |
| Mold Shrinkage | ASTM D 955 | | 0.006–0.007 in/in |
| Water Absorption | ASTM D 570 | 24 hrs at<br>73°F | 0.07% weight |

**PREVEX PMA Molding Grade**

| Properties | Test Methods | Test Conditions | Metric Units |
|---|---|---|---|
| **Mechanical** | | | |
| Impact Resistance<br>Izod Impact<br>3.2mm bar, notched<br>reverse notched | ASTM D 256<br>Method A | 23°C | 270 J/m |
| Gardner Impact<br>3.2mm specimen | ASTM D 3029<br>Method G | 23°C | 27 J |
| Dynatup, Total Energy | | 23°C | |
| Tensile Strength<br>3.2mm thickness | ASTM D 638 | 23°C | 50 MPa |
| Tensile Modulus<br>3.2mm thickness | ASTM D 638 | 23°C | 2.4 GPa |
| Flexural Yield Strength<br>3.2 × 13 × 130mm bar | ASTM D 790 | 23°C | 83 MPa |
| Flexural Modulus<br>3.2 × 13 × 130mm bar | ASTM D 638 | 23°C | 2.5 GPa |
| Rockwell Hardness* | ASTM D 785 | | 114 R |

*Compression molded. All other specimens injection molded.

*(continued)*

**PREVEX PMA Molding Grade *(continued)***

| Properties | Test Methods | Test Conditions | Metric Units |
|---|---|---|---|
| **Thermal** | | | |
| Deflection Temperature under load 13 × 13 × 130mm bar | ASTM D 648 | 1820 kPa<br>440 kPa | 118°C |
| Relative Thermal Index<br>Without Impact<br>With Impact | UL Standard 746 B | | |
| Vicat Softening Temperature | ASTM D 1525 | | 136°C |
| Coefficient of Linear Thermal Expansion | ASTM D 696 | −30°C to +30°C | 7.7 $10^{-5}$ cm/cm/°C |
| Thermal Conductivity | ASTM C 177 | | W/km |
| **Flame-Retardant** | | | |
| Flame Class Rating | UL Standard 94 | 1.6mm thickness | 94HB |
| **Electrical** | | | |
| Hot Wire Ignition | UL Standard 746 A | 1.6mm<br>3.2mm | seconds to ignite |
| High Amperage Arc Ignition | UL Standard 746 A | 1.6mm<br>3.2mm | arcs |
| High Voltage Tracking Rate | UL Standard 746 A | 1.6mm<br>3.2mm | in/min |
| Arc Resistance | ASTM D 495 | 1.6mm<br>3.2mm | seconds |
| Comparative Tracking Index | UL Standard 746 A | 1.6mm<br>3.2mm | volts |
| Dielectric Strength | ASTM D 149 | 23°C | kV/mm |
| Dissipation Factor | ASTM D 150 | 23°C<br>100 cycles | |
| Dielectric Constant | ASTM D 150 | 23°C<br>100 cycles | |
| Volume Resistivity | ASTM D 257 | 23°C | $10^9$ ohm/cm |
| **Other Physical Properties** | | | |
| Specific Gravity* | ASTM D 792 Method A | 23°C | 1.06 |
| Mold Shrinkage | ASTM D 955 | | 0.006–0.007 cm/cm |
| Water Absorption | ASTM D 570 | 24 hrs at 23°C | 0.07% weight |

*Compression molded. All other specimens injection molded.

# Borg-Warner Chemicals Inc.

## Contact Address:

Borg-Warner Chemicals Inc.
International Center
Parkersburg, WV 26102

Phone: 1 (304) 424-5411

# PREVEX PQ1

## Introduction:

Prevex polymers, offered by Borg-Warner Chemicals, Inc., are engineering thermoplastics that balance high impact resistance with excellent performance at elevated temperatures. These resins provide a unique combination of characteristics including Underwriters' Laboratory (UL) flammability classification, Relative Thermal Index (RTI) up to 110°C, electrical properties for direct support, and excellent dimensional stability. Easy flow, minimum drying and excellent thermal stability during molding combine for good processability leading to consistent part production.

To meet the range of design and processing demands presented by today's technology, Prevex polymers are available in numerous grades for injection and structural-foam molding as well as extrusion processing. New grades are being developed to expand the range of options already available from Borg-Warner Chemicals.

## Grade Characteristics:

Prevex PQ1 is an extrusion grade developed for high temperature sheet and profile applications. PQ1 has a 265°F DTUL and very low moisture absorption.

## Product Data:

**PREVEX PQ1 Extrusion Grade**

| Properties | Test Methods | Test Conditions | English Units |
|---|---|---|---|
| **Mechanical** | | | |
| Impact Resistance<br>Izod Impact<br>0.125″ bar, notched<br>reverse notched | ASTM D 256<br>Method A | 73°F | 5.0 ft-lb/in |
| Gardner Impact<br>0.125″ specimen | ASTM D 3029<br>Method G | 73°F | 18 ft-lb |
| Dynatup, Total Energy | | 73°F | |

*(continued)*

NOTE – Information provided in this book has been condensed. The publisher recommends that the product manufacturer be contacted, at the address or phone number provided, for more complete information on product safety, processing parameters, and applications not provided in this information sheet.

PREVEX PQ1 Extrusion Grade *(continued)*

| Properties | Test Methods | Test Conditions | English Units |
|---|---|---|---|
| **Mechanical** | | | |
| Tensile Strength 0.125″ thickness | ASTM D 638 | 73°F | 7,800 psi |
| Tensile Modulus 0.125″ thickness | ASTM D 638 | 73°F | 3.4 $10^5$ psi |
| Flexural Yield Strength 0.125 × 0.5 × 5.0″ bar | ASTM D 790 | 73°F | 12,500 psi |
| Flexural Modulus 0.125 × 0.5 × 5.0″ bar | ASTM D 638 | 73°F | 3.6 $10^5$ psi |
| Rockwell Hardness | ASTM D 785 | | 118 R |
| **Thermal** | | | |
| Deflection Temperature under load 0.5 × 0.5 × 5.0″ bar | ASTM D 648 | 264 psi<br>64 psi | 265°F |
| Relative Thermal Index<br>Without Impact<br>With Impact | UL Standard 746 B | | <br>221°F<br>194°F |
| Vicat Softening Temperature | ASTM D 1525 | | |
| Coefficient of Linear Thermal Expansion | ASTM D 696 | −22°F to 86°F | 3.3 $10^{-5}$ in/in/°F |
| Thermal Conductivity | ASTM C 177 | | BTU in/hr ft²°F |
| **Flame-Retardant** | | | |
| Flame Class Rating | UL Standard 94 | 0.062″ thickness | 94HB |
| **Electrical** | | | |
| Hot Wire Ignition | UL Standard 746 A | 0.062″<br>0.125″ | 120 seconds to ignite |
| High Amperage Arc Ignition | UL Standard 746 A | 0.062″<br>0.125″ | 200+ arcs |
| High Voltage Tracking Rate | UL Standard 746 A | 0.062″<br>0.125″ | 15.8 in/min |
| Arc Resistance | ASTM D 495 | 0.062″<br>0.125″ | 66 seconds |
| Comparative Tracking Index | UL Standard 746 A | 0.062″<br>0.125″ | 470 volts |
| Dielectric Strength | ASTM D 149 | 73°F | 758 volts/mil |

*(continued)*

**PREVEX PQ1 Extrusion Grade *(continued)***

| Properties | Test Methods | Test Conditions | English Units |
|---|---|---|---|
| **Electrical** | | | |
| Dissipation Factor | ASTM D 150 | 73°F<br>100 cycles | 0.0002 |
| Dielectric Constant | ASTM D 150 | 73°F<br>100 cycles | 2.51 |
| Volume Resistivity | ASTM D 257 | 73°F | 16.4 $10^9$ ohm/cm |
| **Other Physical Properties** | | | |
| Specific Gravity* | ASTM D 792<br>Method A | 73°F | 1.06 |
| Mold Shrinkage | ASTM D 955 | | |
| Water Absorption | ASTM D 570 | 24 hrs at<br>73°F | 0.06% weight |

**PREVEX PQ1 Extrusion Grade**

| Properties | Test Methods | Test Conditions | Metric Units |
|---|---|---|---|
| **Mechanical** | | | |
| Impact Resistance<br>Izod Impact<br>3.2mm bar, notched<br>reverse notched | ASTM D 256<br>Method A | 23°C | 270 J/m |
| Gardner Impact<br>3.2mm specimen | ASTM D 3029<br>Method G | 23°C | 24 J |
| Dynatup, Total Energy | | 23°C | |
| Tensile Strength<br>3.2mm thickness | ASTM D 638 | 23°C | 54 MPa |
| Tensile Modulus<br>3.2mm thickness | ASTM D 638 | 23°C | 2.3 GPa |
| Flexural Yield Strength<br>3.2 × 13 × 130mm bar | ASTM D 790 | 23°C | 86 MPa |
| Flexural Modulus<br>3.2 × 13 × 130mm bar | ASTM D 638 | 23°C | 2.5 GPa |
| Rockwell Hardness* | ASTM D 785 | | 118 R |

*Compression molded. All other specimens injection molded.

*(continued)*

**PREVEX PQ1 Extrusion Grade *(continued)***

| Properties | Test Methods | Test Conditions | Metric Units |
|---|---|---|---|
| **Thermal** | | | |
| Deflection Temperature under load 13 × 13 × 130mm bar | ASTM D 648 | 1820 kPa<br>440 kPa | 129°C |
| Relative Thermal Index<br>Without Impact<br>With Impact | UL Standard 746 B | | 105°C<br>90°C |
| Vicat Softening Temperature | ASTM D 1525 | | |
| Coefficient of Linear Thermal Expansion | ASTM D 696 | −30°C to +30°C | 5.9 $10^{-5}$ cm/cm/°C |
| Thermal Conductivity | ASTM C 177 | | W/km |
| **Flame-Retardant** | | | |
| Flame Class Rating | UL Standard 94 | 1.6mm thickness | 94HB |
| **Electrical** | | | |
| Hot Wire Ignition | UL Standard 746 A | 1.6mm<br>3.2mm | 120 seconds to ignite |
| High Amperage Arc Ignition | UL Standard 746 A | 1.6mm<br>3.2mm | 200+ arcs |
| High Voltage Tracking Rate | UL Standard 746 A | 1.6mm<br>3.2mm | 40.1 in/min |
| Arc Resistance | ASTM D 495 | 1.6mm<br>3.2mm | 66 seconds |
| Comparative Tracking Index | UL Standard 746 A | 1.6mm<br>3.2mm | 470 volts |
| Dielectric Strength | ASTM D 149 | 23°C | 29.8 kV/mm |
| Dissipation Factor | ASTM D 150 | 23°C<br>100 cycles | 0.0002 |
| Dielectric Constant | ASTM D 150 | 23°C<br>100 cycles | 2.51 |
| Volume Resistivity | ASTM D 257 | 23°C | 16.4 $10^{9}$ ohm/cm |
| **Other Physical Properties** | | | |
| Specific Gravity* | ASTM D 792 Method A | 23°C | 1.06 |
| Mold Shrinkage | ASTM D 955 | | |
| Water Absorption | ASTM D 570 | 24 hrs at 23°C | 0.06% weight |

*Compression molded. All other specimens injection molded.

# Borg-Warner Chemicals Inc.

## Contact Address:

Borg-Warner Chemicals Inc.
International Center
Parkersburg, WV 26102

Phone: 1 (304) 424-5411

# PREVEX PQA

## Introduction:

Prevex polymers, offered by Borg-Warner Chemicals, Inc., are engineering thermoplastics that balance high impact resistance with excellent performance at elevated temperatures. These resins provide a unique combination of characteristics including Underwriters' Laboratory (UL) flammability classification, Relative Thermal Index (RTI) up to 110°C, electrical properties for direct support, and excellent dimensional stability. Easy flow, minimum drying and excellent thermal stability during molding combine for good processability leading to consistent part production.

To meet the range of design and processing demands presented by today's technology, Prevex polymers are available in numerous grades for injection and structural-foam molding as well as extrusion processing. New grades are being developed to expand the range of options already available from Borg-Warner Chemicals.

## Grade Characteristics:

Prevex PQA, a general purpose grade, is ideal for portable appliances, power tools, hot water appliances and industrial products. With a 265°F DTUL, Prevex PQA provides outstanding dimensional stability and toughness.

## Product Data:

**PREVEX PQA Molding Grade**

| Properties | Test Methods | Test Conditions | English Units |
|---|---|---|---|
| **Mechanical** | | | |
| Impact Resistance<br>Izod Impact<br>0.125″ bar, notched<br>reverse notched | ASTM D 256<br>Method A | 73°F | 5.0 ft-lb/in |
| Gardner Impact<br>0.125″ specimen | ASTM D 3029<br>Method G | 73°F | 18 ft-lb |
| Dynatup, Total Energy | | 73°F | |

*(continued)*

NOTE – Information provided in this book has been condensed. The publisher recommends that the product manufacturer be contacted, at the address or phone number provided, for more complete information on product safety, processing parameters, and applications not provided in this information sheet.

**PREVEX PQA Molding Grade *(continued)***

| Properties | Test Methods | Test Conditions | English Units |
|---|---|---|---|
| **Mechanical** | | | |
| Tensile Strength<br>0.125″ thickness | ASTM D 638 | 73°F | 8,400 psi |
| Tensile Modulus<br>0.125″ thickness | ASTM D 638 | 73°F | 3.7 $10^5$ psi |
| Flexural Yield Strength<br>0.125 × 0.5 × 5.0″ bar | ASTM D 790 | 73°F | 12,500 psi |
| Flexural Modulus<br>0.125 × 0.5 × 5.0″ bar | ASTM D 638 | 73°F | 3.6 $10^5$ psi |
| Rockwell Hardness | ASTM D 785 | | 115 R |
| **Thermal** | | | |
| Deflection Temperature under load<br>0.5 × 0.5 × 5.0″ bar | ASTM D 648 | 264 psi<br>64 psi | 265°F |
| Relative Thermal Index<br>Without Impact<br>With Impact | UL Standard 746 B | | 221°F<br>194°F |
| Vicat Softening Temperature | ASTM D 1525 | | 302°F |
| Coefficient of Linear Thermal Expansion | ASTM D 696 | −22°F to 86°F | 3.7 $10^{-5}$ in/in/°F |
| Thermal Conductivity | ASTM C 177 | | BTU in/hr ft²°F |
| **Flame-Retardant** | | | |
| Flame Class Rating | UL Standard 94 | 0.062″ thickness | 94HB |
| **Electrical** | | | |
| Hot Wire Ignition | UL Standard 746 A | 0.062″<br>0.125″ | 70 seconds<br>70 to ignite |
| High Amperage Arc Ignition | UL Standard 746 A | 0.062″<br>0.125″ | 200+ arcs<br>200+ arcs |
| High Voltage Tracking Rate | UL Standard 746 A | 0.062″<br>0.125″ | 15.8 in/min<br>15.8 in/min |
| Arc Resistance | ASTM D 495 | 0.062″<br>0.125″ | 68 seconds<br>123 seconds |
| Comparative Tracking Index | UL Standard 746 A | 0.062″<br>0.125″ | 470 volts<br>470 volts |
| Dielectric Strength | ASTM D 149 | 73°F | volts/mil |

*(continued)*

PREVEX PQA Molding Grade *(continued)*

| Properties | Test Methods | Test Conditions | English Units |
|---|---|---|---|
| **Electrical** | | | |
| Dissipation Factor | ASTM D 150 | 73°F<br>100 cycles | |
| Dielectric Constant | ASTM D 150 | 73°F<br>100 cycles | |
| Volume Resistivity | ASTM D 257 | 73°F | $10^9$ ohm/cm |
| **Other Physical Properties** | | | |
| Specific Gravity* | ASTM D 792<br>Method A | 73°F | 1.06 |
| Mold Shrinkage | ASTM D 955 | | 0.006–0.007 in/in |
| Water Absorption | ASTM D 570 | 24 hrs at<br>73°F | 0.06% weight |

PREVEX PQA Molding Grade

| Properties | Test Methods | Test Conditions | Metric Units |
|---|---|---|---|
| **Mechanical** | | | |
| Impact Resistance<br>Izod Impact<br>3.2mm bar, notched<br>reverse notched | ASTM D 256<br>Method A | 23°C | 270 J/m |
| Gardner Impact<br>3.2mm specimen | ASTM D 3029<br>Method G | 23°C | 25 J |
| Dynatup, Total Energy | | 23°C | |
| Tensile Strength<br>3.2mm thickness | ASTM D 638 | 23°C | 58 MPa |
| Tensile Modulus<br>3.2mm thickness | ASTM D 638 | 23°C | 2.5 GPa |
| Flexural Yield Strength<br>3.2 × 13 × 130mm bar | ASTM D 790 | 23°C | 86 MPa |
| Flexural Modulus<br>3.2 × 13 × 130mm bar | ASTM D 638 | 23°C | 2.5 GPa |
| Rockwell Hardness* | ASTM D 785 | | 115 R |

*Compression molded. All other specimens injection molded.

*(continued)*

**PREVEX PQA Molding Grade *(continued)***

| Properties | Test Methods | Test Conditions | Metric Units |
|---|---|---|---|
| **Thermal** | | | |
| Deflection Temperature under load 13 × 13 × 130mm bar | ASTM D 648 | 1820 kPa<br>440 kPa | 129°C |
| Relative Thermal Index<br>Without Impact<br>With Impact | UL Standard 746 B | | <br>105°C<br>90°C |
| Vicat Softening Temperature | ASTM D 1525 | | 150°C |
| Coefficient of Linear Thermal Expansion | ASTM D 696 | −30°C to +30°C | 6.6 $10^{-5}$ cm/cm/°C |
| Thermal Conductivity | ASTM C 177 | | W/km |
| **Flame-Retardant** | | | |
| Flame Class Rating | UL Standard 94 | 1.6mm thickness | 94HB |
| **Electrical** | | | |
| Hot Wire Ignition | UL Standard 746 A | 1.6mm<br>3.2mm | 70 seconds<br>70 to ignite |
| High Amperage Arc Ignition | UL Standard 746 A | 1.6mm<br>3.2mm | 200+ arcs<br>200+ arcs |
| High Voltage Tracking Rate | UL Standard 746 A | 1.6mm<br>3.2mm | 40 in/min<br>40 in/min |
| Arc Resistance | ASTM D 495 | 1.6mm<br>3.2mm | 68 seconds<br>123 seconds |
| Comparative Tracking Index | UL Standard 746 A | 1.6mm<br>3.2mm | 470 volts<br>470 volts |
| Dielectric Strength | ASTM D 149 | 23°C | kV/mm |
| Dissipation Factor | ASTM D 150 | 23°C<br>100 cycles | |
| Dielectric Constant | ASTM D 150 | 23°C<br>100 cycles | |
| Volume Resistivity | ASTM D 257 | 23°C | $10^9$ ohm/cm |
| **Other Physical Properties** | | | |
| Specific Gravity* | ASTM D 792 Method A | 23°C | 1.06 |
| Mold Shrinkage | ASTM D 955 | | 0.006–0.007 cm/cm |
| Water Absorption | ASTM D 570 | 24 hrs at 23°C | 0.06% weight |

*Compression molded. All other specimens injection molded.

# Borg-Warner Chemicals Inc.

## Contact Address:

Borg-Warner Chemicals Inc.
International Center
Parkersburg, WV 26102

Phone: 1 (304) 424-5411

# PREVEX V2A

## Introduction:

Prevex polymers, offered by Borg-Warner Chemicals, Inc., are engineering thermoplastics that balance high impact resistance with excellent performance at elevated temperatures. These resins provide a unique combination of characteristics including Underwriters' Laboratory (UL) flammability classification, Relative Thermal Index (RTI) up to 110°C, electrical properties for direct support, and excellent dimensional stability. Easy flow, minimum drying and excellent thermal stability during molding combine for good processability leading to consistent part production.

To meet the range of design and processing demands presented by today's technology, Prevex polymers are available in numerous grades for injection and structural-foam molding as well as extrusion processing. New grades are being developed to expand the range of options already available from Borg-Warner Chemicals.

## Grade Characteristics:

Prevex V2A is a 20% glass-reinforced, flame-retardant grade that combines superior modulus with the flame-retardant properties required for connectors, business equipment, motor housings, terminal blocks and electrical applications.

## Product Data:

**PREVEX V2A Glass Reinforced Grade**

| Properties | Test Methods | Test Conditions | English Units |
|---|---|---|---|
| **Mechanical** | | | |
| Impact Resistance<br>Izod Impact<br>0.125″ bar, notched<br>reverse notched | ASTM D 256<br>Method A | 73°F | 1.8 ft-lb/in<br>5.5 ft-lb/in |
| Gardner Impact<br>0.125″ specimen | ASTM D 3029<br>Method G | 73°F | |
| Dynatup, Total Energy | | 73°F | |

*(continued)*

NOTE—Information provided in this book has been condensed. The publisher recommends that the product manufacturer be contacted, at the address or phone number provided, for more complete information on product safety, processing parameters, and applications not provided in this information sheet.

PREVEX V2A Glass Reinforced Grade *(continued)*

| Properties | Test Methods | Test Conditions | English Units |
|---|---|---|---|
| **Mechanical** | | | |
| Tensile Strength<br>0.125″ thickness | ASTM D 638 | 73°F | 14,000 psi |
| Tensile Modulus<br>0.125″ thickness | ASTM D 638 | 73°F | 7.0 $10^5$ psi |
| Flexural Yield Strength<br>0.125 × 0.5 × 5.0″ bar | ASTM D 790 | 73°F | 21,000 psi |
| Flexural Modulus<br>0.125 × 0.5 × 5.0″ bar | ASTM D 638 | 73°F | 7.5 $10^5$ psi |
| Rockwell Hardness | ASTM D 785 | | |
| **Thermal** | | | |
| Deflection Temperature under load<br>0.5 × 0.5 × 5.0″ bar | ASTM D 648 | 264 psi<br>64 psi | 280°F |
| Relative Thermal Index<br>Without Impact<br>With Impact | UL Standard 746 B | | |
| Vicat Softening Temperature | ASTM D 1525 | | |
| Coefficient of Linear Thermal Expansion | ASTM D 696 | −22°F to 86°F | $10^{-5}$ in/in/°F |
| Thermal Conductivity | ASTM C 177 | | BTU in/hr ft²°F |
| **Flame-Retardant** | | | |
| Flame Class Rating | UL Standard 94 | 0.062″ thickness | 94V-1[1] |
| **Electrical** | | | |
| Hot Wire Ignition | UL Standard 746 A | 0.062″<br>0.125″ | seconds to ignite |
| High Amperage Arc Ignition | UL Standard 746 A | 0.062″<br>0.125″ | arcs |
| High Voltage Tracking Rate | UL Standard 746 A | 0.062″<br>0.125″ | in/min |
| Arc Resistance | ASTM D 495 | 0.062″<br>0.125″ | seconds |
| Comparative Tracking Index | UL Standard 746 A | 0.062″<br>0.125″ | volts |
| Dielectric Strength | ASTM D 149 | 73°F | volts/mil |

[1]BWC Test based on UL Standard 94 test methods at 0.062 inch; UL classification pending.

*(continued)*

**PREVEX V2A Glass Reinforced Grade (*continued*)**

| Properties | Test Methods | Test Conditions | English Units |
|---|---|---|---|
| **Electrical** | | | |
| Dissipation Factor | ASTM D 150 | 73°F<br>100 cycles | |
| Dielectric Constant | ASTM D 150 | 73°F<br>100 cycles | |
| Volume Resistivity | ASTM D 257 | 73°F | $10^9$ ohm/cm |
| **Other Physical Properties** | | | |
| Specific Gravity[2] | ASTM D 792<br>Method A | 73°F | 1.30 |
| Mold Shrinkage | ASTM D 955 | | 0.002–0.004 in/in |
| Water Absorption | ASTM D 570 | 24 hrs at<br>73°F | % weight |

[2]Compression molded. All other specimens injection molded.

**PREVEX V2A Glass Reinforced Grade**

| Properties | Test Methods | Test Conditions | Metric Units |
|---|---|---|---|
| **Mechanical** | | | |
| Impact Resistance<br>Izod Impact<br>3.2mm bar, notched<br>reverse notched | ASTM D 256<br>Method A | 23°C | 96 J/m<br>294 J/m |
| Gardner Impact<br>3.2mm specimen | ASTM D 3029<br>Method G | 23°C | |
| Dynatup, Total Energy | | 23°C | |
| Tensile Strength<br>3.2mm thickness | ASTM D 638 | 23°C | 96 MPa |
| Tensile Modulus<br>3.2mm thickness | ASTM D 638 | 23°C | 4.8 GPa |
| Flexural Yield Strength<br>3.2 × 13 × 130mm bar | ASTM D 790 | 23°C | 145 MPa |
| Flexural Modulus<br>3.2 × 13 × 130mm bar | ASTM D 638 | 23°C | 5.2 GPa |
| Rockwell Hardness* | ASTM D 785 | | |

*Compression molded. All other specimens injection molded.

*(continued)*

## PREVEX V2A Glass Reinforced Grade *(continued)*

| Properties | Test Methods | Test Conditions | Metric Units |
|---|---|---|---|
| **Thermal** | | | |
| Deflection Temperature under load 13 × 13 × 130mm bar | ASTM D 648 | 1820 kPa<br>440 kPa | 138°C |
| Relative Thermal Index<br>Without Impact<br>With Impact | UL Standard 746 B | | |
| Vicat Softening Temperature | ASTM D 1525 | | |
| Coefficient of Linear Thermal Expansion | ASTM D 696 | −30°C to +30°C | $10^{-5}$ cm/cm/°C |
| Thermal Conductivity | ASTM C 177 | | W/km |
| **Flame-Retardant** | | | |
| Flame Class Rating | UL Standard 94 | 1.6mm thickness | 94V-1[1] |
| **Electrical** | | | |
| Hot Wire Ignition | UL Standard 746 A | 1.6mm<br>3.2mm | seconds to ignite |
| High Amperage Arc Ignition | UL Standard 746 A | 1.6mm<br>3.2mm | arcs |
| High Voltage Tracking Rate | UL Standard 746 A | 1.6mm<br>3.2mm | in/min |
| Arc Resistance | ASTM D 495 | 1.6mm<br>3.2mm | seconds |
| Comparative Tracking Index | UL Standard 746 A | 1.6mm<br>3.2mm | volts |
| Dielectric Strength | ASTM D 149 | 23°C | kV/mm |
| Dissipation Factor | ASTM D 150 | 23°C<br>100 cycles | |
| Dielectric Constant | ASTM D 150 | 23°C<br>100 cycles | |
| Volume Resistivity | ASTM D 257 | 23°C | $10^9$ ohm/cm |
| **Other Physical Properties** | | | |
| Specific Gravity[2] | ASTM D 792 Method A | 23°C | 1.30 |
| Mold Shrinkage | ASTM D 955 | | 0.002–0.004 cm/cm |
| Water Absorption | ASTM D 570 | 24 hrs at 23°C | % weight |

[1]BWC Test based on UL Standard 94 test methods at 1.6 mm; UL classification pending.
[2]Compression molded. All other specimens injection molded.

# Borg-Warner Chemicals Inc.

## Contact Address:

Borg-Warner Chemicals Inc.
International Center
Parkersburg, WV 26102

Phone: 1 (304) 424-5411

## PREVEX V3A

### Introduction:

Prevex polymers, offered by Borg-Warner Chemicals, Inc., are engineering thermoplastics that balance high impact resistance with excellent performance at elevated temperatures. These resins provide a unique combination of characteristics including Underwriters' Laboratory (UL) flammability classification, Relative Thermal Index (RTI) up to 110°C, electrical properties for direct support, and excellent dimensional stability. Easy flow, minimum drying and excellent thermal stability during molding combine for good processability leading to consistent part production.

To meet the range of design and processing demands presented by today's technology, Prevex polymers are available in numerous grades for injection and structural-foam molding as well as extrusion processing. New grades are being developed to expand the range of options already available from Borg-Warner Chemicals.

### Grade Characteristics:

Prevex V3A is a 30% glass-reinforced, flame-retardant grade with the highest flexural modulus of all the Prevex flame-retardant grades (1,000,000 psi). It's an outstanding choice for applications in electrical markets and industrial control products.

### Product Data:

**PREVEX V3A Glass Reinforced Grade**

| Properties | Test Methods | Test Conditions | English Units |
|---|---|---|---|
| **Mechanical** | | | |
| Impact Resistance<br>Izod Impact<br>0.125″ bar, notched<br>reverse notched | ASTM D 256<br>Method A | 73°F | 1.8 ft-lb/in<br>5.5 ft-lb/in |
| Gardner Impact<br>0.125″ specimen | ASTM D 3029<br>Method G | 73°F | |
| Dynatup, Total Energy | | 73°F | |

*(continued)*

NOTE – Information provided in this book has been condensed. The publisher recommends that the product manufacturer be contacted, at the address or phone number provided, for more complete information on product safety, processing parameters, and applications not provided in this information sheet.

**PREVEX V3A Glass Reinforced Grade *(continued)***

| Properties | Test Methods | Test Conditions | English Units |
|---|---|---|---|
| **Mechanical** | | | |
| Tensile Strength<br>0.125″ thickness | ASTM D 638 | 73°F | 16,000 psi |
| Tensile Modulus<br>0.125″ thickness | ASTM D 638 | 73°F | 10.0 $10^5$ psi |
| Flexural Yield Strength<br>0.125 × 0.5 × 5.0″ bar | ASTM D 790 | 73°F | 23,000 psi |
| Flexural Modulus<br>0.125 × 0.5 × 5.0″ bar | ASTM D 638 | 73°F | 11.0 $10^5$ psi |
| Rockwell Hardness | ASTM D 785 | | |
| **Thermal** | | | |
| Deflection Temperature under load<br>0.5 × 0.5 × 5.0″ bar | ASTM D 648 | 264 psi<br>64 psi | 285°F |
| Relative Thermal Index<br>Without Impact<br>With Impact | UL Standard 746 B | | |
| Vicat Softening Temperature | ASTM D 1525 | | |
| Coefficient of Linear Thermal Expansion | ASTM D 696 | −22°F to 86°F | $10^{-5}$ in/in/°F |
| Thermal Conductivity | ASTM C 177 | | BTU in/hr ft²°F |
| **Flame-Retardant** | | | |
| Flame Class Rating | UL Standard 94 | 0.062″ thickness | 94V-1[1] |
| **Electrical** | | | |
| Hot Wire Ignition | UL Standard 746 A | 0.062″<br>0.125″ | seconds to ignite |
| High Amperage Arc Ignition | UL Standard 746 A | 0.062″<br>0.125″ | arcs |
| High Voltage Tracking Rate | UL Standard 746 A | 0.062″<br>0.125″ | in/min |
| Arc Resistance | ASTM D 495 | 0.062″<br>0.125″ | seconds |
| Comparative Tracking Index | UL Standard 746 A | 0.062″<br>0.125″ | volts |
| Dielectric Strength | ASTM D 149 | 73°F | volts/mil |

[1]BWC Test based on UL Standard 94 test methods at 0.062 inch; UL classification pending.

*(continued)*

### PREVEX V3A Glass Reinforced Grade *(continued)*

| Properties | Test Methods | Test Conditions | English Units |
|---|---|---|---|
| **Electrical** | | | |
| Dissipation Factor | ASTM D 150 | 73°F<br>100 cycles | |
| Dielectric Constant | ASTM D 150 | 73°F<br>100 cycles | |
| Volume Resistivity | ASTM D 257 | 73°F | $10^9$ ohm/cm |
| **Other Physical Properties** | | | |
| Specific Gravity[2] | ASTM D 792<br>Method A | 73°F | 1.36 |
| Mold Shrinkage | ASTM D 955 | | 0.001–0.003 in/in |
| Water Absorption | ASTM D 570 | 24 hrs at<br>73°F | % weight |

[2]Compression molded. All other specimens injection molded.

### PREVEX V3A Glass Reinforced Grade

| Properties | Test Methods | Test Conditions | Metric Units |
|---|---|---|---|
| **Mechanical** | | | |
| Impact Resistance<br>Izod Impact<br>3.2mm bar, notched<br>reverse notched | ASTM D 256<br>Method A | 23°C | 96 J/m<br>294 J/m |
| Gardner Impact<br>3.2mm specimen | ASTM D 3029<br>Method G | 23°C | |
| Dynatup, Total Energy | | 23°C | |
| Tensile Strength<br>3.2mm thickness | ASTM D 638 | 23°C | 110 MPa |
| Tensile Modulus<br>3.2mm thickness | ASTM D 638 | 23°C | 6.9 GPa |
| Flexural Yield Strength<br>3.2 × 13 × 130mm bar | ASTM D 790 | 23°C | 158 MPa |
| Flexural Modulus<br>3.2 × 13 × 130mm bar | ASTM D 638 | 23°C | 7.6 GPa |
| Rockwell Hardness* | ASTM D 785 | | |

*Compression molded. All other specimens injection molded.

*(continued)*

**PREVEX V3A Glass Reinforced Grade *(continued)***

| Properties | Test Methods | Test Conditions | Metric Units |
|---|---|---|---|
| **Thermal** | | | |
| Deflection Temperature under load 13 × 13 × 130mm bar | ASTM D 648 | 1820 kPa<br>440 kPa | 141 °C |
| Relative Thermal Index<br>Without Impact<br>With Impact | UL Standard 746 B | | |
| Vicat Softening Temperature | ASTM D 1525 | | |
| Coefficient of Linear Thermal Expansion | ASTM D 696 | −30°C to +30°C | $10^{-5}$ cm/cm/°C |
| Thermal Conductivity | ASTM C 177 | | W/km |
| **Flame-Retardant** | | | |
| Flame Class Rating | UL Standard 94 | 1.6mm thickness | 94V-1[1] |
| **Electrical** | | | |
| Hot Wire Ignition | UL Standard 746 A | 1.6mm<br>3.2mm | seconds to ignite |
| High Amperage Arc Ignition | UL Standard 746 A | 1.6mm<br>3.2mm | arcs |
| High Voltage Tracking Rate | UL Standard 746 A | 1.6mm<br>3.2mm | in/min |
| Arc Resistance | ASTM D 495 | 1.6mm<br>3.2mm | seconds |
| Comparative Tracking Index | UL Standard 746 A | 1.6mm<br>3.2mm | volts |
| Dielectric Strength | ASTM D 149 | 23°C | kV/mm |
| Dissipation Factor | ASTM D 150 | 23°C<br>100 cycles | |
| Dielectric Constant | ASTM D 150 | 23°C<br>100 cycles | |
| Volume Resistivity | ASTM D 257 | 23°C | $10^{9}$ ohm/cm |
| **Other Physical Properties** | | | |
| Specific Gravity[2] | ASTM D 792 Method A | 23°C | 1.36 |
| Mold Shrinkage | ASTM D 955 | | 0.001–0.003 cm/cm |
| Water Absorption | ASTM D 570 | 24 hrs at 23°C | % weight |

[1]BWC Test based on UL Standard 94 test methods at 1.6 mm; UL classification pending.
[2]Compression molded. All other specimens injection molded.

# Borg-Warner Chemicals Inc.

## Contact Address:

Borg-Warner Chemicals Inc.
International Center
Parkersburg, WV 26102

Phone: 1 (304) 424-5411

## PREVEX VF1

### Introduction:

Prevex polymers, offered by Borg-Warner Chemicals, Inc., are engineering thermoplastics that balance high impact resistance with excellent performance at elevated temperatures. These resins provide a unique combination of characteristics including Underwriters' Laboratory (UL) flammability classification, Relative Thermal Index (RTI) up to 110°C, electrical properties for direct support, and excellent dimensional stability. Easy flow, minimum drying and excellent thermal stability during molding combine for good processability leading to consistent part production.

To meet the range of design and processing demands presented by today's technology, Prevex polymers are available in numerous grades for injection and structural-foam molding as well as extrusion processing. New grades are being developed to expand the range of options already available from Borg-Warner Chemicals.

### Grade Characteristics:

Prevex VF1 is an extrusion grade developed for sheet and profile applications requiring both flame retardance and high impact strength. VF1 is UL classified 94V-0 with a 20 ft.-lb. Gardner impact and can be used for snap-on insulation or housings used in business machines or telecommunications equipment.

### Product Data:

**PREVEX VF1 Extrusion Grade**

| Properties | Test Methods | Test Conditions | English Units |
|---|---|---|---|
| **Mechanical** | | | |
| Impact Resistance<br>Izod Impact<br>0.125″ bar, notched<br>reverse notched | ASTM D 256<br>Method A | 73°F | 6.0 ft-lb/in |
| Gardner Impact<br>0.125″ specimen | ASTM D 3029<br>Method G | 73°F | 20 ft-lb |
| Dynatup, Total Energy | | 73°F | |

*(continued)*

NOTE—Information provided in this book has been condensed. The publisher recommends that the product manufacturer be contacted, at the address or phone number provided, for more complete information on product safety, processing parameters, and applications not provided in this information sheet.

**PREVEX VF1 Extrusion Grade *(continued)***

| Properties | Test Methods | Test Conditions | English Units |
|---|---|---|---|
| **Mechanical** | | | |
| Tensile Strength 0.125″ thickness | ASTM D 638 | 73°F | 6,500 psi |
| Tensile Modulus 0.125″ thickness | ASTM D 638 | 73°F | 3.3 $10^5$ psi |
| Flexural Yield Strength 0.125 × 0.5 × 5.0″ bar | ASTM D 790 | 73°F | 8,300 psi |
| Flexural Modulus 0.125 × 0.5 × 5.0″ bar | ASTM D 638 | 73°F | 3.4 $10^5$ psi |
| Rockwell Hardness | ASTM D 785 | | 113 R |
| **Thermal** | | | |
| Deflection Temperature under load 0.5 × 0.5 × 5.0″ bar | ASTM D 648 | 264 psi<br>64 psi | 185°F |
| Relative Thermal Index<br>Without Impact<br>With Impact | UL Standard 746 B | | <br>122°F<br>122°F |
| Vicat Softening Temperature | ASTM D 1525 | | |
| Coefficient of Linear Thermal Expansion | ASTM D 696 | −22°F to 86°F | 4.1 $10^{-5}$ in/in/°F |
| Thermal Conductivity | ASTM C 177 | | BTU in/hr ft$^2$°F |
| **Flame-Retardant** | | | |
| Flame Class Rating | UL Standard 94 | 0.062″ thickness | 94V-0 |
| **Electrical** | | | |
| Hot Wire Ignition | UL Standard 746 A | 0.062″<br>0.125″ | seconds to ignite |
| High Amperage Arc Ignition | UL Standard 746 A | 0.062″<br>0.125″ | arcs |
| High Voltage Tracking Rate | UL Standard 746 A | 0.062″<br>0.125″ | in/min |
| Arc Resistance | ASTM D 495 | 0.062″<br>0.125″ | seconds |
| Comparative Tracking Index | UL Standard 746 A | 0.062″<br>0.125″ | volts |
| Dielectric Strength | ASTM D 149 | 73°F | volts/mil |

*(continued)*

**PREVEX VF1 Extrusion Grade *(continued)***

| Properties | Test Methods | Test Conditions | English Units |
|---|---|---|---|
| **Electrical** | | | |
| Dissipation Factor | ASTM D 150 | 73°F 100 cycles | |
| Dielectric Constant | ASTM D 150 | 73°F 100 cycles | |
| Volume Resistivity | ASTM D 257 | 73°F | $10^9$ ohm/cm |
| **Other Physical Properties** | | | |
| Specific Gravity* | ASTM D 792 Method A | 73°F | 1.06 |
| Mold Shrinkage | ASTM D 955 | | |
| Water Absorption | ASTM D 570 | 24 hrs at 73°F | 0.06% weight |

**PREVEX VF1 Extrusion Grade**

| Properties | Test Methods | Test Conditions | Metric Units |
|---|---|---|---|
| **Mechanical** | | | |
| Impact Resistance<br>Izod Impact<br>3.2mm bar, notched<br>reverse notched | ASTM D 256 Method A | 23°C | 320 J/m |
| Gardner Impact 3.2mm specimen | ASTM D 3029 Method G | 23°C | 27 J |
| Dynatup, Total Energy | | 23°C | |
| Tensile Strength 3.2mm thickness | ASTM D 638 | 23°C | 45 MPa |
| Tensile Modulus 3.2mm thickness | ASTM D 638 | 23°C | 2.3 GPa |
| Flexural Yield Strength 3.2 × 13 × 130mm bar | ASTM D 790 | 23°C | 57 MPa |
| Flexural Modulus 3.2 × 13 × 130mm bar | ASTM D 638 | 23°C | 2.3 GPa |
| Rockwell Hardness* | ASTM D 785 | | 113 R |

*Compression molded. All other specimens injection molded.

*(continued)*

**PREVEX VF1 Extrusion Grade *(continued)***

| Properties | Test Methods | Test Conditions | Metric Units |
|---|---|---|---|
| **Thermal** | | | |
| Deflection Temperature under load 13 × 13 × 130mm bar | ASTM D 648 | 1820 kPa<br>440 kPa | 85°C |
| Relative Thermal Index<br>Without Impact<br>With Impact | UL Standard 746 B | | <br>50°C<br>50°C |
| Vicat Softening Temperature | ASTM D 1525 | | |
| Coefficient of Linear Thermal Expansion | ASTM D 696 | −30°C to +30°C | 7.4 $10^{-5}$ cm/cm/°C |
| Thermal Conductivity | ASTM C 177 | | W/km |
| **Flame-Retardant** | | | |
| Flame Class Rating | UL Standard 94 | 1.6mm thickness | 94V-0 |
| **Electrical** | | | |
| Hot Wire Ignition | UL Standard 746 A | 1.6mm<br>3.2mm | seconds to ignite |
| High Amperage Arc Ignition | UL Standard 746 A | 1.6mm<br>3.2mm | arcs |
| High Voltage Tracking Rate | UL Standard 746 A | 1.6mm<br>3.2mm | in/min |
| Arc Resistance | ASTM D 495 | 1.6mm<br>3.2mm | seconds |
| Comparative Tracking Index | UL Standard 746 A | 1.6mm<br>3.2mm | volts |
| Dielectric Strength | ASTM D 149 | 23°C | kV/mm |
| Dissipation Factor | ASTM D 150 | 23°C<br>100 cycles | |
| Dielectric Constant | ASTM D 150 | 23°C<br>100 cycles | |
| Volume Resistivity | ASTM D 257 | 23°C | $10^9$ ohm/cm |
| **Other Physical Properties** | | | |
| Specific Gravity* | ASTM D 792 Method A | 23°C | 1.06 |
| Mold Shrinkage | ASTM D 955 | | |
| Water Absorption | ASTM D 570 | 24 hrs at 23°C | 0.06% weight |

*Compression molded. All other specimens injection molded.

# Borg-Warner Chemicals Inc.

## Contact Address:

Borg-Warner Chemicals Inc.
International Center
Parkersburg, WV 26102

Phone: 1 (304) 424-5411

## PREVEX VK1

### Introduction:

Prevex polymers, offered by Borg-Warner Chemicals, Inc., are engineering thermoplastics that balance high impact resistance with excellent performance at elevated temperatures. These resins provide a unique combination of characteristics including Underwriters' Laboratory (UL) flammability classification, Relative Thermal Index (RTI) up to 110°C, electrical properties for direct support, and excellent dimensional stability. Easy flow, minimum drying and excellent thermal stability during molding combine for good processability leading to consistent part production.

To meet the range of design and processing demands presented by today's technology, Prevex polymers are available in numerous grades for injection and structural-foam molding as well as extrusion processing. New grades are being developed to expand the range of options already available from Borg-Warner Chemicals.

### Grade Characteristics:

Prevex VK1, a flame-retardant grade, is designed for sheet extrusion and profile applications that demand a 94V-0 flame class rating and tolerance for high end-use temperatures. With a DTUL of 225°F, VK1 can be used for internal business machine applications and electrical devices as well as enclosures.

### Product Data:

**PREVEX VK1 Extrusion Grade**

| Properties | Test Methods | Test Conditions | English Units |
|---|---|---|---|
| **Mechanical** | | | |
| Impact Resistance<br>Izod Impact<br>0.125″ bar, notched<br>reverse notched | ASTM D 256<br>Method A | 73°F | 5.5 ft-lb/in |
| Gardner Impact<br>0.125″ specimen | ASTM D 3029<br>Method G | 73°F | 15 ft-lb |
| Dynatup, Total Energy | | 73°F | |

*(continued)*

NOTE—Information provided in this book has been condensed. The publisher recommends that the product manufacturer be contacted, at the address or phone number provided, for more complete information on product safety, processing parameters, and applications not provided in this information sheet.

**PREVEX VK1 Extrusion Grade *(continued)***

| Properties | Test Methods | Test Conditions | English Units |
|---|---|---|---|
| **Mechanical** | | | |
| Tensile Strength 0.125″ thickness | ASTM D 638 | 73°F | 8,000 psi |
| Tensile Modulus 0.125″ thickness | ASTM D 638 | 73°F | 3.6 $10^5$ psi |
| Flexural Yield Strength 0.125 × 0.5 × 5.0″ bar | ASTM D 790 | 73°F | 13,600 psi |
| Flexural Modulus 0.125 × 0.5 × 5.0″ bar | ASTM D 638 | 73°F | 3.7 $10^5$ psi |
| Rockwell Hardness | ASTM D 785 | | 120 R |
| **Thermal** | | | |
| Deflection Temperature under load 0.5 × 0.5 × 5.0″ bar | ASTM D 648 | 264 psi<br>64 psi | 225°F |
| Relative Thermal Index<br>Without Impact<br>With Impact | UL Standard 746 B | | <br>203°F<br>176°F |
| Vicat Softening Temperature | ASTM D 1525 | | 250°F |
| Coefficient of Linear Thermal Expansion | ASTM D 696 | −22°F to 86°F | 3.8 $10^{-5}$ in/in/°F |
| Thermal Conductivity | ASTM C 177 | | BTU in/hr ft²°F |
| **Flame-Retardant** | | | |
| Flame Class Rating | UL Standard 94 | 0.062″ thickness | 94V-0 |
| **Electrical** | | | |
| Hot Wire Ignition | UL Standard 746 A | 0.062″<br>0.125″ | 98 seconds to ignite |
| High Amperage Arc Ignition | UL Standard 746 A | 0.062″<br>0.125″ | 200+ arcs |
| High Voltage Tracking Rate | UL Standard 746 A | 0.062″<br>0.125″ | 15 in/min |
| Arc Resistance | ASTM D 495 | 0.062″<br>0.125″ | 63 seconds |
| Comparative Tracking Index | UL Standard 746 A | 0.062″<br>0.125″ | 460 volts |
| Dielectric Strength | ASTM D 149 | 73°F | volts/mil |

*(continued)*

**PREVEX VK1 Extrusion Grade *(continued)***

| Properties | Test Methods | Test Conditions | English Units |
|---|---|---|---|
| **Electrical** | | | |
| Dissipation Factor | ASTM D 150 | 73°F<br>100 cycles | |
| Dielectric Constant | ASTM D 150 | 73°F<br>100 cycles | |
| Volume Resistivity | ASTM D 257 | 73°F | 3.5[1] $10^{9}$ ohm/cm |
| **Other Physical Properties** | | | |
| Specific Gravity[2] | ASTM D 792<br>Method A | 73°F | 1.09 |
| Mold Shrinkage | ASTM D 955 | | |
| Water Absorption | ASTM D 570 | 24 hrs at<br>73°F | 0.06% weight |

[1]Compression molded. All other specimens injection molded.
[2]$10^{14}$ meg ohms.

**PREVEX VK1 Extrusion Grade**

| Properties | Test Methods | Test Conditions | Metric Units |
|---|---|---|---|
| **Mechanical** | | | |
| Impact Resistance<br>Izod Impact<br>3.2mm bar, notched<br>reverse notched | ASTM D 256<br>Method A | 23°C | 290 J/m |
| Gardner Impact<br>3.2mm specimen | ASTM D 3029<br>Method G | 23°C | 20 J |
| Dynatup, Total Energy | | 23°C | |
| Tensile Strength<br>3.2mm thickness | ASTM D 638 | 23°C | 55 MPa |
| Tensile Modulus<br>3.2mm thickness | ASTM D 638 | 23°C | 2.5 GPa |
| Flexural Yield Strength<br>3.2 × 13 × 130mm bar | ASTM D 790 | 23°C | 94 MPa |
| Flexural Modulus<br>3.2 × 13 × 130mm bar | ASTM D 638 | 23°C | 2.5 GPa |
| Rockwell Hardness* | ASTM D 785 | | 120 R |

*Compression molded. All other specimens injection molded.

*(continued)*

**PREVEX VK1 Extrusion Grade *(continued)***

| Properties | Test Methods | Test Conditions | Metric Units |
|---|---|---|---|
| **Thermal** | | | |
| Deflection Temperature under load 13 × 13 × 130mm bar | ASTM D 648 | 1820 kPa<br>440 kPa | 107°C |
| Relative Thermal Index<br>Without Impact<br>With Impact | UL Standard 746 B | | <br>95°C<br>80°C |
| Vicat Softening Temperature | ASTM D 1525 | | 121°C |
| Coefficient of Linear Thermal Expansion | ASTM D 696 | −30°C to +30°C | 6.8 $10^{-5}$ cm/cm/°C |
| Thermal Conductivity | ASTM C 177 | | W/km |
| **Flame-Retardant** | | | |
| Flame Class Rating | UL Standard 94 | 1.6mm thickness | 94V-0 |
| **Electrical** | | | |
| Hot Wire Ignition | UL Standard 746 A | 1.6mm<br>3.2mm | 98 seconds to ignite |
| High Amperage Arc Ignition | UL Standard 746 A | 1.6mm<br>3.2mm | 200+ arcs |
| High Voltage Tracking Rate | UL Standard 746 A | 1.6mm<br>3.2mm | 38 in/min |
| Arc Resistance | ASTM D 495 | 1.6mm<br>3.2mm | 63 seconds |
| Comparative Tracking Index | UL Standard 746 A | 1.6mm<br>3.2mm | 460 volts |
| Dielectric Strength | ASTM D 149 | 23°C | kV/mm |
| Dissipation Factor | ASTM D 150 | 23°C<br>100 cycles | |
| Dielectric Constant | ASTM D 150 | 23°C<br>100 cycles | |
| Volume Resistivity | ASTM D 257 | 23°C | 3.5[1] $10^9$ ohm/cm |
| **Other Physical Properties** | | | |
| Specific Gravity[2] | ASTM D 792 Method A | 23°C | 1.09 |
| Mold Shrinkage | ASTM D 955 | | |
| Water Absorption | ASTM D 570 | 24 hrs at 23°C | 0.06% weight |

[1]Compression molded. All other specimens injection molded.
[2]$10^{14}$ meg ohms.

# Borg-Warner Chemicals Inc.

## Contact Address:

Borg-Warner Chemicals Inc.
International Center
Parkersburg, WV 26102

Phone: 1 (304) 424-5411

## PREVEX VKA

### Introduction:

Prevex polymers, offered by Borg-Warner Chemicals, Inc., are engineering thermoplastics that balance high impact resistance with excellent performance at elevated temperatures. These resins provide a unique combination of characteristics including Underwriters' Laboratory (UL) flammability classification, Relative Thermal Index (RTI) up to 110°C, electrical properties for direct support, and excellent dimensional stability. Easy flow, minimum drying and excellent thermal stability during molding combine for good processability leading to consistent part production.

To meet the range of design and processing demands presented by today's technology, Prevex polymers are available in numerous grades for injection and structural-foam molding as well as extrusion processing. New grades are being developed to expand the range of options already available from Borg-Warner Chemicals.

### Grade Characteristics:

Prevex VKA is a grade with 94V-0 classification (at 0.062 in.) for thin wall designs that also offers excellent electrical properties. Prevex VKA, with a 95°C RTI is suitable for electronic enclosures, timers, outlet boxes, solenoids, and many internal appliance components.

### Product Data:

**PREVEX VKA Molding Grade**

| Properties | Test Methods | Test Conditions | English Units |
|---|---|---|---|
| **Mechanical** | | | |
| Impact Resistance<br>Izod Impact<br>0.125″ bar, notched<br>reverse notched | ASTM D 256<br>Method A | 73°F | 5.5 ft-lb/in |
| Gardner Impact<br>0.125″ specimen | ASTM D 3029<br>Method G | 73°F | 15 ft-lb |

*(continued)*

NOTE—Information provided in this book has been condensed. The publisher recommends that the product manufacturer be contacted, at the address or phone number provided, for more complete information on product safety, processing parameters, and applications not provided in this information sheet.

PREVEX VKA Molding Grade *(continued)*

| Properties | Test Methods | Test Conditions | English Units |
|---|---|---|---|
| **Mechanical** | | | |
| Dynatup, Total Energy | | 73°F | |
| Tensile Strength<br>0.125″ thickness | ASTM D 638 | 73°F | 8,000 psi |
| Tensile Modulus<br>0.125″ thickness | ASTM D 638 | 73°F | 3.6 $10^5$ psi |
| Flexural Yield Strength<br>0.125 × 0.5 × 5.0″ bar | ASTM D 790 | 73°F | 13,600 psi |
| Flexural Modulus<br>0.125 × 0.5 × 5.0″ bar | ASTM D 638 | 73°F | 3.7 $10^5$ psi |
| Rockwell Hardness | ASTM D 785 | | 120 R |
| **Thermal** | | | |
| Deflection Temperature under load<br>0.5 × 0.5 × 5.0″ bar | ASTM D 648 | 264 psi<br>64 psi | 225°F |
| Relative Thermal Index<br>Without Impact<br>With Impact | UL Standard<br>746 B | | <br>203°F<br>176°F |
| Vicat Softening Temperature | ASTM D 1525 | | 250°F |
| Coefficient of Linear Thermal Expansion | ASTM D 696 | −22°F to 86°F | 3.8 $10^{-5}$ in/in/°F |
| Thermal Conductivity | ASTM C 177 | | BTU in/hr ft²°F |
| **Flame-Retardant** | | | |
| Flame Class Rating | UL Standard<br>94 | 0.062″<br>thickness | 94V-0 |
| **Electrical** | | | |
| Hot Wire Ignition | UL Standard<br>746 A | 0.062″<br>0.125″ | 98 seconds<br>120 to ignite |
| High Amperage Arc Ignition | UL Standard<br>746 A | 0.062″<br>0.125″ | 200+ arcs<br>200+ arcs |
| High Voltage Tracking Rate | UL Standard<br>746 A | 0.062″<br>0.125″ | 15.0 in/min<br>15.0 in/min |
| Arc Resistance | ASTM D 495 | 0.062″<br>0.125″ | 63 seconds<br>94 seconds |
| Comparative Tracking Index | UL Standard<br>746 A | 0.062″<br>0.125″ | 460 volts<br>460 volts |
| Dielectric Strength | ASTM D 149 | 73°F | volts/mil |

*(continued)*

### PREVEX VKA Molding Grade *(continued)*

| Properties | Test Methods | Test Conditions | English Units |
|---|---|---|---|
| **Electrical** | | | |
| Dissipation Factor | ASTM D 150 | 73°F<br>100 cycles | |
| Dielectric Constant | ASTM D 150 | 73°F<br>100 cycles | |
| Volume Resistivity | ASTM D 257 | 73°F | $10^9$ ohm/cm |
| **Other Physical Properties** | | | |
| Specific Gravity* | ASTM D 792<br>Method A | 73°F | 1.09 |
| Mold Shrinkage | ASTM D 955 | | 0.006–0.007 in/in |
| Water Absorption | ASTM D 570 | 24 hrs at<br>73°F | 0.06% weight |

### PREVEX VKA Molding Grade

| Properties | Test Methods | Test Conditions | Metric Units |
|---|---|---|---|
| **Mechanical** | | | |
| Impact Resistance<br>Izod Impact<br>3.2mm bar, notched<br>reverse notched | ASTM D 256<br>Method A | 23°C | 290 J/m |
| Gardner Impact<br>3.2mm specimen | ASTM D 3029<br>Method G | 23°C | 20 J |
| Dynatup, Total Energy | | 23°C | |
| Tensile Strength<br>3.2mm thickness | ASTM D 638 | 23°C | 55 MPa |
| Tensile Modulus<br>3.2mm thickness | ASTM D 638 | 23°C | 2.5 GPa |
| Flexural Yield Strength<br>3.2 × 13 × 130mm bar | ASTM D 790 | 23°C | 94 MPa |
| Flexural Modulus<br>3.2 × 13 × 130mm bar | ASTM D 638 | 23°C | 2.5 GPa |
| Rockwell Hardness* | ASTM D 785 | | 120 R |

*Compression molded. All other specimens injection molded.

*(continued)*

## PREVEX VKA Molding Grade *(continued)*

| Properties | Test Methods | Test Conditions | Metric Units |
|---|---|---|---|
| **Thermal** | | | |
| Deflection Temperature under load 13 × 13 × 130mm bar | ASTM D 648 | 1820 kPa<br>440 kPa | 107°C |
| Relative Thermal Index<br>Without Impact<br>With Impact | UL Standard 746 B | | <br>95°C<br>80°C |
| Vicat Softening Temperature | ASTM D 1525 | | 121°C |
| Coefficient of Linear Thermal Expansion | ASTM D 696 | −30°C to +30°C | 6.8 $10^{-5}$ cm/cm/°C |
| Thermal Conductivity | ASTM C 177 | | W/km |
| **Flame-Retardant** | | | |
| Flame Class Rating | UL Standard 94 | 1.6mm thickness | 94V-0 |
| **Electrical** | | | |
| Hot Wire Ignition | UL Standard 746 A | 1.6mm<br>3.2mm | 98 seconds<br>120 to ignite |
| High Amperage Arc Ignition | UL Standard 746 A | 1.6mm<br>3.2mm | 200+ arcs<br>200+ arcs |
| High Voltage Tracking Rate | UL Standard 746 A | 1.6mm<br>3.2mm | 38 in/min<br>38 in/min |
| Arc Resistance | ASTM D 495 | 1.6mm<br>3.2mm | 63 seconds<br>94 seconds |
| Comparative Tracking Index | UL Standard 746 A | 1.6mm<br>3.2mm | 460 volts<br>460 volts |
| Dielectric Strength | ASTM D 149 | 23°C | kV/mm |
| Dissipation Factor | ASTM D 150 | 23°C<br>100 cycles | |
| Dielectric Constant | ASTM D 150 | 23°C<br>100 cycles | |
| Volume Resistivity | ASTM D 257 | 23°C | $10^9$ ohm/cm |
| **Other Physical Properties** | | | |
| Specific Gravity* | ASTM D 792 Method A | 23°C | 1.09 |
| Mold Shrinkage | ASTM D 955 | | 0.006–0.007 cm/cm |
| Water Absorption | ASTM D 570 | 24 hrs at 23°C | 0.06% weight |

*Compression molded. All other specimens injection molded.

# Borg-Warner Chemicals Inc.

## Contact Address:

Borg-Warner Chemicals Inc.
International Center
Parkersburg, WV 26102

Phone: 1 (304) 424-5411

## PREVEX VJB

### Introduction:

Prevex polymers, offered by Borg-Warner Chemicals, Inc., are engineering thermoplastics that balance high impact resistance with excellent performance at elevated temperatures. These resins provide a unique combination of characteristics including Underwriters' Laboratory (UL) flammability classification, Relative Thermal Index (RTI) up to 110°C, electrical properties for direct support, and excellent dimensional stability. Easy flow, minimum drying and excellent thermal stability during molding combine for good processability leading to consistent part production.

To meet the range of design and processing demands presented by today's technology, Prevex polymers are available in numerous grades for injection and structural-foam molding as well as extrusion processing. New grades are being developed to expand the range of options already available from Borg-Warner Chemicals.

### Grade Characteristics:

Prevex VJB is a flame-retardant grade (UL 94V-1) which provides both heat and impact resistance. VJB exhibits a Deflection Temperature Under Load (DTUL) of 212°F. Its toughness and dimensional stability are required for business machine housings and telecommunications equipment.

### Product Data:

**PREVEX VJB Molding Grade**

| Properties | Test Methods | Test Conditions | English Units |
|---|---|---|---|
| **Mechanical** | | | |
| Impact Resistance<br>Izod Impact<br>0.125″ bar, notched<br>reverse notched | ASTM D 256<br>Method A | 73°F | 5.0 ft-lb/in |
| Gardner Impact<br>0.125″ specimen | ASTM D 3029<br>Method G | 73°F | 16 ft-lb |
| Dynatup, Total Energy | | 73°F | |

*(continued)*

NOTE—Information provided in this book has been condensed. The publisher recommends that the product manufacturer be contacted, at the address or phone number provided, for more complete information on product safety, processing parameters, and applications not provided in this information sheet.

**PREVEX VJB Molding Grade *(continued)***

| Properties | Test Methods | Test Conditions | English Units |
|---|---|---|---|
| **Mechanical** | | | |
| Tensile Strength<br>0.125″ thickness | ASTM D 638 | 73°F | 5,400 psi |
| Tensile Modulus<br>0.125″ thickness | ASTM D 638 | 73°F | 3.6 $10^5$ psi |
| Flexural Yield Strength<br>0.125 × 0.5 × 5.0″ bar | ASTM D 790 | 73°F | 11,000 psi |
| Flexural Modulus<br>0.125 × 0.5 × 5.0″ bar | ASTM D 638 | 73°F | 3.6 $10^5$ psi |
| Rockwell Hardness | ASTM D 785 | | 113 R |
| **Thermal** | | | |
| Deflection Temperature under load<br>0.5 × 0.5 × 5.0″ bar | ASTM D 648 | 264 psi<br>64 psi | 212°F |
| Relative Thermal Index<br>Without Impact<br>With Impact | UL Standard 746 B | | |
| Vicat Softening Temperature | ASTM D 1525 | | 237°F |
| Coefficient of Linear Thermal Expansion | ASTM D 696 | −22°F to 86°F | 3.8 $10^{-5}$ in/in/°F |
| Thermal Conductivity | ASTM C 177 | | BTU in/hr ft²°F |
| **Flame-Retardant** | | | |
| Flame Class Rating | UL Standard 94 | 0.062″ thickness | 94V-1 |
| **Electrical** | | | |
| Hot Wire Ignition | UL Standard 746 A | 0.062″<br>0.125″ | 45 seconds to ignite |
| High Amperage Arc Ignition | UL Standard 746 A | 0.062″<br>0.125″ | 200+ arcs |
| High Voltage Tracking Rate | UL Standard 746 A | 0.062″<br>0.125″ | 5.0 in/min |
| Arc Resistance | ASTM D 495 | 0.062″<br>0.125″ | 64 seconds |
| Comparative Tracking Index | UL Standard 746 A | 0.062″<br>0.125″ | 575 volts<br>575 volts |
| Dielectric Strength | ASTM D 149 | 73°F | volts/mil |

*(continued)*

PREVEX VJB Molding Grade *(continued)*

| Properties | Test Methods | Test Conditions | English Units |
|---|---|---|---|
| **Electrical** | | | |
| Dissipation Factor | ASTM D 150 | 73°F<br>100 cycles | |
| Dielectric Constant | ASTM D 150 | 73°F<br>100 cycles | |
| Volume Resistivity | ASTM D 257 | 73°F | $10^9$ ohm/cm |
| **Other Physical Properties** | | | |
| Specific Gravity* | ASTM D 792<br>Method A | 73°F | 1.06 |
| Mold Shrinkage | ASTM D 955 | | 0.006–0.008 in/in |
| Water Absorption | ASTM D 570 | 24 hrs at<br>73°F | 0.06% weight |

PREVEX VJB Molding Grade

| Properties | Test Methods | Test Conditions | Metric Units |
|---|---|---|---|
| **Mechanical** | | | |
| Impact Resistance<br>Izod Impact<br>3.2mm bar, notched<br>reverse notched | ASTM D 256<br>Method A | 23°C | 270 J/m |
| Gardner Impact<br>3.2mm specimen | ASTM D 3029<br>Method G | 23°C | 22 J |
| Dynatup, Total Energy | | 23°C | |
| Tensile Strength<br>3.2mm thickness | ASTM D 638 | 23°C | 37 MPa |
| Tensile Modulus<br>3.2mm thickness | ASTM D 638 | 23°C | 2.5 GPa |
| Flexural Yield Strength<br>3.2 × 13 × 130mm bar | ASTM D 790 | 23°C | 76 MPa |
| Flexural Modulus<br>3.2 × 13 × 130mm bar | ASTM D 638 | 23°C | 2.5 GPa |
| Rockwell Hardness* | ASTM D 785 | | 113 R |

*Compression molded. All other specimens injection molded.

*(continued)*

**PREVEX VJB Molding Grade *(continued)***

| Properties | Test Methods | Test Conditions | Metric Units |
|---|---|---|---|
| **Thermal** | | | |
| Deflection Temperature under load 13 × 13 × 130mm bar | ASTM D 648 | 1820 kPa<br>440 kPa | 100°C |
| Relative Thermal Index<br>Without Impact<br>With Impact | UL Standard 746 B | | |
| Vicat Softening Temperature | ASTM D 1525 | | 114°C |
| Coefficient of Linear Thermal Expansion | ASTM D 696 | −30°C to +30°C | 6.8 $10^{-5}$ cm/cm/°C |
| Thermal Conductivity | ASTM C 177 | | W/km |
| **Flame-Retardant** | | | |
| Flame Class Rating | UL Standard 94 | 1.6mm thickness | 94V-1 |
| **Electrical** | | | |
| Hot Wire Ignition | UL Standard 746 A | 1.6mm<br>3.2mm | 45 seconds to ignite |
| High Amperage Arc Ignition | UL Standard 746 A | 1.6mm<br>3.2mm | 200+ arcs |
| High Voltage Tracking Rate | UL Standard 746 A | 1.6mm<br>3.2mm | 12.7 in/min |
| Arc Resistance | ASTM D 495 | 1.6mm<br>3.2mm | 64 seconds |
| Comparative Tracking Index | UL Standard 746 A | 1.6mm<br>3.2mm | 575 volts |
| Dielectric Strength | ASTM D 149 | 23°C | kV/mm |
| Dissipation Factor | ASTM D 150 | 23°C<br>100 cycles | |
| Dielectric Constant | ASTM D 150 | 23°C<br>100 cycles | |
| Volume Resistivity | ASTM D 257 | 23°C | $10^9$ ohm/cm |
| **Other Physical Properties** | | | |
| Specific Gravity* | ASTM D 792 Method A | 23°C | 1.06 |
| Mold Shrinkage | ASTM D 955 | | 0.006–0.008 cm/cm |
| Water Absorption | ASTM D 570 | 24 hrs at 23°C | 0.06% weight |

*Compression molded. All other specimens injection molded.

# Borg-Warner Chemicals Inc.

## Contact Address:

Borg-Warner Chemicals Inc.
International Center
Parkersburg, WV 26102

Phone: 1 (304) 424-5411

## PREVEX VM1

### Introduction:

Prevex polymers, offered by Borg-Warner Chemicals, Inc., are engineering thermoplastics that balance high impact resistance with excellent performance at elevated temperatures. These resins provide a unique combination of characteristics including Underwriters' Laboratory (UL) flammability classification, Relative Thermal Index (RTI) up to 110°C, electrical properties for direct support, and excellent dimensional stability. Easy flow, minimum drying and excellent thermal stability during molding combine for good processability leading to consistent part production.

To meet the range of design and processing demands presented by today's technology, Prevex polymers are available in numerous grades for injection and structural-foam molding as well as extrusion processing. New grades are being developed to expand the range of options already available from Borg-Warner Chemicals.

### Grade Characteristics:

Prevex VM1 is a flame-retardant grade that also exhibits high heat resistance. Prevex VM1 is classified 94V-1 and has a 90°C RTI. It is suitable for applications that require dimensional stability and load-bearing capabilities such as load centers, enclosures, panel boards, and other indirect contacting devices.

### Product Data:

**PREVEX VM1 Extrusion Grade**

| Properties | Test Methods | Test Conditions | English Units |
|---|---|---|---|
| **Mechanical** | | | |
| Impact Resistance<br>Izod Impact<br>0.125″ bar, notched<br>reverse notched | ASTM D 256<br>Method A | 73°F | 5.0 ft-lb/in |
| Gardner Impact<br>0.125″ specimen | ASTM D 3029<br>Method G | 73°F | 18 ft-lb |
| Dynatup, Total Energy | | 73°F | |

*(continued)*

NOTE—Information provided in this book has been condensed. The publisher recommends that the product manufacturer be contacted, at the address or phone number provided, for more complete information on product safety, processing parameters, and applications not provided in this information sheet.

**PREVEX VM1 Extrusion Grade *(continued)***

| Properties | Test Methods | Test Conditions | English Units |
|---|---|---|---|
| **Mechanical** | | | |
| Tensile Strength<br>0.125″ thickness | ASTM D 638 | 73°F | 7,700 psi |
| Tensile Modulus<br>0.125″ thickness | ASTM D 638 | 73°F | 3.3 $10^5$ psi |
| Flexural Yield Strength<br>0.125 × 0.5 × 5.0″ bar | ASTM D 790 | 73°F | 14,200 psi |
| Flexural Modulus<br>0.125 × 0.5 × 5.0″ bar | ASTM D 638 | 73°F | 3.6 $10^5$ psi |
| Rockwell Hardness | ASTM D 785 | | 118 R |
| **Thermal** | | | |
| Deflection Temperature under load<br>0.5 × 0.5 × 5.0″ bar | ASTM D 648 | 264 psi<br>64 psi | 240°F |
| Relative Thermal Index<br>Without Impact<br>With Impact | UL Standard 746 B | | |
| Vicat Softening Temperature | ASTM D 1525 | | 279°F |
| Coefficient of Linear Thermal Expansion | ASTM D 696 | −22°F to 86°F | 3.7 $10^{-5}$ in/in/°F |
| Thermal Conductivity | ASTM C 177 | | BTU in/hr ft²°F |
| **Flame-Retardant** | | | |
| Flame Class Rating | UL Standard 94 | 0.062″ thickness | 94V-1[1] |
| **Electrical** | | | |
| Hot Wire Ignition | UL Standard 746 A | 0.062″<br>0.125″ | seconds to ignite |
| High Amperage Arc Ignition | UL Standard 746 A | 0.062″<br>0.125″ | arcs |
| High Voltage Tracking Rate | UL Standard 746 A | 0.062″<br>0.125″ | in/min |
| Arc Resistance | ASTM D 495 | 0.062″<br>0.125″ | seconds |
| Comparative Tracking Index | UL Standard 746 A | 0.062″<br>0.125″ | volts |
| Dielectric Strength | ASTM D 149 | 73°F | volts/mil |

[1]BWC Test based on UL Standard 94 test methods at 0.062 inch; UL classification pending.

*(continued)*

PREVEX VM1 Extrusion Grade *(continued)*

| Properties | Test Methods | Test Conditions | English Units |
|---|---|---|---|
| **Electrical** | | | |
| Dissipation Factor | ASTM D 150 | 73°F 100 cycles | |
| Dielectric Constant | ASTM D 150 | 73°F 100 cycles | |
| Volume Resistivity | ASTM D 257 | 73°F | $10^9$ ohm/cm |
| **Other Physical Properties** | | | |
| Specific Gravity[2] | ASTM D 792 Method A | 73°F | 1.08 |
| Mold Shrinkage | ASTM D 955 | | |
| Water Absorption | ASTM D 570 | 24 hrs at 73°F | 0.06% weight |

[2]Compression molded. All other specimens injection molded.

PREVEX VM1 Extrusion Grade

| Properties | Test Methods | Test Conditions | Metric Units |
|---|---|---|---|
| **Mechanical** | | | |
| Impact Resistance<br>Izod Impact<br>3.2mm bar, notched<br>reverse notched | ASTM D 256 Method A | 23°C | 270 J/m |
| Gardner Impact 3.2mm specimen | ASTM D 3029 Method G | 23°C | 24 J |
| Dynatup, Total Energy | | 23°C | |
| Tensile Strength 3.2mm thickness | ASTM D 638 | 23°C | 53 MPa |
| Tensile Modulus 3.2mm thickness | ASTM D 638 | 23°C | 2.3 GPa |
| Flexural Yield Strength 3.2 × 13 × 130mm bar | ASTM D 790 | 23°C | 98 MPa |
| Flexural Modulus 3.2 × 13 × 130mm bar | ASTM D 638 | 23°C | 2.5 GPa |
| Rockwell Hardness* | ASTM D 785 | | 118 R |

*Compression molded. All other specimens injection molded.

*(continued)*

## PREVEX VM1 Extrusion Grade *(continued)*

| Properties | Test Methods | Test Conditions | Metric Units |
|---|---|---|---|
| **Thermal** | | | |
| Deflection Temperature under load 13 × 13 × 130mm bar | ASTM D 648 | 1820 kPa<br>440 kPa | 116°C |
| Relative Thermal Index<br>Without Impact<br>With Impact | UL Standard 746 B | | |
| Vicat Softening Temperature | ASTM D 1525 | | 137°C |
| Coefficient of Linear Thermal Expansion | ASTM D 696 | −30°C to +30°C | 6.6 $10^{-5}$ cm/cm/°C |
| Thermal Conductivity | ASTM C 177 | | W/km |
| **Flame-Retardant** | | | |
| Flame Class Rating | UL Standard 94 | 1.6mm thickness | 94V-1[1] |
| **Electrical** | | | |
| Hot Wire Ignition | UL Standard 746 A | 1.6mm<br>3.2mm | seconds to ignite |
| High Amperage Arc Ignition | UL Standard 746 A | 1.6mm<br>3.2mm | arcs |
| High Voltage Tracking Rate | UL Standard 746 A | 1.6mm<br>3.2mm | in/min |
| Arc Resistance | ASTM D 495 | 1.6mm<br>3.2mm | seconds |
| Comparative Tracking Index | UL Standard 746 A | 1.6mm<br>3.2mm | volts |
| Dielectric Strength | ASTM D 149 | 23°C | kV/mm |
| Dissipation Factor | ASTM D 150 | 23°C<br>100 cycles | |
| Dielectric Constant | ASTM D 150 | 23°C<br>100 cycles | |
| Volume Resistivity | ASTM D 257 | 23°C | $10^9$ ohm/cm |
| **Other Physical Properties** | | | |
| Specific Gravity[2] | ASTM D 792 Method A | 23°C | 1.08 |
| Mold Shrinkage | ASTM D 955 | | |
| Water Absorption | ASTM D 570 | 24 hrs at 23°C | 0.06% weight |

[1]BWC Test based on UL Standard 94 test methods at 1.6 mm; UL classification pending.
[2]Compression molded. All other specimens injection molded.

# Borg-Warner Chemicals Inc.

## Contact Address:

Borg-Warner Chemicals Inc.
International Center
Parkersburg, WV 26102

Phone: 1 (304) 424-5411

## PREVEX VQ1

### Introduction:

Prevex polymers, offered by Borg-Warner Chemicals, Inc., are engineering thermoplastics that balance high impact resistance with excellent performance at elevated temperatures. These resins provide a unique combination of characteristics including Underwriters' Laboratory (UL) flammability classification, Relative Thermal Index (RTI) up to 110°C, electrical properties for direct support, and excellent dimensional stability. Easy flow, minimum drying and excellent thermal stability during molding combine for good processability leading to consistent part production.

To meet the range of design and processing demands presented by today's technology, Prevex polymers are available in numerous grades for injection and structural-foam molding as well as extrusion processing. New grades are being developed to expand the range of options already available from Borg-Warner Chemicals.

### Grade Characteristics:

Prevex VQ1 is an extrusion grade developed for high temperature sheet and profile applications requiring flame retardance and good electrical properties. VQ1 is classified 94V-1 and has a 105°C RTI.

### Product Data:

**PREVEX VQ1 Extrusion Grade**

| Properties | Test Methods | Test Conditions | English Units |
|---|---|---|---|
| **Mechanical** | | | |
| Impact Resistance<br>Izod Impact<br>0.125″ bar, notched<br>reverse notched | ASTM D 256<br>Method A | 73°F | 5.0 ft-lb/in |
| Gardner Impact<br>0.125″ specimen | ASTM D 3029<br>Method G | 73°F | 16 ft-lb |
| Dynatup, Total Energy | | 73°F | |
| Tensile Strength<br>0.125″ thickness | ASTM D 638 | 73°F | 9,000 psi |

*(continued)*

NOTE—Information provided in this book has been condensed. The publisher recommends that the product manufacturer be contacted, at the address or phone number provided, for more complete information on product safety, processing parameters, and applications not provided in this information sheet.

PREVEX VQ1 Extrusion Grade *(continued)*

| Properties | Test Methods | Test Conditions | English Units |
|---|---|---|---|
| **Mechanical** | | | |
| Tensile Modulus 0.125″ thickness | ASTM D 638 | 73°F | 3.6 $10^5$ psi |
| Flexural Yield Strength 0.125 × 0.5 × 5.0″ bar | ASTM D 790 | 73°F | 14,000 psi |
| Flexural Modulus 0.125 × 0.5 × 5.0″ bar | ASTM D 638 | 73°F | 3.8 $10^5$ psi |
| Rockwell Hardness | ASTM D 785 | | 120 R |
| **Thermal** | | | |
| Deflection Temperature under load 0.5 × 0.5 × 5.0″ bar | ASTM D 648 | 264 psi<br>64 psi | 265°F |
| Relative Thermal Index<br>Without Impact<br>With Impact | UL Standard 746 B | | <br>230°F<br>221°F |
| Vicat Softening Temperature | ASTM D 1525 | | |
| Coefficient of Linear Thermal Expansion | ASTM D 696 | −22°F to 86°F | 3.3 $10^{-5}$ in/in/°F |
| Thermal Conductivity | ASTM C 177 | | BTU in/hr ft²°F |
| **Flame-Retardant** | | | |
| Flame Class Rating | UL Standard 94 | 0.062″ thickness | |
| **Electrical** | | | |
| Hot Wire Ignition | UL Standard 746 A | 0.062″<br>0.125″ | 120 seconds to ignite |
| High Amperage Arc Ignition | UL Standard 746 A | 0.062″<br>0.125″ | 200+ arcs<br>200+ arcs |
| High Voltage Tracking Rate | UL Standard 746 A | 0.062″<br>0.125″ | 18.1 in/min |
| Arc Resistance | ASTM D 495 | 0.062″<br>0.125″ | 68 seconds |
| Comparative Tracking Index | UL Standard 746 A | 0.062″<br>0.125″ | 460 volts |
| Dielectric Strength | ASTM D 149 | 73°F | 744 volts/mil |

*(continued)*

**PREVEX VQ1 Extrusion Grade *(continued)***

| Properties | Test Methods | Test Conditions | English Units |
|---|---|---|---|
| **Electrical** | | | |
| Dissipation Factor | ASTM D 150 | 73°F<br>100 cycles | 0.0023 |
| Dielectric Constant | ASTM D 150 | 73°F<br>100 cycles | 2.41 |
| Volume Resistivity | ASTM D 257 | 73°F | 3.21 $10^9$ ohm/cm |
| **Other Physical Properties** | | | |
| Specific Gravity* | ASTM D 792<br>Method A | 73°F | 1.06 |
| Mold Shrinkage | ASTM D 955 | | |
| Water Absorption | ASTM D 570 | 24 hrs at<br>73°F | 0.06% weight |

**PREVEX VQ1 Extrusion Grade**

| Properties | Test Methods | Test Conditions | Metric Units |
|---|---|---|---|
| **Mechanical** | | | |
| Impact Resistance<br>Izod Impact<br>3.2mm bar, notched<br>reverse notched | ASTM D 256<br>Method A | 23°C | 270 J/m |
| Gardner Impact<br>3.2mm specimen | ASTM D 3029<br>Method G | 23°C | 22 J |
| Dynatup, Total Energy | | 23°C | |
| Tensile Strength<br>3.2mm thickness | ASTM D 638 | 23°C | 62 MPa |
| Tensile Modulus<br>3.2mm thickness | ASTM D 638 | 23°C | 2.5 GPa |
| Flexural Yield Strength<br>3.2 × 13 × 130mm bar | ASTM D 790 | 23°C | 96 MPa |
| Flexural Modulus<br>3.2 × 13 × 130mm bar | ASTM D 638 | 23°C | 2.6 GPa |
| Rockwell Hardness* | ASTM D 785 | | 120 R |

*Compression molded. All other specimens injection molded.

*(continued)*

**PREVEX VQ1 Extrusion Grade *(continued)***

| Properties | Test Methods | Test Conditions | Metric Units |
|---|---|---|---|
| **Thermal** | | | |
| Deflection Temperature under load 13 × 13 × 130mm bar | ASTM D 648 | 1820 kPa<br>440 kPa | 129°C |
| Relative Thermal Index<br>Without Impact<br>With Impact | UL Standard 746 B | | <br>110°C<br>105°C |
| Vicat Softening Temperature | ASTM D 1525 | | |
| Coefficient of Linear Thermal Expansion | ASTM D 696 | −30°C to +30°C | 5.9 $10^{-5}$ cm/cm/°C |
| Thermal Conductivity | ASTM C 177 | | W/km |
| **Flame-Retardant** | | | |
| Flame Class Rating | UL Standard 94 | 1.6mm thickness | |
| **Electrical** | | | |
| Hot Wire Ignition | UL Standard 746 A | 1.6mm<br>3.2mm | 120 seconds to ignite |
| High Amperage Arc Ignition | UL Standard 746 A | 1.6mm<br>3.2mm | 200+ arcs |
| High Voltage Tracking Rate | UL Standard 746 A | 1.6mm<br>3.2mm | 46 in/min |
| Arc Resistance | ASTM D 495 | 1.6mm<br>3.2mm | 68 seconds |
| Comparative Tracking Index | UL Standard 746 A | 1.6mm<br>3.2mm | 460 volts |
| Dielectric Strength | ASTM D 149 | 23°C | 29.3 kV/mm |
| Dissipation Factor | ASTM D 150 | 23°C<br>100 cycles | 0.0023 |
| Dielectric Constant | ASTM D 150 | 23°C<br>100 cycles | 2.41 |
| Volume Resistivity | ASTM D 257 | 23°C | 3.21 $10^{9}$ ohm/cm |
| **Other Physical Properties** | | | |
| Specific Gravity* | ASTM D 792 Method A | 23°C | 1.06 |
| Mold Shrinkage | ASTM D 955 | | |
| Water Absorption | ASTM D 570 | 24 hrs at 23°C | 0.06% weight |

*Compression molded. All other specimens injection molded.

# Borg-Warner Chemicals Inc.

## Contact Address:

Borg-Warner Chemicals Inc.
International Center
Parkersburg, WV 26102

Phone: 1 (304) 424-5411

## PREVEX VQA

### Introduction:

Prevex polymers, offered by Borg-Warner Chemicals, Inc., are engineering thermoplastics that balance high impact resistance with excellent performance at elevated temperatures. These resins provide a unique combination of characteristics including Underwriters' Laboratory (UL) flammability classification, Relative Thermal Index (RTI) up to 110°C, electrical properties for direct support, and excellent dimensional stability. Easy flow, minimum drying and excellent thermal stability during molding combine for good processability leading to consistent part production.

To meet the range of design and processing demands presented by today's technology, Prevex polymers are available in numerous grades for injection and structural-foam molding as well as extrusion processing. New grades are being developed to expand the range of options already available from Borg-Warner Chemicals.

### Grade Characteristics:

Prevex VQA is a flame-retardant grade for products requiring good dimensional stability, durability and high heat resistance. This grade has an RTI of 105°C. Its 94V-1 UL classification means it meets the performance requirements for many electrical devices, enclosures, industrial instruments and connectors.

### Product Data:

**PREVEX VQA Molding Grade**

| Properties | Test Methods | Test Conditions | English Units |
|---|---|---|---|
| **Mechanical** | | | |
| Impact Resistance<br>Izod Impact<br>0.125″ bar, notched<br>reverse notched | ASTM D 256<br>Method A | 73°F | 5.0 ft-lb/in |
| Gardner Impact<br>0.125″ specimen | ASTM D 3029<br>Method G | 73°F | 13 ft-lb |
| Dynatup, Total Energy | | 73°F | |

*(continued)*

NOTE—Information provided in this book has been condensed. The publisher recommends that the product manufacturer be contacted, at the address or phone number provided, for more complete information on product safety, processing parameters, and applications not provided in this information sheet.

**PREVEX VQA Molding Grade *(continued)***

| Properties | Test Methods | Test Conditions | English Units |
|---|---|---|---|
| **Mechanical** | | | |
| Tensile Strength 0.125″ thickness | ASTM D 638 | 73°F | 8,850 psi |
| Tensile Modulus 0.125″ thickness | ASTM D 638 | 73°F | 3.9 $10^5$ psi |
| Flexural Yield Strength 0.125 × 0.5 × 5.0″ bar | ASTM D 790 | 73°F | 14,000 psi |
| Flexural Modulus 0.125 × 0.5 × 5.0″ bar | ASTM D 790 | 73°F | 3.8 $10^5$ psi |
| Rockwell Hardness | ASTM D 785 | | 120 R |
| **Thermal** | | | |
| Deflection Temperature under load 0.5 × 0.5 × 5.0″ bar | ASTM D 648 | 264 psi<br>64 psi | 265°F |
| Relative Thermal Index<br>Without Impact<br>With Impact | UL Standard 746 B | | <br>230°F<br>221°F |
| Vicat Softening Temperature | ASTM D 1525 | | 300°F |
| Coefficient of Linear Thermal Expansion | ASTM D 696 | −22°F to 86°F | 3.6 $10^{-5}$ in/in/°F |
| Thermal Conductivity | ASTM C 177 | | BTU in/hr ft²°F |
| **Flame-Retardant** | | | |
| Flame Class Rating | UL Standard 94 | 0.062″ thickness | 94V-1 |
| **Electrical** | | | |
| Hot Wire Ignition | UL Standard 746 A | 0.062″<br>0.125″ | 102 seconds<br>120 to ignite |
| High Amperage Arc Ignition | UL Standard 746 A | 0.062″<br>0.125″ | 200+ arcs<br>200+ arcs |
| High Voltage Tracking Rate | UL Standard 746 A | 0.062″<br>0.125″ | 18.1 in/min<br>18.1 in/min |
| Arc Resistance | ASTM D 495 | 0.062″<br>0.125″ | 66 seconds<br>89 seconds |
| Comparative Tracking Index | UL Standard 746 A | 0.062″<br>0.125″ | 460 volts<br>460 volts |
| Dielectric Strength | ASTM D 149 | 73°F | volts/mil |

*(continued)*

**PREVEX VQA Molding Grade *(continued)***

| *Properties* | *Test Methods* | *Test Conditions* | *English Units* |
|---|---|---|---|
| **Electrical** | | | |
| Dissipation Factor | ASTM D 150 | 73°F<br>100 cycles | |
| Dielectric Constant | ASTM D 150 | 73°F<br>100 cycles | |
| Volume Resistivity | ASTM D 257 | 73°F | $10^9$ ohm/cm |
| **Other Physical Properties** | | | |
| Specific Gravity* | ASTM D 792<br>Method A | 73°F | 1.06 |
| Mold Shrinkage | ASTM D 955 | | 0.006–0.008 in/in |
| Water Absorption | ASTM D 570 | 24 hrs at<br>73°F | 0.06% weight |

**PREVEX VQA Molding Grade**

| **Properties** | **Test Methods** | **Test Conditions** | **Metric Units** |
|---|---|---|---|
| **Mechanical** | | | |
| Impact Resistance<br>Izod Impact<br>3.2mm bar, notched<br>reverse notched | ASTM D 256<br>Method A | 23°C | 270 J/m |
| Gardner Impact<br>3.2mm specimen | ASTM D 3029<br>Method G | 23°C | 17 J |
| Dynatup, Total Energy | | 23°C | |
| Tensile Strength<br>3.2mm thickness | ASTM D 638 | 23°C | 61 MPa |
| Tensile Modulus<br>3.2mm thickness | ASTM D 638 | 23°C | 2.7 GPa |
| Flexural Yield Strength<br>3.2 × 13 × 130mm bar | ASTM D 790 | 23°C | 96 MPa |
| Flexural Modulus<br>3.2 × 13 × 130mm bar | ASTM D 638 | 23°C | 2.6 GPa |
| Rockwell Hardness* | ASTM D 785 | | 120 R |

*Compression molded. All other specimens injection molded.

*(continued)*

PREVEX VQA Molding Grade *(continued)*

| Properties | Test Methods | Test Conditions | Metric Units |
|---|---|---|---|
| **Thermal** | | | |
| Deflection Temperature under load 13 × 13 × 130mm bar | ASTM D 648 | 1820 kPa<br>440 kPa | 129°C |
| Relative Thermal Index<br>Without Impact<br>With Impact | UL Standard 746 B | | <br>110°C<br>105°C |
| Vicat Softening Temperature | ASTM D 1525 | | 149°C |
| Coefficient of Linear Thermal Expansion | ASTM D 696 | −30°C to +30°C | 6.4 $10^{-5}$ cm/cm/°C |
| Thermal Conductivity | ASTM C 177 | | W/km |
| **Flame-Retardant** | | | |
| Flame Class Rating | UL Standard 94 | 1.6mm thickness | 94V-1 |
| **Electrical** | | | |
| Hot Wire Ignition | UL Standard 746 A | 1.6mm<br>3.2mm | 102 seconds<br>120 to ignite |
| High Amperage Arc Ignition | UL Standard | 1.6mm<br>3.2mm | 200+ arcs<br>200+ arcs |
| High Voltage Tracking Rate | UL Standard 746 A | 1.6mm<br>3.2mm | 46 in/min<br>46 in/min |
| Arc Resistance | ASTM D 495 | 1.6mm<br>3.2mm | 66 seconds<br>89 seconds |
| Comparative Tracking Index | UL Standard 746 A | 1.6mm<br>3.2mm | 460 volts<br>460 volts |
| Dielectric Strength | ASTM D 149 | 23°C | kV/mm |
| Dissipation Factor | ASTM D 150 | 23°C<br>100 cycles | |
| Dielectric Constant | ASTM D 150 | 23°C<br>100 cycles | |
| Volume Resistivity | ASTM D 257 | 23°C | $10^9$ ohm/cm |
| **Other Physical Properties** | | | |
| Specific Gravity* | ASTM D 792 Method A | 23°C | 1.06 |
| Mold Shrinkage | ASTM D 955 | | 0.006–0.008 cm/cm |
| Water Absorption | ASTM D 570 | 24 hrs at 23°C | 0.06% weight |

*Compression molded. All other specimens injection molded.

# BP Performance Polymers Inc.

Contact Address:

BP Performance Polymers Inc.
Newberg Road
P.O. Box 400
Hackettstown, NJ 07840

Phone: 1 (201) 852-1100

## TPR® 451 (BP PPI Code: 65-451-04)

Product Data:

**TPR® 451 (BP PPI Code: 65-451-04)**

| Properties[1] | Unit | Typical Value | Test Method |
|---|---|---|---|
| **Typical** | | | |
| Tensile Strength at Break[a] | psi<br>MPa | 1400<br>9.7 | ASTM D-412 |
| Elongation at Break[a] | | 550% | ASTM D-412 |
| Flex Modulus[b] | psi<br>MPa | 41,000<br>283 | ASTM D-790 |
| 100% Modulus[a] | psi<br>MPa | | ASTM D-412 |
| Hardness (not cross linked)[c] | D | 38 | ASTM D-2240 |
| Tear Resistance[a] | PLi | 400 | ASTM D-CZK |
| Compression Set | | | ASTM D-395 |
| Brittle Point | °F<br>°C | < −85<br>< −65 | |
| Viscosity | pascal seconds | | |
| Surface resistivity | ohm/cm | | |

[1]Obtained testing 80 mil molded plaques. Should not be treated as specification values.
[a]Die C, die cut, 20″/min
[b]Method I, Procedure B, secant at 0.1″ deflection, 1/8″ × 3″ specimen, die cut, 0.5″/min
[c]3 second delay .250″ plied sample

NOTE—Information provided in this book has been condensed. The publisher recommends that the product manufacturer be contacted, at the address or phone number provided, for more complete information on product safety, processing parameters, and applications not provided in this information sheet.

# BP Performance Polymers Inc.

## Contact Address:

BP Performance Polymers Inc.
Newberg Road
P.O. Box 400
Hackettstown, NJ 07840

Phone: 1 (201) 852-1100

## TPR® 538E Black (BP PPI Code: 65-132-04)

### Description:

A non-conductive olefin based thermoplastic elastomer.

### Application:

Suitable for weatherstrip, seals, gaskets, automotive, sheet and fiilm applications.

### Features:

- intermediate modulus
- extrusion
- paintable after treatment with an approved surface primer
- approved under Ford Engineering Specification Materials Number ESB-M2D246-A2 and Fisher Body Materials Specification FBMS-2-25, Type I

**TPR® 538E Black (BP PPI Code: 65-132-04)**

| Properties[1] | Unit | Typical Value | Test Method |
|---|---|---|---|
| **Typical** | | | |
| Specific Gravity | | 0.91 | ASTM D-792 |
| Hardness | Shore D | 42 | ASTM D-2240 |
| Tensile Strength | psi<br>MPa | 1600<br>11.0 | ASTM D-412 |
| Elongation | % | 500 | ASTM D-412 |
| Flexural Modulus | psi<br>MPa | 45,000<br>310 | ASTM D-790 |

[1]Obtained testing 125 mil molded plaques. Should not be treated as specification values.

*(continued)*

NOTE – Information provided in this book has been condensed. The publisher recommends that the product manufacturer be contacted, at the address or phone number provided, for more complete information on product safety, processing parameters, and applications not provided in this information sheet.

**TPR® 538E Black (BP PPI Code: 65-132-04) *(continued)***

| Properties[1] | Unit | Typical Value | Test Method |
|---|---|---|---|
| **Typical** | | | |
| Tear Strength | PLi<br>kN/m | 535<br>93.7 | ASTM D-624 |
| Surface Resistivity | ohm/cm | $>10^{16}$ | ASTM D-257 |
| Brittle Point | °F<br>°C | $<-100$<br>$<-73$ | ASTM D-746<br>Method I<br>Procedure B |
| Melt Viscosity | pascal<br>seconds | 425 | |

[1]Obtained testing 80 mil molded plaques. Should not be treated as specification values.

# BP Performance Polymers Inc.

**Contact Address:**

BP Performance Polymers Inc.
Newberg Road
P.O. Box 400
Hackettstown, NJ 07840

Phone: 1 (201) 852-1100

## TPR® 538R (BP PPI Code: 65-131-14)

**Product Data:**

**TPR® 538R (BP PPI Code: 65-131-14)**

| Properties | Unit | Typical Value | Test Method |
|---|---|---|---|
| **Typical** | | | |
| Tensile Strength at Break[a] | psi<br>MPa | | ASTM D-412 |
| Elongation at Break[a] | % | 300 | ASTM D-412 |
| Flex Modulus[b] | psi<br>MPa | 30,000<br>206.9 | ASTM D-790 |
| 100% Modulus[a] | psi<br>MPa | 725<br>5.0 | ASTM D-412 |
| Hardness (not cross linked)[c] | D | 35 | ASTM D-2240 |
| Tear Resistance[a] | PLi | 400 | ASTM D-CZK |
| Compression Set | | | ASTM D-395 |
| Brittle Point | °F<br>°C | < −100<br>< −73 | |
| Viscosity | pascal seconds | | |
| Surface resistivity | ohm/cm | | |

[a]Die C, die cut, 20″/min
[b]Method I, Procedure B, secant at 0.1″ deflection, 1/8″ × 3″ specimen, die cut, 0.5″/min
[c]3 second delay .250″ plied sample

NOTE – Information provided in this book has been condensed. The publisher recommends that the product manufacturer be contacted, at the address or phone number provided, for more complete information on product safety, processing parameters, and applications not provided in this information sheet.

# BP Performance Polymers Inc.

## Contact Address:

BP Performance Polymers Inc.
Newberg Road
P.O. Box 400
Hackettstown, NJ 07840

Phone: 1 (201) 852-1100

## TPR® 0720 (BP PPI Code: 65-100-04)

### Product Data:

**TPR® 0720 (BP PPI Code: 65-100-04)**

| Properties | Unit | Typical Value | Test Method |
|---|---|---|---|
| **Typical** | | | |
| Tensile Strength at Break[a] | psi<br>MPa | 1900<br>13.1 | ASTM D-412 |
| Elongation at Break | % | 350 | ASTM D-412 |
| Flex Modulus[b] | psi<br>MPa | 97,000<br>669 | ASTM D-790 |
| 100% Modulus[a] | psi<br>MPa | | ASTM D-412 |
| Hardness (not cross linked)[c] | D | 57 | ASTM D-2240 |
| Tear Resistance[a] | PLi | 700 | ASTM D-CZK |
| Compression Set | | | ASTM D-395 |
| Brittle Point | °F<br>°C | $< -40$<br>$< -40$ | |
| Viscosity | pascal seconds | | |
| Surface resistivity | ohm/cm | $>10^{16}$ | |

[a]Die C, die cut, 20″/min
[b]Method I, Procedure B, secant at 0.1″ deflection, 1/8″ × 3″ specimen, die cut, 0.5″/min
[c]3 second delay .250″ plied sample

NOTE – Information provided in this book has been condensed. The publisher recommends that the product manufacturer be contacted, at the address or phone number provided, for more complete information on product safety, processing parameters, and applications not provided in this information sheet.

# BP Performance Polymers Inc.

## Contact Address:

BP Performance Polymers Inc.
Newberg Road
P.O. Box 400
Hackettstown, NJ 07840

Phone: 1 (201) 852-1100

---

## TPR® 0720 Natural (BP PPI Code: 65-100-14)

### Description:

Olefin based thermoplastic elastomer.

### Application:

Suitable for automotive sheet and film applications.

### Features:

- high modulus
- extrusion or injection molding
- paintable after treatment with an approved surface primer

**TPR® 0720 Natural (BP PPI Code: 65-100-14)**

| Properties[1] | Unit | Typical Value | Test Method |
|---|---|---|---|
| **Typical** | | | |
| Specific Gravity | | 0.91 | ASTM D-792 |
| Hardness | Shore D | 57 | ASTM D-2240 |
| Tensile Strength | psi<br>MPa | 1900<br>13.1 | ASTM D-412 |
| Elongation | % | 350 | ASTM D-412 |
| Flexural Modulus | psi<br>MPa | 97,000<br>669 | ASTM D-790 |

[1]Obtained testing 125 mil molded plaques. Should not be treated as specification values.

*(continued)*

NOTE—Information provided in this book has been condensed. The publisher recommends that the product manufacturer be contacted, at the address or phone number provided, for more complete information on product safety, processing parameters, and applications not provided in this information sheet.

**TPR® 0720 Natural (BP PPI Code: 65-100-14) *(continued)***

| **Properties**[1] | **Unit** | **Typical Value** | **Test Method** |
|---|---|---|---|
| **Typical** | | | |
| Tear Strength | | | ASTM D-624 |
| | PLi | 700 | |
| | kN/m | 122.6 | |
| Surface Resistivity | ohm/cm | $>10^{16}$ | ASTM D-257 |
| Brittle Point | | | ASTM D-746 |
| | °F | $<-40$ | Method I |
| | °C | $<-40$ | Procedure B |

[1]Obtained testing 125 mil molded plaques. Should not be treated as specification values.

# BP Performance Polymers Inc.

Contact Address:

BP Performance Polymers Inc.
Newberg Road
P.O. Box 400
Hackettstown, NJ 07840

Phone: 1 (201) 852-1100

## TPR® 0721 (BP PPI Code: 65-100-04)

Product Data:

**TPR® 0721 (BP PPI Code: 65-100-04)**

| Properties | Unit | Typical Value | Test Method |
|---|---|---|---|
| **Typical** | | | |
| Tensile Strength at Break[a] | psi<br>MPa | 1900<br>13.1 | ASTM D-412 |
| Elongation at Break[a] | % | 350 | ASTM D-412 |
| Flex Modulus[b] | psi<br>MPa | 97,000<br>669 | ASTM D-790 |
| 100% Modulus[a] | psi<br>MPa | | ASTM D-412 |
| Hardness (not cross linked)[c] | D | 57 | ASTM D-2240 |
| Tear Resistance[a] | PLi | 700 | ASTM D-CZK |
| Compression Set | | | ASTM D-395 |
| Brittle Point | °F<br>°C | $<-58$<br>$<-50$ | |
| Viscosity | pascal seconds | | |
| Surface resistivity | ohm/cm | $>10^{16}$ | |

[a]Die C, die cut, 20″/min
[b]Method I, Procedure B, secant at 0.1″ deflection, 1/8″ × 3″ specimen, die cut, 0.5″/min
[c]3 second delay .250″ plied sample

NOTE – Information provided in this book has been condensed. The publisher recommends that the product manufacturer be contacted, at the address or phone number provided, for more complete information on product safety, processing parameters, and applications not provided in this information sheet.

# BP Performance Polymers Inc.

## Contact Address:

BP Performance Polymers Inc.
Newberg Road
P.O. Box 400
Hackettstown, NJ 07840

Phone: 1 (201) 852-1100

## TPR® 791 (BP PPI Code: 95-791-114)

### Product Data:

**TPR® 791 (BP PPI Code: 95-791-114)**

| Properties | Unit | Typical Value | Test Method |
|---|---|---|---|
| **Typical** | | | |
| Tensile Strength at Break[a] | psi<br>MPa | 1600<br>11 | ASTM D-412 |
| Elongation at Break[a] | % | 525 | ASTM D-790 |
| Flex Modulus[b] | psi<br>MPa | | ASTM D-412 |
| 100% Modulus[a] | psi<br>MPa | 1100<br>7.6 | ASTM D-412 |
| Hardness (not cross linked)[c] | A | 90 | ASTM D-2240 |
| Tear Resistance[a] | PLi | | ASTM D-CZK |
| Compression Set | | | ASTM D-395 |
| Brittle Point | °F<br>°C | | |
| Viscosity | pascal seconds | | |
| Surface resistivity | ohm/cm | | |

[a]Die C, die cut, 20″/min
[b]Method I, Procedure B, secant at 0.1″ deflection, 1/8″ × 3″ specimen, die cut, 0.5″/min
[c]3 second delay .250″ plied sample

NOTE – Information provided in this book has been condensed. The publisher recommends that the product manufacturer be contacted, at the address or phone number provided, for more complete information on product safety, processing parameters, and applications not provided in this information sheet.

# BP Performance Polymers Inc.

Contact Address:

BP Performance Polymers Inc.
Newberg Road
P.O. Box 400
Hackettstown, NJ 07840

Phone: 1 (201) 852-1100

## TPR® 904C (BP PPI Code: 65-460-04)

Product Data:

**TPR® 904C (BP PPI Code: 65-460-04)**

| Properties | Unit | Typical Value | Test Method |
|---|---|---|---|
| **Typical** | | | |
| Tensile Strength at Break[a] | psi<br>MPa | 1400<br>9.7 | ASTM D-412 |
| Elongation at Break[a] | % | 120 | ASTM D-412 |
| Flex Modulus[b] | psi<br>MPa | 23,000<br>157 | ASTM D-790 |
| 100% Modulus[a] | psi<br>MPa | | ASTM D-412 |
| Hardness (not cross linked)[c] | D | 40 | ASTM D-2240 |
| Tear Resistance[a] | PLi | 430 | ASTM D-CZK |
| Compression Set | | | ASTM D-395 |
| Brittle Point | °F<br>°C | < −58<br>< −50 | |
| Viscosity | pascal seconds | | |
| Surface resistivity | ohm/cm | | |

[a]Die C, die cut, 20″/min
[b]Method I, Procedure B, secant at 0.1″ deflection, 1/8″ × 3″ specimen, die cut, 0.5″/min
[c]3 second delay .250″ plied sample

NOTE – Information provided in this book has been condensed. The publisher recommends that the product manufacturer be contacted, at the address or phone number provided, for more complete information on product safety, processing parameters, and applications not provided in this information sheet.

# BP Performance Polymers Inc.

## Contact Address:

BP Performance Polymers Inc.
Newberg Road
P.O. Box 400
Hackettstown, NJ 07840

Phone: 1 (201) 852-1100

## TPR® 905 (BP PPI Code: 65-465-04)

### Product Data:

**TPR® 905 (BP PPI Code: 65-465-04)**

| Properties | Unit | Typical Value | Test Method |
|---|---|---|---|
| **Typical** | | | |
| Tensile Strength at Break[a] | psi<br>MPa | 1400<br>9.7 | ASTM D-412 |
| Elongation at Break[a] | % | 350 | ASTM D-412 |
| Flex Modulus[b] | psi<br>MPa | 60,000<br>414 | ASTM D-790 |
| 100% Modulus[a] | psi<br>MPa | | ASTM D-412 |
| Hardness (not cross linked)[c] | D | 35 | ASTM D-2240 |
| Tear Resistance[a] | PLi | 480 | ASTM D-CZK |
| Compression Set | | | ASTM D-395 |
| Brittle Point | °F<br>°C | < −100<br>< −73 | |
| Viscosity | pascal seconds | 350 | |
| Surface resistivity | ohm/cm | | |

[a]Die C, die cut, 20″/min
[b]Method I, Procedure B, secant at 0.1″ deflection, 1/8″ × 3″ specimen, die cut, 0.5″/min
[c]3 second delay .250″ plied sample

NOTE – Information provided in this book has been condensed. The publisher recommends that the product manufacturer be contacted, at the address or phone number provided, for more complete information on product safety, processing parameters, and applications not provided in this information sheet.

# BP Performance Polymers Inc.

## Contact Address:

BP Performance Polymers Inc.
Newberg Road
P.O. Box 400
Hackettstown, NJ 07840

Phone: 1 (201) 852-1100

## TPR® 905 Black (BP PPI Code: 65-465-04)

### Description:

A non-conductive olefin based thermoplastic elastomer.

### Application:

Suitable for automotive, caps and closure applications.

### Features:

- intermediate modulus
- injection molding
- paintable after treatment with an approved surface primer
- approved as TPE 905B under Fisher Body Materials Specification (FBMS) 2-25, Type I

**TPR® 905 Black (BP PPI Code: 65-465-04)**

| Properties[1] | Unit | Typical Value | Test Method |
|---|---|---|---|
| **Typical** | | | |
| Specific Gravity | | 0.89 | ASTM D-792 |
| Hardness | Shore A | 35 | ASTM D-2240 |
| Tensile Strength | psi<br>MPa | 1400<br>9.7 | ASTM D-412 |
| Elongation | % | 350 | ASTM D-412 |
| Flexural Modulus | psi<br>MPa | 60,000<br>414 | ASTM D-790 |

[1]Obtained testing 125 mil molded plaques. Should not be treated as specification values.

*(continued)*

NOTE – Information provided in this book has been condensed. The publisher recommends that the product manufacturer be contacted, at the address or phone number provided, for more complete information on product safety, processing parameters, and applications not provided in this information sheet.

### TPR® 905 Black (BP PPI Code: 65-465-04) *(continued)*

| Properties[1] | Unit | Typical Value | Test Method |
|---|---|---|---|
| **Typical** | | | |
| Tear Strength | PLi<br>kN/m | 480<br>84.1 | ASTM D-624 |
| Brittle Point | °F<br>°C | < −100<br>< −73 | ASTM D-746<br>Method I<br>Procedure B |
| Melt Viscosity | pascal seconds | 370 | |

[1]Obtained testing 125 mil molded plaques. Should not be treated as specification values.

# BP Performance Polymers Inc.

Contact Address:

BP Performance Polymers Inc.
Newberg Road
P.O. Box 400
Hackettstown, NJ 07840

Phone: 1 (201) 852-1100

## TPR® 1291 (BP PPI Code: 65-105-04)

Product Data:

**TPR® 1291 (BP PPI Code: 65-105-04)**

| Properties | Unit | Typical Value | Test Method |
|---|---|---|---|
| **Typical** | | | |
| Tensile Strength at Break[a] | psi<br>MPa | 1200<br>83 | ASTM D-412 |
| Elongation at Break[a] | % | 400 | ASTM D-412 |
| Flex Modulus[b] | psi<br>MPa | 35,000 | ASTM D-790 |
| 100% Modulus[a] | psi<br>MPa | 950<br>6.6 | ASTM D-412 |
| Hardness (not cross linked)[c] | D | 35 | ASTM D-2240 |
| Tear Resistance[a] | PLi | 300 | ASTM D-CZK |
| Compression Set | | | ASTM D-395 |
| Brittle Point | °F<br>°C | | |
| Viscosity | pascal seconds | 275 | |
| Surface resistivity | ohm/cm | | |

[a]Die C, die cut, 20″/min
[b]Method I, Procedure B, secant at 0.1″ deflection, 1/8″ × 3″ specimen, die cut, 0.5″/min
[c]3 second delay .250″ plied sample

NOTE – Information provided in this book has been condensed. The publisher recommends that the product manufacturer be contacted, at the address or phone number provided, for more complete information on product safety, processing parameters, and applications not provided in this information sheet.

# BP Performance Polymers Inc.

**Contact Address:**

BP Performance Polymers Inc.
Newberg Road
P.O. Box 400
Hackettstown, NJ 07840

Phone: 1 (201) 852-1100

## TPR® 1500 (BP PPI Code: 65-102-04)

**Product Data:**

**TPR® 1500 (BP PPI Code: 65-102-04)**

| Properties | Unit | Typical Value | Test Method |
|---|---|---|---|
| **Typical** | | | |
| Tensile Strength at Break[a] | psi<br>MPa | 550<br>3.8 | ASTM D-412 |
| Elongation at Break[a] | % | 150 | ASTM D-412 |
| Flex Modulus[b] | psi<br>MPa | 12,000<br>82.8 | ASTM D-790 |
| 100% Modulus[a] | psi<br>MPa | 600<br>3.4 | ASTM D-412 |
| Hardness (not cross linked)[c] | A | 58 | ASTM D-2240 |
| Tear Resistance[a] | PLi | 100 | ASTM D-CZK |
| Compression Set | | 25 | ASTM D-395 |
| Brittle Point | °F<br>°C | < −100<br>< −73 | |
| Viscosity | pascal seconds | | |
| Surface resistivity | ohm/cm | | |

[a]Die C, die cut, 20″/min
[b]Method I, Procedure B, secant at 0.1″ deflection, 1/8″ × 3″ specimen, die cut, 0.5″/min
[c]3 second delay .250″ plied sample

NOTE – Information provided in this book has been condensed. The publisher recommends that the product manufacturer be contacted, at the address or phone number provided, for more complete information on product safety, processing parameters, and applications not provided in this information sheet.

# BP Performance Polymers Inc.

**Contact Address:**

BP Performance Polymers Inc.
Newberg Road
P.O. Box 400
Hackettstown, NJ 07840

Phone: 1 (201) 852-1100

## TPR® 1500 Natural (BP PPI Code: 65-102-04)

**Description:**

Olefin based thermoplastic elastomer.

**Application:**

Suitable for gaskets and toy applications.

**Features:**

- low modulus
- injection molding
- good compression set resistance

**TPR® 1500 Natural (BP PPI Code 65-102-04)**

| Properties[1] | Unit | Typical Value | Test Method |
|---|---|---|---|
| **Typical** | | | |
| Specific Gravity | | 0.89 | ASTM D-792 |
| Hardness | Shore A | 58 | ASTM D-2240 |
| Tensile Strength | | | ASTM D-412 |
| | psi | 550 | |
| | MPa | 3.8 | |
| 100% Modulus | | | ASTM D-412 |
| | psi | 500 | |
| | MPa | 3.4 | |
| Elongation | % | 150 | ASTM D-412 |
| Flexural Modulus | | | ASTM D-790 |
| | psi | 1200 | |
| | MPa | 82.8 | |

*(continued)*

NOTE—Information provided in this book has been condensed. The publisher recommends that the product manufacturer be contacted, at the address or phone number provided, for more complete information on product safety, processing parameters, and applications not provided in this information sheet.

**TPR® 1500 Natural (BP PPI Code 65-102-04) *(continued)***

| Properties[1] | Unit | Typical Value | Test Method |
|---|---|---|---|
| **Typical** | | | |
| Tear Strength | PLi<br>kN/m | 100<br>17.5 | ASTM D-624 |
| Compression Set | % | 25 | ASTM D-395 |
| Brittle Point | °F<br>°C | <−100<br><−73 | ASTM D-746<br>Method I<br>Procedure B |

[1]Obtained testing 125 mil molded plaques. Should not be treated as specification values.

# BP Performance Polymers Inc.

## Contact Address:

BP Performance Polymers Inc.
Newberg Road
P.O. Box 400
Hackettstown, NJ 07840

Phone: 1 (201) 852-1100

## TPR® 1600 (BP PPI Code: 65-100-04)

### Product Data:

**TPR® 1600 (BP PPI Code: 65-100-04)**

| Properties | Unit | Typical Value | Test Method |
|---|---|---|---|
| **Typical** | | | |
| Tensile Strength at Break[a] | psi<br>MPa | 550<br>3.8 | ASTM D-412 |
| Elongation at Break[a] | % | 280 | ASTM D-790 |
| Flex Modulus[b] | psi<br>MPa | 1500<br>10.3 | ASTM D-790 |
| 100% Modulus[a] | psi<br>MPa | | ASTM D-412 |
| Hardness (not cross linked)[c] | A | 65 | ASTM D-2240 |
| Tear Resistance[a] | PLi | 25 | ASTM D-CZK |
| Compression Set | | 37 | ASTM D-395 |
| Brittle Point | °F<br>°C | < −100<br>< −73 | |
| Viscosity | pascal seconds | 400 | |
| Surface resistivity | ohm/cm | | |

[a]Die C, die cut, 20″/min
[b]Method I, Procedure B, secant at 0.1″ deflection, 1/8″ × 3″ specimen, die cut, 0.5″/min
[c]3 second delay .250″ plied sample

NOTE—Information provided in this book has been condensed. The publisher recommends that the product manufacturer be contacted, at the address or phone number provided, for more complete information on product safety, processing parameters, and applications not provided in this information sheet.

# BP Performance Polymers Inc.

## Contact Address:

BP Performance Polymers Inc.
Newberg Road
P.O. Box 400
Hackettstown, NJ 07840

Phone: 1 (201) 852-1100

## TPR® 1600 Natural (BP PPI Code: 65-106-04)

### Description:

Olefin based thermoplastic elastomer.

### Application:

Suitable for weatherstrip, seals, gaskets, grips, bumpers, footers, toys, caps, closures, automotive, sheet film, hose, tubing and mechanical goods applications.

### Features:

- low modulus
- extrusion or injection molding

**TPR® 1600 Natural (BP PPI Code: 65-106-04)**

| Properties[1] | Unit | Typical Value | Test Method |
|---|---|---|---|
| **Typical** | | | |
| Specific Gravity | | 0.89 | ASTM D-792 |
| Hardness | Shore A | 65 | ASTM D-2240 |
| Tensile Strength | | | ASTM D-412 |
| | psi | 550 | |
| | MPa | 3.8 | |
| Elongation | % | 280 | ASTM D-412 |
| Flexural Modulus | | | ASTM D-790 |
| | psi | 1500 | |
| | MPa | 10.3 | |

[1]Obtained testing 125 mil molded plaques. Should not be treated as specification values.

*(continued)*

NOTE – Information provided in this book has been condensed. The publisher recommends that the product manufacturer be contacted, at the address or phone number provided, for more complete information on product safety, processing parameters, and applications not provided in this information sheet.

### TPR® 1600 Natural (BP PPI Code: 65-106-04) *(continued)*

| Properties[1] | Unit | Typical Value | Test Method |
|---|---|---|---|
| **Typical** | | | |
| Tear Strength | PLi<br>kN/m | 25<br>4.4 | ASTM D-790 |
| Compression Set | % | 37 | ASTM D-395 |
| Brittle Point | °F<br>°C | $<-100$<br>$<-73$ | ASTM D-746<br>Method I<br>Procedure B |
| Melt Viscosity | pascal seconds | 400 | |

[1]Obtained testing 125 mil molded plaques. Should not be treated as specification values.

# BP Performance Polymers Inc.

## Contact Address:

BP Performance Polymers Inc.
Newberg Road
P.O. Box 400
Hackettstown, NJ 07840

Phone: 1 (201) 852-1100

## TPR® 1600E (BP PPI Code: 65-107-04)

### Product Data:

**TPR® 1600E (BP PPI Code: 65-107-04)**

| Properties | Unit | Typical Value | Test Method |
|---|---|---|---|
| **Typical** | | | |
| Tensile Strength at Break[a] | psi<br>MPa | 550<br>3.8 | ASTM D-412 |
| Elongation at Break[a] | % | 350 | ASTM D-412 |
| Flex Modulus[b] | psi<br>MPa | 1500<br>10.3 | ASTM D-790 |
| 100% Modulus[a] | psi<br>MPa | 430<br>3.0 | ASTM D-412 |
| Hardness (not cross linked)[c] | A | 65 | ASTM D-2240 |
| Tear Resistance[a] | PLi | 25 | ASTM D-CZK |
| Compression Set | | | ASTM D-395 |
| Brittle Point | °F<br>°C | $<-100$<br>$<-73$ | |
| Viscosity | pascal seconds | 400 | |
| Surface resistivity | ohm/cm | | |

[a]Die C, die cut, 20″/min
[b]Method I, Procedure B, secant at 0.1″ deflection, 1/8″ × 3″ specimen, die cut, 0.5″/min
[c]3 second delay .250″ plied sample

NOTE—Information provided in this book has been condensed. The publisher recommends that the product manufacturer be contacted, at the address or phone number provided, for more complete information on product safety, processing parameters, and applications not provided in this information sheet.

# BP Performance Polymers Inc.

Contact Address:

BP Performance Polymers Inc.
Newberg Road
P.O. Box 400
Hackettstown, NJ 07840

Phone: 1 (201) 852-1100

## TPR® 1600M (BP PPI Code: 65-107-14)

Product Data:

**TPR® 1600M (BP PPI Code: 65-107-14)**

| Properties | Unit | Typical Value | Test Method |
|---|---|---|---|
| **Typical** | | | |
| Tensile Strength at Break[a] | psi<br>MPa | 550<br>3.8 | ASTM D-412 |
| Elongation at Break[a] | % | 250 | ASTM D-412 |
| Flex Modulus[b] | psi<br>MPa | 1500<br>10.3 | ASTM D-790 |
| 100% Modulus[a] | psi<br>MPa | | ASTM D-412 |
| Hardness (not cross linked)[c] | A | 65 | ASTM D-2240 |
| Tear Resistance[a] | PLi | | ASTM D-CZK |
| Compression Set | | 70 | ASTM D-395 |
| Brittle Point | °F<br>°C | < −100<br>< −73 | |
| Viscosity | pascal seconds | 500 | |
| Surface resistivity | ohm/cm | | |

[a]Die C, die cut, 20″/min
[b]Method I, Procedure B, secant at 0.1″ deflection, 1/8″ × 3″ specimen, die cut, 0.5″/min
[c]3 second delay .250″ plied sample

NOTE—Information provided in this book has been condensed. The publisher recommends that the product manufacturer be contacted, at the address or phone number provided, for more complete information on product safety, processing parameters, and applications not provided in this information sheet.

# BP Performance Polymers Inc.

Contact Address:

BP Performance Polymers Inc.
Newberg Road
P.O. Box 400
Hackettstown, NJ 07840

Phone: 1 (201) 852-1100

## TPR® 1612M (BP PPI Code: 65-107-24)

Product Data:

**TPR® 1612M (BP PPI Code: 65-107-24)**

| Properties | Unit | Typical Value | Test Method |
|---|---|---|---|
| **Typical** | | | |
| Tensile Strength at Break[a] | psi<br>MPa | 580<br>4.0 | ASTM D-412 |
| Elongation at Break[a] | % | 200 | ASTM D-412 |
| Flex Modulus[b] | psi<br>MPa | 1500<br>10.3 | ASTM D-790 |
| 100% Modulus[a] | psi<br>MPa | | ASTM D-412 |
| Hardness (not cross linked)[c] | A | 65 | ASTM D-2240 |
| Tear Resistance[a] | PLi | 140 | ASTM D-CZK |
| Compression Set | | 65 | ASTM D-395 |
| Brittle Point | °F<br>°C | | |
| Viscosity | pascal seconds | 600 | |
| Surface resistivity | ohm/cm | | |

[a]Die C, die cut, 20″/min
[b]Method I, Procedure B, secant at 0.1″ deflection, 1/8″ × 3″ specimen, die cut, 0.5″/min
[c]3 second delay .250″ plied sample

NOTE—Information provided in this book has been condensed. The publisher recommends that the product manufacturer be contacted, at the address or phone number provided, for more complete information on product safety, processing parameters, and applications not provided in this information sheet.

# BP Performance Polymers Inc.

## Contact Address:

BP Performance Polymers Inc.
Newberg Road
P.O. Box 400
Hackettstown, NJ 07840

Phone: 1 (201) 852-1100

## TPR® 1700 (BP PPI Code: 65-110-04)

### Product Data:

**TPR® 1700 (BP PPI Code: 65-110-04)**

| Properties | Unit | Typical Value | Test Method |
|---|---|---|---|
| **Typical** | | | |
| Tensile Strength at Break[a] | psi<br>MPa | 775<br>5.3 | ASTM D-412 |
| Elongation at Break[a] | % | 250 | ASTM D-412 |
| Flex Modulus[b] | psi<br>MPa | 30,000<br>206.9 | ASTM D-790 |
| 100% Modulus[a] | psi<br>MPa | | ASTM D-412 |
| Hardness (not cross linked)[c] | A | 74 | ASTM D-2240 |
| Tear Resistance[a] | PLi | 30 | ASTM D-CZK |
| Compression Set | | 44 | ASTM D-395 |
| Brittle Point | °F<br>°C | < −100<br>< −73 | |
| Viscosity | pascal seconds | 575 | |
| Surface resistivity | ohm/cm | | |

[a]Die C, die cut, 20″/min
[b]Method I, Procedure B, secant at 0.1″ deflection, 1/8″ × 3″ specimen, die cut, 0.5″/min
[c]3 second delay .250″ plied sample

NOTE – Information provided in this book has been condensed. The publisher recommends that the product manufacturer be contacted, at the address or phone number provided, for more complete information on product safety, processing parameters, and applications not provided in this information sheet.

# BP Performance Polymers Inc.

## Contact Address:

BP Performance Polymers Inc.
Newberg Road
P.O. Box 400
Hackettstown, NJ 07840

Phone: 1 (201) 852-1100

## TPR® 1700 Natural (BP PPI Code: 65-110-04)

### Description:

Olefin based thermoplastic elastomer.

### Application:

Suitable for weatherstrip, seals, gaskets, grips, bumpers, footers, toys, caps, closures and mechanical goods applications.

### Features:

- low modulus
- extrusion or injection molding

**TPR® 1700 Natural (BP PPI Code: 65-110-04)**

| Properties[1] | Unit | Typical Value | Test Method |
|---|---|---|---|
| **Typical** | | | |
| Specific Gravity | | 0.89 | ASTM D-792 |
| Hardness | Shore A | 74 | ASTM D-2240 |
| Tensile Strength | | | ASTM D-412 |
| | psi | 775 | |
| | MPa | 5.3 | |
| Elongation | % | 250 | ASTM D-412 |
| Flexural Modulus | | | ASTM D-790 |
| | psi | 3000 | |
| | MPa | 206.9 | |
| Tear Strength | | | ASTM D-790 |
| | PLi | 30 | |
| | kN/m | 5.3 | |

[1]Obtained testing 80 mil molded plaques. Should not be treated as specification values.

*(continued)*

NOTE—Information provided in this book has been condensed. The publisher recommends that the product manufacturer be contacted, at the address or phone number provided, for more complete information on product safety, processing parameters, and applications not provided in this information sheet.

**TPR® 1700 Natural (BP PPI Code: 65-110-04) *(continued)***

| Properties[1] | Unit | Typical Value | Test Method |
|---|---|---|---|
| **Typical** | | | |
| Brittle Point | °F<br>°C | < −100<br>< −73 | ASTM D-746<br>Method I<br>Procedure B |
| Melt Viscosity | pascal seconds | 575 | |
| Compression Set | % | 44 | ASTM D-395 |

[1]Obtained testing 80 mil molded plaques. Should not be treated as specification values.

# BP Performance Polymers Inc.

## Contact Address:

BP Performance Polymers Inc.
Newberg Road
P.O. Box 400
Hackettstown, NJ 07840

Phone: 1 (201) 852-1100

## TPR® 1710D (BP PPI Code: 65-112-04)

### Product Data:

**TPR® 1710D (BP PPI Code: 65-112-04)**

| Properties | Unit | Typical Value | Test Method |
|---|---|---|---|
| **Typical** | | | |
| Tensile Strength at Break[a] | psi<br>MPa | 800<br>5.5 | ASTM D-412 |
| Elongation at Break[a] | % | 490 | ASTM D-412 |
| Flex Modulus[b] | psi<br>MPa | 15,000<br>103.4 | ASTM D-790 |
| 100% Modulus[a] | psi<br>MPa | | ASTM D-412 |
| Hardness (not cross linked)[c] | A | 84 | ASTM D-2240 |
| Tear Resistance[a] | PLi | 225 | ASTM D-CZK |
| Compression Set | | | ASTM D-395 |
| Brittle Point | °F<br>°C | | |
| Viscosity | pascal seconds | | |
| Surface resistivity | ohm/cm | | |

[a]Die C, die cut, 20″/min
[b]Method I, Procedure B, secant at 0.1″ deflection, 1/8″ × 3″ specimen, die cut, 0.5″/min
[c]3 second delay .250″ plied sample

NOTE—Information provided in this book has been condensed. The publisher recommends that the product manufacturer be contacted, at the address or phone number provided, for more complete information on product safety, processing parameters, and applications not provided in this information sheet.

# BP Performance Polymers Inc.

## Contact Address:

BP Performance Polymers Inc.
Newberg Road
P.O. Box 400
Hackettstown, NJ 07840

Phone: 1 (201) 852-1100

## TPR® 1712 (BP PPI Code: 65-110-14)

## Product Data:

**TPR® 1712 (BP PPI Code: 65-110-14)**

| Properties | Unit | Typical Value | Test Method |
|---|---|---|---|
| **Typical** | | | |
| Tensile Strength at Break[a] | psi<br>MPa | 600<br>4.1 | ASTM D-412 |
| Elongation at Break[a] | % | 250 | ASTM D-412 |
| Flex Modulus[b] | psi<br>MPa | 3000<br>20.7 | ASTM D-790 |
| 100% Modulus[a] | psi<br>MPa | | ASTM D-412 |
| Hardness (not cross linked)[c] | A | 73 | ASTM D-2240 |
| Tear Resistance[a] | PLi | 30 | ASTM D-CZK |
| Compression Set | | 50 | ASTM D-395 |
| Brittle Point | °F<br>°C | | |
| Viscosity | pascal seconds | | |
| Surface resistivity | ohm/cm | | |

[a]Die C, die cut, 20″/min
[b]Method I, Procedure B, secant at 0.1″ deflection, 1/8″ × 3″ specimen, die cut, 0.5″/min
[c]3 second delay .250″ plied sample

NOTE—Information provided in this book has been condensed. The publisher recommends that the product manufacturer be contacted, at the address or phone number provided, for more complete information on product safety, processing parameters, and applications not provided in this information sheet.

# BP Performance Polymers Inc.

**Contact Address:**

BP Performance Polymers Inc.
Newberg Road
P.O. Box 400
Hackettstown, NJ 07840

Phone: 1 (201) 852-1100

## TPR® (BP PPI Code: 65-114-04)

**Product Data:**

**TPR® 1812 (BP PPI Code: 65-114-04)**

| Properties | Unit | Typical Value | Test Method |
|---|---|---|---|
| **Typical** | | | |
| Tensile Strength at Break[a] | psi<br>MPa | 850<br>5.9 | ASTM D-412 |
| Elongation at Break[a] | % | 350 | ASTM D-412 |
| Flex Modulus[b] | psi<br>MPa | 15,000<br>103.4 | ASTM D-790 |
| 100% Modulus[a] | psi<br>MPa | 650<br>4.5 | ASTM D-412 |
| Hardness (not cross linked)[c] | A | 84 | ASTM D-2240 |
| Tear Resistance[a] | PLi | 220 | ASTM D-CZK |
| Compression Set | | | ASTM D-395 |
| Brittle Point | °F<br>°C | | |
| Viscosity | pascal seconds | | |
| Surface resistivity | ohm/cm | | |

[a]Die C, die cut, 20″/min
[b]Method I, Procedure B, secant at 0.1″ deflection, 1/8″ × 3″ specimen, die cut, 0.5″/min
[c]3 second delay .250″ plied sample

NOTE – Information provided in this book has been condensed. The publisher recommends that the product manufacturer be contacted, at the address or phone number provided, for more complete information on product safety, processing parameters, and applications not provided in this information sheet.

# BP Performance Polymers Inc.

## Contact Address:

BP Performance Polymers Inc.
Newberg Road
P.O. Box 400
Hackettstown, NJ 07840

Phone: 1 (201) 852-1100

## TPR® 1830R (BP PPI Code: 65-114-14)

### Product Data:

**TPR® 1830R (BP PPI Code: 65-114-14)**

| Properties | Unit | Typical Value | Test Method |
|---|---|---|---|
| **Typical** | | | |
| Tensile Strength at Break[a] | psi<br>MPa | 850<br>5.9 | ASTM D-412 |
| Elongation at Break[a] | % | 350 | ASTM D-412 |
| Flex Modulus[b] | psi<br>MPa | 15,000<br>103.4 | ASTM D-790 |
| 100% Modulus[a] | psi<br>MPa | 650<br>4.5 | ASTM D-412 |
| Hardness (not cross linked)[c] | A | 87 | ASTM D-2240 |
| Tear Resistance[a] | PLi | 220 | ASTM D-CZK |
| Compression Set | | | ASTM D-395 |
| Brittle Point | °F<br>°C | | |
| Viscosity | pascal seconds | | |
| Surface resistivity | ohm/cm | | |

[a]Die C, die cut, 20″/min
[b]Method I, Procedure B, secant at 0.1″ deflection, 1/8″ × 3″ specimen, die cut, 0.5″/min
[c]3 second delay .250″ plied sample

NOTE – Information provided in this book has been condensed. The publisher recommends that the product manufacturer be contacted, at the address or phone number provided, for more complete information on product safety, processing parameters, and applications not provided in this information sheet.

# BP Performance Polymers Inc.

Contact Address:

BP Performance Polymers Inc.
Newberg Road
P.O. Box 400
Hackettstown, NJ 07840

Phone: 1 (201) 852-1100

## TPR® 1900 (BP PPI Code: 65-115-04)

Product Data:

**TPR® 1900 (BP PPI Code: 65-115-04)**

| Properties | Unit | Typical Value | Test Method |
|---|---|---|---|
| **Typical** | | | |
| Tensile Strength at Break[a] | psi<br>MPa | 950<br>6.5 | ASTM D-412 |
| Elongation at Break[a] | % | 425 | ASTM D-412 |
| Flex Modulus[b] | psi<br>MPa | 27,000<br>186.2 | ASTM D-790 |
| 100% Modulus[a] | psi<br>MPa | | ASTM D-412 |
| Hardness (not cross linked)[c] | D | 30 | ASTM D-2240 |
| Tear Resistance[a] | PLi | 380 | ASTM D-CZK |
| Compression Set | | | ASTM D-395 |
| Brittle Point | °F<br>°C | < −100<br>< −73 | |
| Viscosity | pascal seconds | 440 | |
| Surface resistivity | ohm/cm | | |

[a]Die C, die cut, 20″/min
[b]Method I, Procedure B, secant at 0.1″ deflection, 1/8″ × 3″ specimen, die cut, 0.5″/min
[c]3 second delay .250″ plied sample

NOTE—Information provided in this book has been condensed. The publisher recommends that the product manufacturer be contacted, at the address or phone number provided, for more complete information on product safety, processing parameters, and applications not provided in this information sheet.

# BP Performance Polymers Inc.

Contact Address:

BP Performance Polymers Inc.
Newberg Road
P.O. Box 400
Hackettstown, NJ 07840

Phone: 1 (201) 852-1100

## TPR® 1900 Natural (BP PPI Code: 65-115-04)

Description:

Olefin based thermoplastic elastomer.

Application:

Suitable for weatherstrip, seals, gaskets, wheels, caps, closures, film and sheet applications.

Features:

- medium modulus
- extrusion or injection molding
- recognized under UL Component Plastics Program (UL yellow card)

**TPR® 1900 Natural (BP PPI Code: 65-115-04)**

| Properties[1] | Unit | Typical Value | Test Method |
|---|---|---|---|
| **Typical** | | | |
| Specific Gravity | | 0.89 | ASTM D-792 |
| Hardness | Shore D | 30 | ASTM D-2240 |
| Tensile Strength | | | ASTM D-412 |
| | psi | 950 | |
| | MPa | 6.5 | |
| Elongation | % | 425 | ASTM D-412 |
| Flexural Modulus | | | ASTM D-790 |
| | psi | 27,000 | |
| | MPa | 186.2 | |

[1]Obtained testing 80 mil molded plaques. Should not be treated as specification values.

*(continued)*

NOTE—Information provided in this book has been condensed. The publisher recommends that the product manufacturer be contacted, at the address or phone number provided, for more complete information on product safety, processing parameters, and applications not provided in this information sheet.

**TPR® 1900 Natural (BP PPI Code: 65-115-04) *(continued)***

| Properties[1] | Unit | Typical Value | Test Method |
|---|---|---|---|
| **Typical** | | | |
| Tear Strength | PLi<br>kN/m | 380<br>66.5 | ASTM D-624 |
| Brittle Point | °F<br>°C | $< -100$<br>$< -73$ | ASTM D-746<br>Method I<br>Procedure B |
| Melt Viscosity | pascal seconds | 440 | |

[1]Obtained testing 80 mil molded plaques. Should not be treated as specification values.

# BP Performance Polymers Inc.

**Contact Address:**

BP Performance Polymers Inc.
Newberg Road
P.O. Box 400
Hackettstown, NJ 07840

Phone: 1 (201) 852-1100

## TPR® 2800 White (BP PPI Code: 65-112-14)

**Description:**

Olefin based thermoplastic elastomer.

**Application:**

Suitable for plug and connector applications.

**Features:**

- low modulus
- injection molding
- easy flow
- recognized under UL Component Plastics Program (UL yellow card)

**TPR® 2800 White (BP PPI Code: 65-112-14)**

| Properties[1] | Unit | Typical Value | Test Method |
|---|---|---|---|
| **Typical** | | | |
| Specific Gravity | | 1.04 | ASTM D-792 |
| Hardness | Shore A | 84 | ASTM D-2240 |
| Tensile Strength | | | ASTM D-412 |
| | psi | 800 | |
| | MPa | 5.2 | |
| Elongation | % | 500 | ASTM D-412 |
| Flexural Modulus | | | ASTM D-790 |
| | psi | 13,000 | |
| | MPa | 89.7 | |

[1]Obtained testing 80 mil molded plaques. Should not be treated as specification values.

*(continued)*

NOTE—Information provided in this book has been condensed. The publisher recommends that the product manufacturer be contacted, at the address or phone number provided, for more complete information on product safety, processing parameters, and applications not provided in this information sheet.

**TPR® 2800 White (BP PPI Code: 65-112-14) *(continued)***

| Properties[1] | Unit | Typical Value | Test Method |
|---|---|---|---|
| **Typical** | | | |
| Tear Strength | PLi<br>kN/m | 225<br>39.4 | ASTM D-624 |
| Brittle Point | °F<br>°C | < −75<br>< −59 | ASTM D-746<br>Method I<br>Procedure B |
| Melt Viscosity | pascal seconds | 190 | |

[1]Obtained testing 80 mil molded plaques. Should not be treated as specification values.

# BP Performance Polymers Inc.

**Contact Address:**

BP Performance Polymers Inc.
Newberg Road
P.O. Box 400
Hackettstown, NJ 07840

Phone: 1 (201) 852-1100

## TPR® 2900 (BP PPI Code: 65-112-04)

**Product Data:**

**TPR® 2900 (BP PPI Code: 65-112-04)**

| Properties | Unit | Typical Value | Test Method |
|---|---|---|---|
| **Typical** | | | |
| Tensile Strength at Break[a] | psi<br>MPa | 800<br>5.2 | ASTM D-412 |
| Elongation at Break[a] | % | 500 | ASTM D-412 |
| Flex Modulus[b] | psi<br>MPa | 113,000<br>206.9 | ASTM D-790 |
| 100% Modulus[a] | psi<br>MPa | | ASTM D-412 |
| Hardness (not cross linked)[c] | A | 74 | ASTM D-2240 |
| Tear Resistance[a] | PLi | 30 | ASTM D-CZK |
| Compression Set | | 44 | ASTM D-395 |
| Brittle Point | °F<br>°C | < −100<br>< −73 | |
| Viscosity | pascal seconds | 575 | |
| Surface resistivity | ohm/cm | | |

[a]Die C, die cut, 20″/min
[b]Method I, Procedure B, secant at 0.1″ deflection, 1/8″ × 3″ specimen, die cut, 0.5″/min
[c]3 second delay .250″ plied sample

NOTE—Information provided in this book has been condensed. The publisher recommends that the product manufacturer be contacted, at the address or phone number provided, for more complete information on product safety, processing parameters, and applications not provided in this information sheet.

# BP Performance Polymers Inc.

## Contact Address:

BP Performance Polymers Inc.
Newberg Road
P.O. Box 400
Hackettstown, NJ 07840

Phone: 1 (201) 852-1100

## TPR® 2900 (BP PPI Code: 65-115-14)

### Product Data:

**TPR® 2900 (BP PPI Code: 65-115-14)**

| Properties | Unit | Typical Value | Test Method |
|---|---|---|---|
| **Typical** | | | |
| Tensile Strength at Break[a] | psi<br>MPa | 2900<br>1000 | ASTM D-412 |
| Elongation at Break[a] | % | 500 | ASTM D-412 |
| Flex Modulus[b] | psi<br>MPa | 32,000<br>221 | ASTM D-790 |
| 100% Modulus[a] | psi<br>MPa | | ASTM D-412 |
| Hardness (not cross linked)[c] | A | 89 | ASTM D-2240 |
| Tear Resistance[a] | PLi | 270 | ASTM D-CZK |
| Compression Set | | | ASTM D-395 |
| Brittle Point | °F<br>°C | < −65<br>< −43 | |
| Viscosity | pascal seconds | 140 | |
| Surface resistivity | ohm/cm | | |

[a]Die C, die cut, 20″/min
[b]Method I, Procedure B, secant at 0.1″ deflection, 1/8″ × 3″ specimen, die cut, 0.5″/min
[c]3 second delay .250″ plied sample

NOTE – Information provided in this book has been condensed. The publisher recommends that the product manufacturer be contacted, at the address or phone number provided, for more complete information on product safety, processing parameters, and applications not provided in this information sheet.

# BP Performance Polymers Inc.

## Contact Address:

BP Performance Polymers Inc.
Newberg Road
P.O. Box 400
Hackettstown, NJ 07840

Phone: 1 (201) 852-1100

## TPR® 2900 Natural (BP PPI Code: 65-115-14)

### Description:

Olefin based thermoplastic elastomer.

### Application:

Suitable for plug and connector applications.

### Features:

- low modulus
- injection molding
- recognized under UL Component Plastics Program (UL yellow card)

**TPR® 2900 Natural (BP PPI Code: 65-115-14)**

| Properties[1] | Unit | Typical Value | Test Method |
|---|---|---|---|
| **Typical** | | | |
| Specific Gravity | | 1.04 | ASTM D-792 |
| Hardness | Shore A | 89 | ASTM D-2240 |
| Tensile Strength | psi<br>MPa | 1000<br>6.9 | ASTM D-412 |
| Elongation | % | 500 | ASTM D-412 |
| Flexural Modulus | psi<br>MPa | 32,000<br>221 | ASTM D-790 |

[1]Obtained testing 80 mil molded plaques. Should not be treated as specification values.

*(continued)*

NOTE—Information provided in this book has been condensed. The publisher recommends that the product manufacturer be contacted, at the address or phone number provided, for more complete information on product safety, processing parameters, and applications not provided in this information sheet.

TPR® 2900 Natural (BP PPI Code: 65-115-14) *(continued)*

| Properties[1] | Unit | Typical Value | Test Method |
|---|---|---|---|
| **Typical** | | | |
| Tear Strength | PLi<br>kN/m | 270<br>47.3 | ASTM D-624 |
| Brittle Point | °F<br>°C | < −55<br>< −48 | ASTM D-746<br>Method I<br>Procedure B |
| Melt Viscosity | pascal seconds | 140 | |

[1]Obtained testing 125 mil molded plaques. Should not be treated as specification values.

# BP Performance Polymers Inc.

**Contact Address:**

BP Performance Polymers Inc.
Newberg Road
P.O. Box 400
Hackettstown, NJ 07840

Phone: 1 (201) 852-1100

## TPR® 3210 (BP PPI Code: 95-626-04)

**Product Data:**

**TPR® 3210 (BP PPI Code: 95-626-04)**

| Properties | Unit | Typical Value | Test Method |
|---|---|---|---|
| **Typical** | | | |
| Tensile Strength at Break[a] | psi<br>MPa | 1300 | ASTM D-412 |
| Elongation at Break[a] | % | 400 | ASTM D-412 |
| Flex Modulus[b] | psi<br>MPa | 35,000 | ASTM D-790 |
| 100% Modulus[a] | psi<br>MPa | | ASTM D-412 |
| Hardness (not cross linked)[c] | D | 34 | ASTM D-2240 |
| Tear Resistance[a] | PLi | 450 | ASTM D-CZK |
| Compression Set | | | ASTM D-395 |
| Brittle Point | °F<br>°C | < −52 | |
| Viscosity | pascal seconds | | |
| Surface resistivity | ohm/cm | | |

[a]Die C, die cut, 20″/min
[b]Method I, Procedure B, secant at 0.1″ deflection, 1/8″ × 3″ specimen, die cut, 0.5″/min
[c]3 second delay .250″ plied sample

NOTE—Information provided in this book has been condensed. The publisher recommends that the product manufacturer be contacted, at the address or phone number provided, for more complete information on product safety, processing parameters, and applications not provided in this information sheet.

# BP Performance Polymers Inc.

## Contact Address:

BP Performance Polymers Inc.
Newberg Road
P.O. Box 400
Hackettstown, NJ 07840

Phone: 1 (201) 852-1100

---

## TPR® 3251 Black (BP PPI Code: 65-406-04)

### Description:

Olefin based thermoplastic elastomer.

### Application:

Suitable for automotive applications.

### Features:

- high modulus
- injection molding

**TPR® 3251 Black (BP PPI Code: 65-406-04)**

| Properties[1] | Unit | Typical Value | Test Method |
|---|---|---|---|
| **Typical** | | | |
| Specific Gravity | | 1.18 | ASTM D-792 |
| Tensile Strength | psi<br>MPa | 1500<br>10.3 | ASTM D-412 |
| Elongation | % | 200 | ASTM D-412 |
| Brittle Point | °F<br>°C | +14<br>−10 | ASTM D-746<br>Method I<br>Procedure B |

[1]Obtained testing 125 mil molded plaques. Should not be treated as specification values.

NOTE—Information provided in this book has been condensed. The publisher recommends that the product manufacturer be contacted, at the address or phone number provided, for more complete information on product safety, processing parameters, and applications not provided in this information sheet.

# BP Performance Polymers Inc.

## Contact Address:

BP Performance Polymers Inc.
Newberg Road
P.O. Box 400
Hackettstown, NJ 07840

Phone: 1 (201) 852-1100

## TPR® 5160 (BP PPI Code: 65-107-34)

### Product Data:

**TPR® 5160 (BP PPI Code: 65-107-34)**

| Properties | Unit | Typical Value | Test Method |
|---|---|---|---|
| **Typical** | | | |
| Tensile Strength at Break[a] | psi<br>MPa | 450<br>3.1 | ASTM D-412 |
| Elongation at Break[a] | % | 500 | ASTM D-412 |
| Flex Modulus[b] | psi<br>MPa | | ASTM D-790 |
| 100% Modulus[a] | psi<br>MPa | 600<br>4.1 | ASTM D-412 |
| Hardness (not cross linked)[c] | A | 63 | ASTM D-2240 |
| Tear Resistance[a] | PLi | | ASTM D-CZK |
| Compression Set | | 70 | ASTM D-395 |
| Brittle Point | °F<br>°C | <−100<br><−73 | |
| Viscosity | pascal seconds | | |
| Surface resistivity | ohm/cm | | |

[a]Die C, die cut, 20″/min
[b]Method I, Procedure B, secant at 0.1″ deflection, 1/8″ × 3″ specimen, die cut, 0.5″/min
[c]3 second delay .250″ plied sample

NOTE—Information provided in this book has been condensed. The publisher recommends that the product manufacturer be contacted, at the address or phone number provided, for more complete information on product safety, processing parameters, and applications not provided in this information sheet.

# BP Performance Polymers Inc.

**Contact Address:**

BP Performance Polymers Inc.
Newberg Road
P.O. Box 400
Hackettstown, NJ 07840

Phone: 1 (201) 852-1100

## TPR® 5190 (BP PPI Code: 65-115-39)

**Product Data:**

**TPR® 5190 (BP PPI Code: 65-115-39)**

| Properties | Unit | Typical Value | Test Method |
|---|---|---|---|
| **Typical** | | | |
| Tensile Strength at Break[a] | psi<br>MPa | 1100<br>7.6 | ASTM D-412 |
| Elongation at Break[a] | % | 450 | ASTM D-412 |
| Flex Modulus[b] | psi<br>MPa | 27,000<br>186.2 | ASTM D-790 |
| 100% Modulus[a] | psi<br>MPa | 630<br>4.3 | ASTM D-412 |
| Hardness (not cross linked)[c] | D | 30 | ASTM D-2240 |
| Tear Resistance[a] | PLi | | ASTM D-CZK |
| Compression Set | | | ASTM D-395 |
| Brittle Point | °F<br>°C | < −100<br>< −73 | |
| Viscosity | pascal seconds | 450 | |
| Surface resistivity | ohm/cm | | |

[a]Die C, die cut, 20″/min
[b]Method I, Procedure B, secant at 0.1″ deflection, 1/8″ × 3″ specimen, die cut, 0.5″/min
[c]3 second delay .250″ plied sample

NOTE – Information provided in this book has been condensed. The publisher recommends that the product manufacturer be contacted, at the address or phone number provided, for more complete information on product safety, processing parameters, and applications not provided in this information sheet.

# BP Performance Polymers Inc.

**Contact Address:**

BP Performance Polymers Inc.
Newberg Road
P.O. Box 400
Hackettstown, NJ 07840

Phone: 1 (201) 852-1100

## TPR® 5381 (BP PPI Code: 65-131-04)

**Product Data:**

**TPR® 5381 (BP PPI Code: 65-131-04)**

| Properties | Unit | Typical Value | Test Method |
|---|---|---|---|
| **Typical** | | | |
| Tensile Strength at Break[a] | psi<br>MPa | 1050<br>7.2 | ASTM D-412 |
| Elongation at Break[a] | % | 250 | ASTM D-412 |
| Flex Modulus[b] | psi<br>MPa | 27,000<br>186.2 | ASTM D-790 |
| 100% Modulus[a] | psi<br>MPa | | ASTM D-412 |
| Hardness (not cross linked)[c] | D | 38 | ASTM D-2240 |
| Tear Resistance[a] | PLi | 360 | ASTM D-CZK |
| Compression Set | | | ASTM D-395 |
| Brittle Point | °F<br>°C | < −100<br>< −73 | |
| Viscosity | pascal seconds | 300 | |
| Surface resistivity | ohm/cm | | |

[a]Die C, die cut, 20″/min
[b]Method I, Procedure B, secant at 0.1″ deflection, 1/8″ × 3″ specimen, die cut, 0.5″/min
[c]3 second delay .250″ plied sample

NOTE – Information provided in this book has been condensed. The publisher recommends that the product manufacturer be contacted, at the address or phone number provided, for more complete information on product safety, processing parameters, and applications not provided in this information sheet.

# BP Performance Polymers Inc.

## Contact Address:

BP Performance Polymers Inc.
Newberg Road
P.O. Box 400
Hackettstown, NJ 07840

Phone: 1 (201) 852-1100

## TPR® 5381 (BP PPI Code: 65-131-24)

### Product Data:

**TPR® 5381 (BP PPI Code: 65-131-24)**

| Properties | Unit | Typical Value | Test Method |
|---|---|---|---|
| **Typical** | | | |
| Tensile Strength at Break[a] | psi<br>MPa | 1100<br>7.6 | ASTM D-412 |
| Elongation at Break[a] | % | 460 | ASTM D-412 |
| Flex Modulus[b] | psi<br>MPa | 30,000<br>206.9 | ASTM D-790 |
| 100% Modulus[a] | psi<br>MPa | 900<br>6.2 | ASTM D-412 |
| Hardness (not cross linked)[c] | D | 31 | ASTM D-2240 |
| Tear Resistance[a] | PLi | 360 | ASTM D-CZK |
| Compression Set | | | ASTM D-395 |
| Brittle Point | °F<br>°C | < −100<br>< −73 | |
| Viscosity | pascal seconds | | |
| Surface resistivity | ohm/cm | | |

[a]Die C, die cut, 20″/min
[b]Method I, Procedure B, secant at 0.1″ deflection, 1/8″ × 3″ specimen, die cut, 0.5″/min
[c]3 second delay .250″ plied sample

NOTE – Information provided in this book has been condensed. The publisher recommends that the product manufacturer be contacted, at the address or phone number provided, for more complete information on product safety, processing parameters, and applications not provided in this information sheet.

# BP Performance Polymers Inc.


Contact Address:

BP Performance Polymers Inc.
Newberg Road
P.O. Box 400
Hackettstown, NJ 07840

Phone: 1 (201) 852-1100


## TPR® 5381 Black (BP PPI Code: 65-131-04)

### Description:

A non-conductive olefin based thermoplastic elastomer.

### Application:

Suitable for wheels and automotive applications.

### Features:

- intermediate modulus
- injection molding
- paintable after treatment with an approved surface primer
- approved under Chrysler Corporation Automotive Specification Materials Number MSDC-243-CPN-2390 and AMC Specification Materials Number AMC-5090

**TPR® 5381 Black (BP PPI Code: 65-131-04)**

| Properties[1] | Unit | Typical Value | Test Method |
|---|---|---|---|
| **Typical** | | | |
| Specific Gravity | | 0.90 | ASTM D-792 |
| Hardness | Shore D | 39 | ASTM D-2240 |
| Tensile Strength | psi<br>MPa | 1050<br>7.2 | ASTM D-412 |
| Elongation | % | 250 | ASTM D-412 |
| Flexural Modulus | psi<br>MPa | 27,000<br>186.2 | ASTM D-790 |

[1]Obtained testing 80 mil molded plaques. Should not be treated as specification values.

*(continued)*

NOTE—Information provided in this book has been condensed. The publisher recommends that the product manufacturer be contacted, at the address or phone number provided, for more complete information on product safety, processing parameters, and applications not provided in this information sheet.

**TPR® 5381 Black (BP PPI Code: 65-131-04) *(continued)***

| Properties[1] | Unit | Typical Value | Test Method |
|---|---|---|---|
| **Typical** | | | |
| Tear Strength | PLi<br>kN/m | 360<br>63.0 | ASTM D-624 |
| Brittle Point | °F<br>°C | < −100<br>< −73 | ASTM D-746<br>Method I<br>Procedure B |
| Melt Viscosity | pascal seconds | 300 | |

[1]Obtained testing 80 mil molded plaques. Should not be treated as specification values.

# BP Performance Polymers Inc.

Contact Address:

BP Performance Polymers Inc.
Newberg Road
P.O. Box 400
Hackettstown, NJ 07840

Phone: 1 (201) 852-1100

## TPR® 8781 (BP PPI Code: 65-102-14)

Product Data:

**TPR® 8781 (BP PPI Code: 65-102-14)**

| Properties | Unit | Typical Value | Test Method |
|---|---|---|---|
| **Typical** | | | |
| Tensile Strength at Break[a] | psi<br>MPa | 550<br>3.8 | ASTM D-412 |
| Elongation at Break[a] | % | 150 | ASTM D-412 |
| Flex Modulus[b] | psi<br>MPa | | ASTM D-790 |
| 100% Modulus[a] | psi<br>MPa | 500<br>3.4 | ASTM D-412 |
| Hardness (not cross linked)[c] | A | 55 | ASTM D-2240 |
| Tear Resistance[a] | PLi | 100 | ASTM D-CZK |
| Compression Set | | 45 | ASTM D-395 |
| Brittle Point | °F<br>°C | < −100<br>< −73 | |
| Viscosity | pascal seconds | 350 | |
| Surface resistivity | ohm/cm | | |

[a]Die C, die cut, 20″/min
[b]Method I, Procedure B, secant at 0.1″ deflection, 1/8″ × 3″ specimen, die cut, 0.5″/min
[c]3 second delay .250″ plied sample

NOTE—Information provided in this book has been condensed. The publisher recommends that the product manufacturer be contacted, at the address or phone number provided, for more complete information on product safety, processing parameters, and applications not provided in this information sheet.

# BP Performance Polymers Inc.

Contact Address:

BP Performance Polymers Inc.
Newberg Road
P.O. Box 400
Hackettstown, NJ 07840

Phone: 1 (201) 852-1100

## TPR® 8800 (BP PPI Code: 65-141-14)

Product Data:

**TPR® 8800 (BP PPI Code: 65-141-14)**

| Properties | Unit | Typical Value | Test Method |
|---|---|---|---|
| **Typical** | | | |
| Tensile Strength at Break[a] | psi<br>MPa | 1330<br>9.2 | ASTM D-412 |
| Elongation at Break[a] | % | 630 | ASTM D-412 |
| Flex Modulus[b] | psi<br>MPa | 25,000<br>172.4 | ASTM D-790 |
| 100% Modulus[a] | psi<br>MPa | | ASTM D-412 |
| Hardness (not cross linked)[c] | D | 36 | ASTM D-2240 |
| Tear Resistance[a] | PLi | | ASTM D-CZK |
| Compression Set | | | ASTM D-395 |
| Brittle Point | °F<br>°C | | |
| Viscosity | pascal seconds | 200 | |
| Surface resistivity | ohm/cm | | |

[a]Die C, die cut, 20″/min
[b]Method I, Procedure B, secant at 0.1″ deflection, 1/8″ × 3″ specimen, die cut, 0.5″/min
[c]3 second delay .250″ plied sample

NOTE—Information provided in this book has been condensed. The publisher recommends that the product manufacturer be contacted, at the address or phone number provided, for more complete information on product safety, processing parameters, and applications not provided in this information sheet.

# BP Performance Polymers Inc.

## Contact Address:

BP Performance Polymers Inc.
Newberg Road
P.O. Box 400
Hackettstown, NJ 07840

Phone: 1 (201) 852-1100

## TPR® 8951 (BP PPI Code: 65-139-04)

### Product Data:

**TPR® 8951 (BP PPI Code: 65-139-04)**

| Properties | Unit | Typical Value | Test Method |
|---|---|---|---|
| **Typical** | | | |
| Tensile Strength at Break[a] | psi<br>MPa | 950<br>6.5 | ASTM D-412 |
| Elongation at Break[a] | % | 250 | ASTM D-412 |
| Flex Modulus[b] | psi<br>MPa | 25,000<br>172.4 | ASTM D-790 |
| 100% Modulus[a] | psi<br>MPa | | ASTM D-412 |
| Hardness (not cross linked)[c] | D | 31 | ASTM D-2240 |
| Tear Resistance[a] | PLi | 270 | ASTM D-CZK |
| Compression Set | | | ASTM D-395 |
| Brittle Point | °F<br>°C | | |
| Viscosity | pascal seconds | 425 | |
| Surface resistivity | ohm/cm | | |

[a]Die C, die cut, 20″/min
[b]Method I, Procedure B, secant at 0.1″ deflection, 1/8″ × 3″ specimen, die cut, 0.5″/min
[c]3 second delay .250″ plied sample

NOTE—Information provided in this book has been condensed. The publisher recommends that the product manufacturer be contacted, at the address or phone number provided, for more complete information on product safety, processing parameters, and applications not provided in this information sheet.

# BP Performance Polymers Inc.

## Contact Address:

BP Performance Polymers Inc.
Newberg Road
P.O. Box 400
Hackettstown, NJ 07840

Phone: 1 (201) 852-1100

## TPR® 9010 (BP PPI Code: 65-144-04)

### Product Data:

**TPR® 9010 (BP PPI Code: 65-144-04)**

| Properties | Unit | Typical Value | Test Method |
|---|---|---|---|
| **Typical** | | | |
| Tensile Strength at Break[a] | psi<br>MPa | 350<br>2.4 | ASTM D-412 |
| Elongation at Break[a] | % | 450 | ASTM D-412 |
| Flex Modulus[b] | psi<br>MPa | | ASTM D-790 |
| 100% Modulus[a] | psi<br>MPa | | ASTM D-412 |
| Hardness (not cross linked)[c] | A | 63 | ASTM D-2240 |
| Tear Resistance[a] | PLi | 25 | ASTM D-CZK |
| Compression Set | | | ASTM D-395 |
| Brittle Point | °F<br>°C | < −100<br>< −73 | |
| Viscosity | pascal seconds | 350 | |
| Surface resistivity | ohm/cm | | |

[a]Die C, die cut, 20″/min
[b]Method I, Procedure B, secant at 0.1″ deflection, 1/8″ × 3″ specimen, die cut, 0.5″/min
[c]3 second delay .250″ plied sample

NOTE—Information provided in this book has been condensed. The publisher recommends that the product manufacturer be contacted, at the address or phone number provided, for more complete information on product safety, processing parameters, and applications not provided in this information sheet.

# BP Performance Polymers Inc.

## Contact Address:

BP Performance Polymers Inc.
Newberg Road
P.O. Box 400
Hackettstown, NJ 07840

Phone: 1 (201) 852-1100

## TPR® 9180 (BP PPI Code: 65-110-24)

### Product Data:

**TPR® 9180 (BP PPI Code: 65-110-24)**

| Properties | Unit | Typical Value | Test Method |
|---|---|---|---|
| **Typical** | | | |
| Tensile Strength at Break[a] | psi<br>MPa | 750<br>5.2 | ASTM D-412 |
| Elongation at Break[a] | % | 300 | ASTM D-412 |
| Flex Modulus[b] | psi<br>MPa | | ASTM D-790 |
| 100% Modulus[a] | psi<br>MPa | | ASTM D-412 |
| Hardness (not cross linked)[c] | A | 74 | ASTM D-2240 |
| Tear Resistance[a] | PLi | | ASTM D-CZK |
| Compression Set | | 65 | ASTM D-395 |
| Brittle Point | °F<br>°C | | |
| Viscosity | pascal seconds | | |
| Surface resistivity | ohm/cm | 600 | |

[a]Die C, die cut, 20″/min
[b]Method I, Procedure B, secant at 0.1″ deflection, 1/8″ × 3″ specimen, die cut, 0.5″/min
[c]3 second delay .250″ plied sample

NOTE – Information provided in this book has been condensed. The publisher recommends that the product manufacturer be contacted, at the address or phone number provided, for more complete information on product safety, processing parameters, and applications not provided in this information sheet.

# Celanese Engineering Resins Inc.

## Contact Address:

Celanese Engineering Resin Inc.
26 Main Street
Chatham, NJ 07928

Phone: 1 (212) 635-2600

## VANDAR 2000

### Product Data:

**Vandar 2000**

| Vandar (PBT) Property | ASTM Test Method | Units |
|---|---|---|
| **Mechanical** | | |
| Tensile Strength at Yield<br>−40°F<br>73°F | D638 | <br>—<br>4500 psi |
| Elongation<br>−40°F<br>73° | D638 | <br>—<br>100% |
| Flexural Modulus<br>73°F | D790 | <br>200,000 psi |
| Izod Impact Strength (notched)<br>−20°F<br>73°F | D256 | <br>3 ft-lb/in<br>N/B |
| Gardner Impact<br>−20°F<br>73°F | | <br>25 ft-lb<br>23 ft-lb |
| Water Absorption, 73°F<br>24 hr. Immersion | | —<br>— |
| **Thermal** | | |
| H.D.T. @ 264 psi<br>@ 66 psi | D648 | 100°F<br>170°F |
| Vicat. Point | | 390°F |

*(continued)*

NOTE—Information provided in this book has been condensed. The publisher recommends that the product manufacturer be contacted, at the address or phone number provided, for more complete information on product safety, processing parameters, and applications not provided in this information sheet.

**Vandar 2000 *(continued)***

| **Vandar (PBT) Property** | **ASTM Test Method** | **Units** |
|---|---|---|
| **Electrical** (at 73°F and 50% RH) | | |
| UL Rating | | 1/16 in HB |
| **Miscellaneous** | | |
| Appearance opaque | | — |
| Cost/lb | | $1.95 |
| Cost/cu in | | $0.0837 |
| Specific Gravity | | 1.19 |
| Shrinkage: Flow<br>Transverse | 1/8 (in/in) | 0.016<br>0.014 |
| Material Comp. | | unfilled |

Values shown are based on limited testing. These provisional values are not intended for use in establishing maximum, minimum or range for specification purposes.

# Celanese Engineering Resins Inc.

Contact Address:

Celanese Engineering Resin Inc.
26 Main Street
Chatham, NJ 07928

Phone: 1 (212) 635-2600

## VANDAR 2100

Product Data:

**Vandar 2100**

| **Vandar (PBT) Property** | **ASTM Test Method** | **Units** |
|---|---|---|
| **Mechanical** | | |
| Tensile Strength at Yield<br>−40°F<br>73°F | D638 | —<br>6300 psi |
| Elongation<br>−40°F<br>73° | D638 | —<br>150% |
| Flexural Modulus<br>73°F | D790 | 280,000 psi |
| Izod Impact Strength (notched)<br>−20°F<br>73°F | D256 | 3 ft-lb/in<br>N/B |
| Gardner Impact<br>−20°F<br>73°F | | 35 ft-lb<br>30 ft-lb |
| Water Absorption, 73°F<br>24 hr. Immersion | | 0.13% |
| **Thermal** | | |
| H.D.T. @ 264 psi<br>@ 66 psi | D648 | 105°F<br>180°F |
| Vicat. Point | | 398°F |

*(continued)*

NOTE – Information provided in this book has been condensed. The publisher recommends that the product manufacturer be contacted, at the address or phone number provided, for more complete information on product safety, processing parameters, and applications not provided in this information sheet.

**Vandar 2100 *(continued)***

| Vandar (PBT) Property | ASTM Test Method | Units |
|---|---|---|
| **Electrical** (at 73°F and 50% RH) | | |
| UL Rating | | 1/16 in HB |
| **Miscellaneous** | | |
| Appearance opaque | | — |
| Cost/lb | | $1.95 |
| Cost/cu in | | $0.0865 |
| Specific Gravity | | 1.23 |
| Shrinkage: Flow<br>Transverse | 1/8 (in/in) | 0.018<br>0.016 |
| Material Comp. | | unfilled |

Values shown are based on limited testing. These provisional values are not intended for use in establishing maximum, minimum or range for specification purposes.

# Celanese Engineering Resins Inc.

## Contact Address:

Celanese Engineering Resin Inc.
26 Main Street
Chatham, NJ 07928

Phone: 1 (212) 635-2600

## VANDAR 2122

### Product Data:

**Vandar 2122**

| Vandar (PBT) Property | ASTM Test Method | Units |
|---|---|---|
| **Mechanical** | | |
| Tensile Strength at Yield<br>−40°F<br>73°F | D638 | <br>—<br>6400 psi |
| Elongation<br>−40°F<br>73° | D638 | <br>—<br>50% |
| Flexural Modulus<br>73°F | D790 | 325,000 psi |
| Izod Impact Strength (notched)<br>−20°F<br>73°F | D256 | <br>1 ft-lb/in<br>3 ft-lb/in |
| Gardner Impact<br>−20°F<br>73°F | | <br>26 ft-lb<br>26 ft-lb |
| Water Absorption, 73°F<br>24 hr. Immersion | | 0.13% |
| **Thermal** | | |
| H.D.T. @ 264 psi<br>@ 66 psi | D648 | 120°F<br>190°F |
| Vicat. Point | | 420°F |

*(continued)*

NOTE – Information provided in this book has been condensed. The publisher recommends that the product manufacturer be contacted, at the address or phone number provided, for more complete information on product safety, processing parameters, and applications not provided in this information sheet.

Vandar 2122 *(continued)*

| Vandar (PBT) Property | ASTM Test Method | Units |
|---|---|---|
| **Electrical** (at 73°F and 50% RH) | | |
| UL Rating | | 1/16 in HB |
| **Miscellaneous** | | |
| Appearance opaque | | — |
| Cost/lb | | $1.90 |
| Cost/cu in | | $0.0891 |
| Specific Gravity | | 1.30 |
| Shrinkage: Flow<br>Transverse | 1/8 (in/in) | 0.013<br>0.015 |
| Material Comp. | | mineral filled |

Values shown are based on limited testing. These provisional values are not intended for use in establishing maximum, minimum or range for specification purposes.

# Celanese Engineering Resins Inc.

**Contact Address:**

Celanese Engineering Resin Inc.
26 Main Street
Chatham, NJ 07928

Phone: 1 (212) 635-2600

## VANDAR 4602Z

**Product Data:**

**Vandar 4602Z**

| Vandar (PBT) Property | ASTM Test Method | Units |
|---|---|---|
| **Mechanical** | | |
| Tensile Strength at Yield<br>−40°F<br>73°F | D638 | <br>—<br>5800 psi |
| Elongation<br>−40°F<br>73° | D638 | <br>—<br>270% |
| Flexural Modulus<br>73°F | D790 | <br>220,000 psi |
| Izod Impact Strength (notched)<br>−20°F<br>73°F | D256 | <br>—<br>N/B |
| Gardner Impact<br>−20°F<br>73°F | | <br>—<br>N/B |
| Water Absorption, 73°F<br>24 hr. Immersion | | <br>0.10% |
| **Thermal** | | |
| H.D.T. @ 264 psi<br>@ 66 psi | D648 | 125°F*<br>270°F* |
| Vicat. Point | | — |

*annealed @121°C

*(continued)*

NOTE—Information provided in this book has been condensed. The publisher recommends that the product manufacturer be contacted, at the address or phone number provided, for more complete information on product safety, processing parameters, and applications not provided in this information sheet.

Vandar 4602Z *(continued)*

| Vandar (PBT) Property | ASTM Test Method | Units |
|---|---|---|
| **Electrical** (at 73°F and 50% RH) | | |
| UL Rating | | 1/16 in HB |
| **Miscellaneous** | | |
| Appearance opaque | | — |
| Cost/lb | | $1.95 |
| Cost/cu in | | $0.0889 |
| Specific Gravity | | 1.25 |
| Shrinkage: Flow<br>Transverse | 1/8 (in/in) | 0.017<br>0.017 |
| Material Comp. | | unfilled |

Values shown are based on limited testing. These provisional values are not intended for use in establishing maximum, minimum or range for specification purposes.

# Celanese Engineering Resins Inc.

## Contact Address:

Celanese Engineering Resin Inc.
26 Main Street
Chatham, NJ 07928

Phone: 1 (212) 635-2600

## VANDAR 4612R

### Product Data:

**Vandar 4612R**

| Vandar (PBT) Property | ASTM Test Method | Units |
|---|---|---|
| **Mechanical** | | |
| Tensile Strength at Yield | D638 | |
| −40°F | | — |
| 73°F | | 7400 psi |
| Elongation | D638 | |
| −40°F | | — |
| 73° | | 6% |
| Flexural Modulus | D790 | |
| 73°F | | 395,000 psi |
| Izod Impact Strength (notched) | D256 | |
| −20°F | | — |
| 73°F | | 3.4 ft-lb/in |
| Gardner Impact | | |
| −20°F | | — |
| 73°F | | 8 ft-lb |
| Water Absorption, 73°F 24 hr. Immersion | | 0.10% |
| **Thermal** | | |
| H.D.T. @ 264 psi | | 170°F |
| @ 66 psi | D648 | 379°F |
| Vicat. Point | | — |

*(continued)*

NOTE—Information provided in this book has been condensed. The publisher recommends that the product manufacturer be contacted, at the address or phone number provided, for more complete information on product safety, processing parameters, and applications not provided in this information sheet.

Vandar 4612R *(continued)*

| Vandar (PBT) Property | ASTM Test Method | Units |
|---|---|---|
| **Electrical** (at 73°F and 50% RH) | | |
| UL Rating | | 1/16 in HB |
| **Miscellaneous** | | |
| Appearance opaque | | — |
| Cost/lb | | $1.85 |
| Cost/cu in | | $0.0867 |
| Specific Gravity | | 1.30 |
| Shrinkage: Flow<br>Transverse | 1/8 (in/in) | 0.007–8<br>0.012–13 |
| Material Comp. | | 7.5% glass |

Values shown are based on limited testing. These provisional values are not intended for use in establishing maximum, minimum or range for specification purposes.

# Celanese Engineering Resins Inc.

Contact Address:

Celanese Engineering Resin Inc.
26 Main Street
Chatham, NJ 07928

Phone: 1 (212) 635-2600

## VANDAR 4632Z

Product Data:

**Vandar 4632Z**

| Vandar (PBT) Property | ASTM Test Method | Units |
|---|---|---|
| **Mechanical** | | |
| Tensile Strength at Yield<br>−40°F<br>73°F | D638 | <br>—<br>9000 psi |
| Elongation<br>−40°F<br>73° | D638 | <br>—<br>5% |
| Flexural Modulus<br>73°F | D790 | <br>550,000 psi |
| Izod Impact Strength (notched)<br>−20°F<br>73°F | D256 | <br>—<br>2.8 ft-lb/in |
| Gardner Impact<br>−20°F<br>73°F | | <br>—<br>— |
| Water Absorption, 73°F<br>24 hr. Immersion | | <br>0.10% |
| **Thermal** | | |
| H.D.T. @ 264 psi<br>@ 66 psi | <br>D648 | 365°F<br>405°F |
| Vicat. Point | | — |

*(continued)*

NOTE – Information provided in this book has been condensed. The publisher recommends that the product manufacturer be contacted, at the address or phone number provided, for more complete information on product safety, processing parameters, and applications not provided in this information sheet.

Vandar 4632Z *(continued)*

| Vandar (PBT) Property | ASTM Test Method | Units |
|---|---|---|
| **Electrical** (at 73°F and 50% RH) | | |
| UL Rating | | 1/16 in HB |
| **Miscellaneous** | | |
| Appearance opaque | | |
| Cost/lb | | $1.85 |
| Cost/cu in | | $0.0894 |
| Specific Gravity | | 1.34 |
| Shrinkage: Flow<br>Transverse | 1/8 (in/in) | 0.005–.006<br>0.012–0.013 |
| Material Comp. | | 15% glass |

Values shown are based on limited testing. These provisional values are not intended for use in establishing maximum, minimum or range for specification purposes.

# Celanese Engineering Resins Inc.


Contact Address:

Celanese Engineering Resin Inc.
26 Main Street
Chatham, NJ 07928

Phone: 1 (212) 635-2600


## VANDAR 4662Z

Product Data:

**Vandar 4662Z**

| Vandar (PBT) Property | ASTM Test Method | Units |
|---|---|---|
| **Mechanical** | | |
| Tensile Strength at Yield<br>−40°F<br>73°F | D638 | <br>—<br>13,000 psi |
| Elongation<br>−40°F<br>73° | D638 | <br>—<br>3.5% |
| Flexural Modulus<br>73°F | D790 | <br>925,000 psi |
| Izod Impact Strength (notched)<br>−20°F<br>73°F | D256 | <br>—<br>4.3 ft-lb/in |
| Gardner Impact<br>−20°F<br>73°F | | <br>—<br>— |
| Water Absorption, 73°F<br>24 hr. Immersion | | <br>0.08% |
| **Thermal** | | |
| H.D.T. @ 264 psi<br>@ 66 psi | D648 | 374°F<br>417°F |
| Vicat. Point | | — |

*(continued)*

NOTE—Information provided in this book has been condensed. The publisher recommends that the product manufacturer be contacted, at the address or phone number provided, for more complete information on product safety, processing parameters, and applications not provided in this information sheet.

**Vandar 4662Z *(continued)***

| Vandar (PBT) Property | ASTM Test Method | Units |
|---|---|---|
| **Electrical** (at 73°F and 50% RH) | | |
| UL Rating | | 1/16 in HB |
| **Miscellaneous** | | |
| Appearance opaque | | — |
| Cost/lb | | $1.85 |
| Cost/cu in | | $0.0981 |
| Specific Gravity | | 1.47 |
| Shrinkage: Flow<br>Transverse | 1/8 (in/in) | 0.003–4<br>0.013–14 |
| Material Comp. | | 30% glass |

Values shown are based on limited testing. These provisional values are not intended for use in establishing maximum, minimum or range for specification purposes.

# Celanese Engineering Resins Inc.

**Contact Address:**

Celanese Engineering Resin Inc.
26 Main Street
Chatham, NJ 07928

Phone: 1 (212) 635-2600

## VANDAR 8000

**Product Data:**

**Vandar 8000**

| Vandar (PBT) Property | ASTM Test Method | Units |
|---|---|---|
| **Mechanical** | | |
| Tensile Strength at Yield | D638 | |
| −40°F | | — |
| 73°F | | 5700 psi |
| Elongation | D638 | |
| −40°F | | — |
| 73° | | 50% |
| Flexural Modulus | D790 | |
| 73°F | | 270,000 psi |
| Izod Impact Strength (notched) | D256 | |
| −20°F | | — |
| 73°F | | 12 ft-lb/in |
| Gardner Impact | | |
| −20°F | | — |
| 73°F | | — |
| Water Absorption, 73°F 24 hr. Immersion | | — |
| **Thermal** | | |
| H.D.T. @ 264 psi | D648 | 125°F |
| @ 66 psi | | — |

*(continued)*

NOTE—Information provided in this book has been condensed. The publisher recommends that the product manufacturer be contacted, at the address or phone number provided, for more complete information on product safety, processing parameters, and applications not provided in this information sheet.

### Vandar 8000 *(continued)*

| Vandar (PBT) Property | ASTM Test Method | Units |
|---|---|---|
| **Electrical** (at 73°F and 50% RH) | | |
| UL Rating | | V@ 1/32″ |
| **Miscellaneous** | | |
| Appearance opaque | | — |
| Cost/lb | | $2.88 |
| Specific Gravity | | |
| Shrinkage: Flow<br>Transverse | 1/8 (in/in) | 0.005–0.006<br>0.012–0.013 |
| Material Comp. | | 15% glass |

Values shown are based on limited testing. These provisional values are not intended for use in establishing maximum, minimum or range for specification purposes.

# Dow Chemical Company

## Contact Address:

Dow Chemical Company
Plastics Group Services
2040 Williard H. Dow Center
Midland, MI 48674

Phone: 1 (800) 258-CHEM

## PULSE 600

### Description:

PULSE 600 Engineering Resin is specifically designed for automotive parts where there is little need for a low gloss surface, but where a superior balance of impact strength, heat distortion, and processability is needed. PULSE 600 Engineering Resin offers excellent processability, high heat distortion, and high impact strength at both room and low temperatures.

### Processing:

Recommended conditions for molding PULSE 600 Engineering Resin are: injection molding temperatures 490°–540°F (255°–282°C); mold temperatures 175°–200°F (79°–93°C); moisture content below .02%; and drying prior to fabrication for 3–4 hours at 200°F (93°C) in a dehumidifying dryer.

### Product Data:

**PULSE 600**

| Property[1] | ASTM | |
|---|---|---|
| **Physical** | | |
| Melt flow rate<br>g/10 min @ 230°C/3.8 kg | D1238 | 5.0 |
| Specific Gravity, g/cc | D792 | 1.12 |
| Water absorption, %<br>24 hrs @ 73°F<br>Equilibrium @ 73°F | D570 | <br>.15<br>.32 |
| Mold shrinkage, in/in | D955 | 0.006 |
| **Thermal** | | |
| Deflection temp, °F @ 264 psi<br>1/8″ thick, unannealed<br>1/8″ thick, annealed | D648 | <br>205<br>225 |
| Coefficient of linear thermal expansion, in/in/°F | D696 | $4.11 \times 10^{-5}$ |

[1]Typical property values; not to be construed as sales specifications.

*(continued)*

NOTE – Information provided in this book has been condensed. The publisher recommends that the product manufacturer be contacted, at the address or phone number provided, for more complete information on product safety, processing parameters, and applications not provided in this information sheet.

**PULSE 600 *(continued)***

| Property[1] | ASTM | |
|---|---|---|
| **Mechanical** | | |
| Tensile yield strength, psi | D638 | 7700 |
| Tensile break strength, psi | D638 | 7000 |
| Elongation, break, % | D638 | 100 |
| Flexural yield strength, psi | D790 | 12,000 |
| Flexural modulus (tangent), $10^5$ psi | D790 | 3.75 |
| Izod impact strength,<br>ft-lb/in notch, 1/8″ thick @ 73°F<br>@ −20°F | D256 | <br>10.0<br>9.0 |
| Rockwell hardness, R | D785 | 110 |
| Instrumented dart impact[2],<br>in-lb, total energy @ 73°<br>@ −20° F | D3763-85 | <br>500<br>500 |
| **Flammability Rating[3]** | | |
| Average extent of burning | D635 | 1″ |

[1]Typical property values; not to be construed as sales specifications.
[2]0.125″ thick, clamp opening 1.5″, 8000 IPM, 0.5″ dart.
[3]These flammability ratings are not intended to reflect hazards presented by these or any other materials under actual fire conditions.

# Dow Chemical Company

## Contact Address:

Dow Chemical Company
Plastics Group Services
2040 Williard H. Dow Center
Midland, MI 48674

Phone: 1 (800) 258-CHEM

## PULSE 700

### Description:

PULSE 700 Engineering Resin is designed specifically for the automotive interior designer. PULSE 700 Engineering Resin offers properties very similar to PULSE 710 while improving on the processability. The combination of excellent processability, low gloss, high heat distortion, and impact strength make PULSE 700 Engineering Resin an excellent choice for interior trim applications.

### Processing:

Recommended conditions for molding PULSE 700 Engineering Resin are: injection molding temperatures 500°–550°F (260°–288°C); mold temperatures 175°–200°F (79°–93°C); moisture content below .02%; and drying prior to fabrication for 3–4 hours at 200°F (93°C) in a dehumidifying dryer.

### Product Data:

**PULSE 700**

| Property[1] | ASTM | |
|---|---|---|
| **Physical** | | |
| Melt flow rate<br>g/10 min @ 230°C/3.8 kg | D1238 | 5.0 |
| Specific Gravity, g/cc | D792 | 1.10 |
| Water absorption, %<br>24 hrs @ 73°F<br>Equilibrium @ 73°F | D570 | <br>.15<br>.32 |
| Mold shrinkage, in/in | D955 | 0.006 |
| **Thermal** | | |
| Deflection temp, °F @ 264 psi<br>1/8″ thick, unannealed<br>1/8″ thick, annealed | D648 | <br>200<br>220 |
| Coefficient of linear thermal expansion,<br>in/in/°F | D696 | $4.11 \times 10^{-5}$ |

[1]Typical property values; not to be construed as sales specifications.

*(continued)*

NOTE–Information provided in this book has been condensed. The publisher recommends that the product manufacturer be contacted, at the address or phone number provided, for more complete information on product safety, processing parameters, and applications not provided in this information sheet.

PULSE 700 *(continued)*

| Property[1] | ASTM | |
|---|---|---|
| **Mechanical** | | |
| Tensile yield strength, psi | D638 | 6500 |
| Tensile break strength, psi | D638 | 6000 |
| Elongation, break, % | D638 | 80 |
| Flexural yield strength, psi | D790 | 10,000 |
| Flexural modulus (tangent), $10^5$ psi | D790 | 3.10 |
| Izod impact strength, ft-lb/in notch, 1/8″ thick @ 73°F<br>@ −20°F | D256 | 9.0<br>5.0 |
| Rockwell hardness, R | D785 | 108 |
| Instrumented dart impact[2], in-lb, total energy @ 73°<br>@ −20° F | D3763-85 | 450<br>300 |
| **Flammability Rating[3]** | | |
| Average extent of burning | D635 | 1″ |

[1]Typical property values; not to be construed as sales specifications.
[2]0.125″ thick, clamp opening 1.5″, 8000 IPM, 0.5″ dart.
[3]These flammability ratings are not intended to reflect hazards presented by these or any other materials under actual fire conditions.

# Dow Chemical Company

## Contact Address:

Dow Chemical Company
Plastics Group Services
2040 Williard H. Dow Center
Midland, MI 48674

Phone: 1 (800) 258-CHEM

## PULSE 710

### Description:

PULSE 710 Engineering Resin offers the automotive interior designer a high heat distortion resin which can provide the surface finish to match today's interior design aesthetics. Excellent processability, low gloss, and high impact strength combine to make PULSE 710 Engineering Resin an excellent choice for interior trim applications.

### Processing:

Recommended conditions for molding PULSE 710 Engineering Resin are: injection molding temperatures 500°–550°F (260°–288°C); mold temperatures 175°–200°F (79°–93°C); moisture content below .02%; and drying prior to fabrication for 3–4 hours at 200°F (93°C) in a dehumidifying dryer.

### Product Data:

**PULSE 710**

| Property[1] | ASTM | |
|---|---|---|
| **Physical** | | |
| Melt flow rate<br>g/10 min @ 230°C/3.8 kg | D1238 | 4.0 |
| Specific Gravity, g/cc | D792 | 1.12 |
| Water absorption, %<br>24 hrs @ 73°F<br>Equilibrium @ 73°F | D570 | <br>.15<br>.32 |
| Mold shrinkage, in/in | D955 | 0.006 |
| **Thermal** | | |
| Deflection temp, °F @ 264 psi<br>1/8″ thick, unannealed<br>1/8″ thick, annealed | D648 | <br>215<br>235 |
| Coefficient of linear thermal expansion,<br>in/in/°F | D696 | $4.11 \times 10^{-5}$ |

[1]Typical property values; not to be construed as sales specifications.

*(continued)*

NOTE—Information provided in this book has been condensed. The publisher recommends that the product manufacturer be contacted, at the address or phone number provided, for more complete information on product safety, processing parameters, and applications not provided in this information sheet.

**PULSE 710 *(continued)***

| Property[1] | ASTM | |
|---|---|---|
| **Mechanical** | | |
| Tensile yield strength, psi | D638 | 7000 |
| Tensile break strength, psi | D638 | 6500 |
| Elongation, break, % | D638 | 60 |
| Flexural yield strength, psi | D790 | 12,000 |
| Flexural modulus (tangent), $10^5$ psi | D790 | 3.50 |
| Izod impact strength,<br>ft-lb/in notch, 1/8″ thick @ 73°F<br>@ −20°F | D256 | <br>10.0<br>9.0 |
| Rockwell hardness, R | D785 | 110 |
| Instrumented dart impact[2],<br>in-lb, total energy @ 73°<br>@ −20° F | D3763-85 | <br>500<br>500 |
| **Flammability Rating[3]** | | |
| Average extent of burning | D635 | 1″ |

[1]Typical property values; not to be construed as sales specifications.
[2]0.125″ thick, clamp opening 1.5″, 8000 IPM, 0.5″ dart.
[3]These flammability ratings are not intended to reflect hazards presented by these or any other materials under actual fire conditions.

# Dow Chemical Company

## Contact Address:

Dow Chemical Company
Plastics Group Services
2040 Williard H. Dow Center
Midland, MI 48674

Phone: 1 (800) 258-CHEM

## PULSE 830

### Description:

PULSE 830 Engineering Resin offers the automotive parts designer superior impact strength while retaining excellent processability. Very high heat distortion and low-temperature (ductile) impact strength make PULSE 830 Engineering Resin an excellent choice for instrument panels and instrument panel components.

### Processing:

Recommended conditions for molding PULSE 830 Engineering Resin are: injection molding melt temperatures 525°-575°F (274°-302°C); mold temperatures 180°-200°F (82°-93°C); moisture content below .02%; and drying prior to fabrication for 3–4 hours at 200°F (93°C) in a dehumidifying dryer.

### Product Data:

**PULSE 830**

| Property[1] | ASTM | |
|---|---|---|
| **Physical** | | |
| Melt flow rate<br>g/10 min @ 230°C/3.8 kg | D1238 | 3.0 |
| Specific Gravity, g/cc | D792 | 1.135 |
| Water absorption, %<br>24 hrs @ 73°F<br>Equilibrium @ 73°F | D570 | <br>.15<br>.32 |
| Mold shrinkage, in/in | D955 | 0.006 |
| **Thermal** | | |
| Deflection temp, °F @ 264 psi<br>1/8″ thick, unannealed<br>1/8″ thick, annealed | D648 | <br>230<br>250 |
| Coefficient of linear thermal expansion,<br>in/in/°F | D696 | $4.11 \times 10^{-5}$ |

[1]Typical property values; not to be construed as sales specifications.

*(continued)*

NOTE – Information provided in this book has been condensed. The publisher recommends that the product manufacturer be contacted, at the address or phone number provided, for more complete information on product safety, processing parameters, and applications not provided in this information sheet.

PULSE 830 *(continued)*

| Property[1] | ASTM | |
|---|---|---|
| **Mechanical** | | |
| Tensile yield strength, psi | D638 | 7500 |
| Tensile break strength, psi | D638 | 6500 |
| Elongation, break, % | D638 | 125 |
| Flexural yield strength, psi | D790 | 12,000 |
| Flexural modulus (tangent), $10^5$ psi | D790 | 3.35 |
| Izod impact strength,<br>ft-lb/in notch, 1/8″ thick @ 73°F<br>@ −20°F | D256 | <br>12.0<br>10 |
| Rockwell hardness, R | D785 | 115 |
| Instrumented dart impact[2],<br>in-lb, total energy @ 73°<br>@ −20° F | D3763-85 | <br>550<br>550 |
| **Flammability Rating[3]** | | |
| Average extent of burning | D635 | 1″ |

[1]Typical property values; not to be construed as sales specifications.
[2]0.125″ thick, clamp opening 1.5″, 8000 IPM, 0.5″ dart.
[3]These flammability ratings are not intended to reflect hazards presented by these or any other materials under actual fire conditions.

# Dow Chemical Company

## Contact Address:

Dow Chemical Company
Plastics Group Services
2040 Williard H. Dow Center
Midland, MI 48674

Phone: 1 (800) 258-CHEM

## PULSE 855

### Description:

PULSE 855 Engineering Resin is designed specifically for automotive instrument panels and instrument panel components. PULSE 855 combines good processability (3.0 MFR) and good impact strength (4.0 Izod) with very high heat distortion and good chemical resistance.

### Processing:

Recommended conditions for molding PULSE 855 Engineering Resin are: injection molding melt temperatures 525°–575°F (274°–302°C); mold temperatures 180°–200°F (82°–93°C); moisture content below .02%; and drying prior to fabrication for 3–4 hours at 200°F (93°C) in a dehumidifying dryer.

### Product Data:

**PULSE 855**

| **Property**[1] | **ASTM** | |
|---|---|---|
| **Physical** | | |
| Melt flow rate<br>g/10 min @ 230°C/3.8 kg | D1238 | 3.0 |
| Specific Gravity, g/cc | D792 | 1.15 |
| Water absorption, %<br>24 hrs @ 73°F<br>Equilibrium @ 73°F | D570 | <br>.15<br>.32 |
| Mold shrinkage, in/in | D955 | 0.005 |
| **Thermal** | | |
| Deflection temp, °F @ 264 psi<br>1/8″ thick, unannealed<br>1/8″ thick, annealed | D648 | <br>250<br>260 |
| Coefficient of linear thermal expansion, in/in/°F | D696 | $3.7 \times 10^{-5}$ |

[1]Typical property values; not to be construed as sales specifications.

*(continued)*

NOTE—Information provided in this book has been condensed. The publisher recommends that the product manufacturer be contacted, at the address or phone number provided, for more complete information on product safety, processing parameters, and applications not provided in this information sheet.

**PULSE 855 *(continued)***

| Property[1] | ASTM | |
|---|---|---|
| **Mechanical** | | |
| Tensile yield strength, psi | D638 | 7400 |
| Tensile break strength, psi | D638 | 5700 |
| Elongation, break, % | D638 | 10 |
| Flexural yield strength, psi | D790 | 14,000 |
| Flexural modulus (tangent), $10^5$ psi | D790 | 4.10 |
| Izod impact strength,<br>ft-lb/in notch, 1/8″ thick @ 73°F<br>@ −20°F | D256 | <br>4.0<br>4.0 |
| Rockwell hardness, R | D785 | 120 |
| Instrumented dart impact[2],<br>in-lb, total energy @ 73°<br>@ −20°F | D3763-85 | <br>200<br>200 |
| **Flammability Rating**[3] | | |
| Average extent of burning | D635 | 1″ |

[1]Typical property values; not to be construed as sales specifications.
[2]0.125″ thick, clamp opening 1.5″, 8000 IPM, 0.5″ dart.
[3]These flammability ratings are not intended to reflect hazards presented by these or any other materials under actual fire conditions.

# Dow Chemical Company

## Contact Address:

Dow Chemical Company
Plastics Group Services
2040 Williard H. Dow Center
Midland, MI 48674

Phone: 1 (800) 258-CHEM

## PULSE 869

### Description:

PULSE 869 Engineering Resin is designed specifically for automotive instrument panels and components. PULSE 869 Engineering Resin combines good flow, good impact strength, and excellent tensile and flexural strength along with very high heat distortion and good chemical resistance.

### Processing:

Recommended conditions for molding PULSE 869 Engineering Resin are: injection molding temperatures 525°–575°F (274°–302°C); mold temperatures 180°–200°F (82°–93°C); moisture content below .02%; and drying prior to fabrication for 3–4 hours at 200°F (93°C) in a dehumidifying dryer.

### Product Data:

**PULSE 869**

| Property[1] | ASTM | |
|---|---|---|
| **Physical** | | |
| Melt flow rate<br>g/10 min @ 230°C/3.8 kg | D1238 | 3.0 |
| Specific Gravity, g/cc | D792 | 1.19 |
| Water absorption, %<br>24 hrs @ 73°F<br>Equilibrium @ 73°F | D570 | <br>.15<br>.32 |
| Mold shrinkage, in/in | D955 | 0.004 |
| **Thermal** | | |
| Deflection temp, °F @ 264 psi<br>1/8″ thick, unannealed<br>1/8″ thick, annealed | D648 | <br>260<br>265 |
| Coefficient of linear thermal expansion,<br>in/in/°F | D696 | $3.3 \times 10^{-5}$ |

[1]Typical property values; not to be construed as sales specifications.

*(continued)*

NOTE–Information provided in this book has been condensed. The publisher recommends that the product manufacturer be contacted, at the address or phone number provided, for more complete information on product safety, processing parameters, and applications not provided in this information sheet.

**PULSE 869 *(continued)***

| Property[1] | ASTM | |
|---|---|---|
| **Mechanical** | | |
| Tensile yield strength, psi | D638 | 7800 |
| Tensile break strength, psi | D638 | 6800 |
| Elongation, break, % | D638 | 6 |
| Flexural yield strength, psi | D790 | 16,000 |
| Flexural modulus (tangent), $10^5$ psi | D790 | 5.00 |
| Izod impact strength,<br>ft-lb/in notch, 1/8″ thick @ 73°F<br>@ −20°F | D256 | <br>3.0<br>3.0 |
| Rockwell hardness, R | D785 | 120 |
| Instrumented dart impact[2],<br>in-lb, total energy @ 73°<br>@ −20°F | D3763-85 | <br>150<br>150 |
| **Flammability Rating[3]** | | |
| Average extent of burning | D635 | 1″ |

[1]Typical property values; not to be construed as sales specifications.
[2]0.125″ thick, clamp opening 1.5″, 8000 IPM, 0.5″ dart.
[3]These flammability ratings are not intended to reflect hazards presented by these or any other materials under actual fire conditions.

# Dow Chemical Company

Contact Address:

Dow Chemical Company
Plastics Group Services
2040 Williard H. Dow Center
Midland, MI 48674

Phone: 1 (800) 258-CHEM

## REACTIVE POLYSTYRENE

Product Information:

In addition to polymer blends and alloys listed, Dow has developed a reactive polystyrene used to chemically bond with other polymers used to compound composite systems, polymer blends, or polymer alloys. By incorporating oxazoline reactive sites along the polystyrene chain, this polymer can graft to dissimilar polymers, and produce polymer blends from previously incompatible polymers. The bonding between reactive polystyrene and co-reactive polymers is fast, and blending can be carried out by melt blending in conventional polymer compounding equipment, such as a twin-screw extruder.

Physical properties are dependent on specific polymers in the blend. For a polyolefin/reactive polystyrene alloy, the following properties were improved—heat distortion under load, tensile strength, modulus, and processability. Only elongation was reduced—from excellent to good.

# E. I. du Pont de Nemours & Co. Inc.

## Contact Address:

E. I. du Pont de Nemours & Co. Inc.
Polymer Products Department
Wilmington, DE 19898

Phone: 1 (302) 999-4592

## DELRIN 100ST

### General Information:

DELRIN is Du Pont's registered trademark for acetyl resins. The 100ST and 500T grades are acetyl-elastomer alloys that provide an extra tough polymer for engineering applications.

DELRIN resins have excellent mechanical properties. Their high strength and inherent lubricity make them natural candidates for mechanical components such as gears, cams, and bearings. They are resistant to chemical attack, and are not subject to build up of carbonates found in hard water. DELRIN resins have excellent electrical insulation properties, and can perform well at moderately elevated temperatures.

### Product Information:

DELRIN 100ST is generally processed by extrusion or by injection molding. It finds applications where the ultimate in impact resistance is needed, however, some reduction in other mechanical properties occurs.

### Product Data:

**DELRIN 100ST**

| Property | ASTM Method | English Units | SI Units |
|---|---|---|---|
| **Strength** | | | |
| Tensile Elongation at Break (0.2 in/min) | D638 | | |
| −68°F | | — | — |
| +73°F | | 200% | >200% |
| +158°F | | >260% | >260% |
| +212°F | | >260% | >260% |
| +250°F | | — | — |
| Tensile Strength (0.2 in/min) | D638 | | |
| −68°F | | — | — |
| +73°F | | 6.5 kpsi | 44.8 MPa |
| +158°F | | 4.7 kpsi | 32.8 MPa |
| +212°F | | 3.4 kpsi | 23.4 MPa |
| +250°F | | — | — |

*(continued)*

NOTE—Information provided in this book has been condensed. The publisher recommends that the product manufacturer be contacted, at the address or phone number provided, for more complete information on product safety, processing parameters, and applications not provided in this information sheet.

**DELRIN 100ST *(continued)***

| Property | ASTM Method | English Units | SI Units |
|---|---|---|---|
| **Strength** | | | |
| Shear Strength<br>+73°F | D732 | <br>5.0 kpsi | <br>34.4 MPa |
| Flexural Yield Strength (0.05 in/min)<br>+73°F | D790 | <br>5.8 kpsi | <br>40.0 MPa |
| **Stiffness & Creep** | | | |
| Modulus of Elasticity (0.2 in/min)<br>+73°F | D638 | <br>190 kpsi | <br>1309 MPa |
| Flexural Modulus (0.05 in/min)<br>−68°F<br>+73°F<br>+158°F<br>+212°F<br>+250°F | D790 | <br>470 kpsi<br>180 kpsi<br>100 kpsi<br>50 kpsi<br>33 kpsi | <br>3240 MPa<br>1240 MPa<br>690 MPa<br>345 MPa<br>227 MPa |
| Compressive Stress (0.05 in/min)<br>+73°F @ 1% Def.<br>+73°F @ 10% Def. | D695 | <br>1.2 kpsi<br>7.6 kpsi | <br>8.3 MPa<br>52.4 MPa |
| Deformation under load<br>2000 psi @ +122°F | D621 | <br>— | <br>— |
| Deflection Temperature<br>264 psi<br>66 psi | D648 | <br>194°F<br>293°F | <br>90.0°C<br>145.0°C |
| Flexural Fatigue Endurance Limit<br>50% RH, +73°F, $10^{6}$ Cycles | D671 | <br>— | <br>— |
| Tensile Impact Strength<br>+73°F | D1822<br>Long | <br>750 ft lb/in² | <br>1577.3 kJ/m² |
| **Toughness** | | | |
| Izod Impact<br>−40°F<br>+73° | D256 | <br>—<br>17 ft lb/in | <br>—<br>907.8 J/m |
| Unnotched Impact<br>+73°F | D256 | >100 ft lb/in<br>(no break) | no break |
| **Thermal** | | | |
| Melting Point (Fisher-Johns) | D2133 | 347°F | 175.0°C |
| Coef. of Linear Thermal Expansion<br>−40 to +85°F<br>+85 to +140°F<br>+140 to 220°F<br>+220 to 300°F | D696 | <br>6.8 $10^{5}$ in/in °F<br>7.0 $10^{5}$ in/in °F<br>7.1 $10^{5}$ in/in °F<br>7.8 $10^{5}$ in/in °F | <br>12.2 $10^{5}$ m/m°C<br>12.6 $10^{5}$ m/m°C<br>12.8 $10^{5}$ m/m°C<br>14.0 $10^{5}$ m/m°C |
| Thermal Conductivity | — | — | — |

*(continued)*

### DELRIN 100ST *(continued)*

| Property | ASTM Method | English Units | SI Units |
|---|---|---|---|
| **Electrical** | | | |
| Volume Resistivity<br>2% water, +73F | D257 | $2.3 \times 10^{14}$ ohm-cm | $2.3 \times 10^{14}$ ohm-cm |
| Dielectric Constant<br>50% RH, +73°F, $10^2$–$10^6$ Hz | D150 | 4.1 | 4.1 |
| Dissipation Factor<br>50% RH, +73°F, $10^6$ Hz | — | 0.007 | 0.007 |
| Dielectric Strength Short Time (90 mils) | D149 | 480 V/mil | 18.9 MV/m |
| Arc Resistance<br>Flame extinguishes when arcing stops (120 mils) | D495 | 120 sec no tracking | 120 sec no tracking |
| **Miscellaneous** | | | |
| Water Absorption, +73°F<br>24 hrs. immersion<br>Equil., 50% RH<br>Equil., immersion | D570 | <br>0.44%<br>—<br>— | <br>0.44%<br>—<br>— |
| Specific Gravity | D792 | 1.34 | 1.34 |
| Rockwell Hardness | D785 | M58, R105 | M58, R105 |
| Combustibility | UL 94 | 94HB | 94 |
| Coef. of Friction (no lubricant)<br>Static<br>Dynamic | Thrust Washer | <br>—<br>0.14 | <br>—<br>0.14 |
| Chemical Resistance | | All resins | |

# E. I. du Pont de Nemours & Co. Inc.

**Contact Address:**

E. I. du Pont de Nemours & Co. Inc.
Polymer Products Department
Wilmington, DE 19898

Phone: 1 (302) 999-4592

## DELRIN 500T

**General Information:**

DELRIN is Du Pont's registered trademark for acetyl resins. The 100ST and 500T grades are acetyl-elastomer alloys that provide an extra tough polymer for engineering applications.

DELRIN resins have excellent mechanical properties. Their high strength and inherent lubricity make them natural candidates for mechanical components such as gears, cams, and bearings. They are resistant to chemical attack, and are not subject to build up of carbonates found in hard water. DELRIN resins have excellent electrical insulation properties, and can perform well at moderately elevated temperatures.

**Product Information:**

DELRIN 500T is generally injection molded. The alloy shows improved wear resistance in service. It finds application where improved impact resistance is required, and other property requirements are those associated with acetyl resins.

**Product Data:**

**DELRIN 500T**

| Property | ASTM Method | English Units | SI Units |
|---|---|---|---|
| **Strength** | | | |
| Tensile Elongation at Break (0.2 in/min) | D638 | | |
| −68°F | | — | — |
| +73°F | | 60% | 60% |
| +158°F | | >260% | >260% |
| +212°F | | >260% | >260% |
| +250°F | | — | — |
| Tensile Strength (0.2 in/min) | D638 | | |
| −68°F | | — | — |
| +73°F | | 8.4 kpsi | 57.9 MPa |
| +158°F | | 6.2 kpsi | 42.7 MPa |
| +212°F | | 5.0 kpsi | 34.4 MPa |
| +250°F | | — | — |

*(continued)*

NOTE—Information provided in this book has been condensed. The publisher recommends that the product manufacturer be contacted, at the address or phone number provided, for more complete information on product safety, processing parameters, and applications not provided in this information sheet.

## DELRIN 500T *(continued)*

| Property | ASTM Method | English Units | SI Units |
|---|---|---|---|
| **Strength** | | | |
| Shear Strength<br>+73°F | D732 | <br>6.4 kpsi | <br>44.1 MPa |
| Flexural Yield Strength (0.05 in/min)<br>+73°F | D790 | <br>10.0 kpsi | <br>69.0 MPa |
| **Stiffness & Creep** | | | |
| Modulus of Elasticity (0.2 in/min)<br>+73°F | D638 | <br>350 kpsi | <br>2410 MPa |
| Flexural Modulus (0.05 in/min)<br>−68°F<br>+73°F<br>+158°F<br>+212°F<br>+250°F | D790 | <br>530 kpsi<br>350 kpsi<br>180 kpsi<br>100 kpsi<br>60 kpsi | <br>3650 MPa<br>2412 MPa<br>1240 MPa<br>690 MPa<br>413 MPa |
| Compressive Stress (0.05 in/min)<br>+73°F @ 1% Def.<br>+73°F @ 10% Def. | D695 | <br>2.3 kpsi<br>11.8 kpsi | <br>15.8 MPa<br>81.3 MPa |
| Deformation under load<br>2000 psi @ +122°F | D621 | <br>— | <br>— |
| Deflection Temperature<br>264 psi<br>66 psi | D648 | <br>212°F<br>336°F | <br>100.0°C<br>168.9°C |
| Flexural Fatigue Endurance Limit<br>50% RH, +73°F, $10^6$ Cycles | D671 | <br>— | <br>— |
| Tensile Impact Strength<br>+73°F | D1822<br>Long | <br>380 ft lb/$in^2$ | <br>799.1 kJ/$m^2$ |
| **Toughness** | | | |
| Izod Impact<br>−40°F<br>+73° | D256 | <br>—<br>2.1 ft lb/in | <br>—<br>112.1 J/m |
| Unnotched Impact<br>+73°F | D256 | 50 ft lb/in<br>(no break) | 2670 J/m |
| **Thermal** | | | |
| Melting Point (Fisher-Johns) | D2133 | 347°F | 175.0°C |
| Coef. of Linear Thermal Expansion<br>−40 to +85°F<br>+85 to +140°F<br>+140 to 220°F<br>+220 to 300°F | D696 | <br>6.1 $10^5$ in/in °F<br>6.8 $10^5$ in/in °F<br>7.4 $10^5$ in/in °F<br>8.5 $10^5$ in/in °F | <br>11.0 $10^5$ m/m°C<br>12.2 $10^5$ m/m°C<br>13.3 $10^5$ m/m°C<br>15.3 $10^5$ m/m°C |
| Thermal Conductivity | — | — | — |

*(continued)*

DELRIN 500T *(continued)*

| Property | ASTM Method | English Units | SI Units |
|---|---|---|---|
| **Electrical** | | | |
| Volume Resistivity<br>2% water, +73F | D257 | $2.0 \times 10^{14}$ ohm-cm | $2.0 \times 10^{14}$ ohm-cm |
| Dielectric Constant<br>50% RH, +73°F, $10^2$–$10^6$ Hz | D150 | 3.6 | 3.6 |
| Dissipation Factor<br>50% RH, +73°F, $10^6$ Hz | — | — | — |
| Dielectric Strength Short Time (90 mils) | D149 | 400 V/mil | 15.8 MV/m |
| Arc Resistance<br>Flame extinguishes when arcing stops (120 mils) | D495 | 120 sec<br>no tracking | 120 sec<br>no tracking |
| **Miscellaneous** | | | |
| Water Absorption, +73°F<br>24 hrs. immersion<br>Equil., 50% RH<br>Equil., immersion | D570 | 0.30%<br>—<br>— | 0.30%<br>—<br>— |
| Specific Gravity | D792 | 1.39 | 1.39 |
| Rockwell Hardness | D785 | M79, R117 | M79, R117 |
| Combustibility | UL 94 | 94HB | 94HB |
| Coef. of Friction (no lubricant)<br>Static<br>Dynamic | Thrust Washer | —<br>0.17 | —<br>0.17 |
| Chemical Resistance | | All resins | |

# E. I. du Pont de Nemours & Co. Inc.

## Contact Address:

E. I. du Pont de Nemours & Co. Inc.
Polymer Products Department
Wilmington, DE 19898

Phone: 1 (302) 999-4592

# MINLON 12T

## General Information:

MINLON is Du Pont's trademark for engineering thermoplastics that combine Du Pont's nylon with mineral, or a mineral/glass reinforced resin. The word "reinforced" is important. Unlike filled resin systems, the mineral or mineral/glass reinforcement is chemically bonded to the polymer binder; combining nylon's inherent strength, toughness, and chemical resistance with the improved stiffness and dimensional stability characteristic of filled resin systems.

## Product Information:

MINLON 12T is of particular interest because the polymeric binder is a nylon alloy that provides a super tough resin with outstanding dimensional stability, and good stiffness, while offering superior toughness and impact strength.

## Product Data:

**MINLON 12T**

| Property | ASTM Method | English Units | SI Units |
|---|---|---|---|
| **Strength** | | | |
| Tensile Strength | D638 | | |
| 73°F (23°), DAM 66 psi | | 11,500 psi | 79.4 MPa |
| 73°F (23°C), 50% RH 264 psi | | 8,800 psi | 60.7 MPa |
| Elongation at Break | D638 | | |
| 73°F (23°), DAM | | 20% | |
| 73°F (23°C), 50% RH | | 45% | |
| **Stiffness & Creep** | | | |
| Flexural Modulus | D790 | | |
| 73°F (23°), DAM | | 665,000 psi | 4589 MPa |
| 73°F (23°C), 50% RH | | 255,000 psi | 1760 MPa |

*(continued)*

NOTE – Information provided in this book has been condensed. The publisher recommends that the product manufacturer be contacted, at the address or phone number provided, for more complete information on product safety, processing parameters, and applications not provided in this information sheet.

MINLON 12T *(continued)*

| Property | ASTM Method | English Units | SI Units |
|---|---|---|---|
| **Stiffness & Creep** | | | |
| Heat Deflection Temp. | D648 | | |
| 73°F (23°), DAM 66 psi | | 436°F | 225°C |
| 73°F (23°C), 50% RH 264 psi | | 347°F | 175°C |
| Izod Impact Strength | D256 | | |
| 73°F (23°), DAM | | 2.4 ft lb/in. | 128 J/m |
| 73°F (23°C), 50% RH | | 3.5 ft lb/in | 187 J/m |
| **Toughness** | | | |
| Rockwell Hardness | D785 | R120 | |
| Coeff. of Linear Thermal Expansion 73°F (23°C) | | $3 \times 10^{-6}$ in/in/°F | $5.4 \times 10^{-5}$ m/m/°C |
| Specific Gravity | D792 | 1.42 | |
| Melt Point (Fisher Johns) | D789 | 493°F | 256°C |
| Mold Shrinkage (Flow) | | 1.0% | |

# E. I. du Pont de Nemours & Co. Inc.

## Contact Address:

E. I. du Pont de Nemours & Co. Inc.
Polymer Products Department
Wilmington, DE 19898

Phone: 1 (302) 999-4592

## ZYTEL

### General Information:

ZYTEL is Du Pont's registered trademark for nylon based polymer alloys.

### Product Information:

The ZYTEL 408 series of resins are impact modified 66 nylons offering good processability and stiffness, as well as superior toughness, particularly under conditions of low moisture content and low temperatures. Available grades include:

| | | |
|---|---|---|
| ZYTEL 408 | L NC-10 | General purpose natural color. |
| ZYTEL 408 | HS NC-10 | Heat stabilized. |
| ZYTEL 408 | L NC-10 FC | For food and potable water contact. |
| ZYTEL 408 | HS BK-09 | Heat stabilized black, containing a black dye colorant. |
| ZYTEL 408 | Custom colors | Fully compounded custom colors. |

The good melt flow properties of ZYTEL resins produce higher yields where there is difficulty in filling the mold cavity. This includes molding parts having different wall thicknesses and having thin sections or especially long narrow cavities applications include hardware, gears, fasteners, and helmet straps. ZYTEL 408 resins have been extruded to form tubing, rods, and cable jacketing.

*(continued)*

NOTE—Information provided in this book has been condensed. The publisher recommends that the product manufacturer be contacted, at the address or phone number provided, for more complete information on product safety, processing parameters, and applications not provided in this information sheet.

### Typical Properties—Family of ZYTEL® 408 Resins

| Property | ASTM Method | Units | ZYTEL 408 L NC-10<br>ZYTEL 408 L NC-10 FC<br>ZYTEL 408 HS L NC-10 DAM | 50% RH |
|---|---|---|---|---|
| **Strength** | | | | |
| Tensile Strength | | | | |
| −40°F | D638 | psi | 15,100 | 13,100 |
| −40°C | | MPa | 104.1 | 90.3 |
| 73°F | | psi | 9,000 | 7,500 |
| 23°C | | MPa | 62.1 | 51.7 |
| 170°F | | psi | 6,300 | 5,000 |
| 77°C | | MPa | 43.4 | 34.5 |
| 250°F | | psi | 4,600 | 4,000 |
| 121°C | | MPa | 31.7 | 27.6 |
| Yield Strength | | | | |
| −40°F | D638 | psi | — | — |
| −40°C | | MPa | — | — |
| 73°F | | psi | 8,800 | 7,500 |
| 23°C | | MPa | 60.7 | 51.7 |
| 170°F | | psi | 4,900 | 3,800 |
| 77°C | | MPa | 33.8 | 26.2 |
| 250°F | | psi | 3,300 | 3,000 |
| 121°C | | MPa | 22.8 | 20.7 |
| Elongation at Break | | | | |
| −40°F (−40°C) | D638 | % | — | 20 |
| 73°F (23°C) | | % | 80 | 270 |
| 170°F (77°C) | | % | 210 | ≥300 |
| 250°F (121°C) | | % | ≥300 | ≥300 |
| **Stiffness & Creep** | | | | |
| Flexural Modulus | | | | |
| −40°F | D790 | psi | 410,000 | 480,000 |
| −40°C | | MPa | 2,827 | 3,309 |
| 73°F | | psi | 285,000 | 160,000 |
| 23°C | | MPa | 1,965 | 1,103 |
| 170°F | | psi | 80,000 | 60,000 |
| 77°C | | MPa | 552 | 414 |
| 250°F | | psi | 50,000 | 50,000 |
| 121°C | | MPa | 345 | 345 |

*(continued)*

Typical Properties—Family of ZYTEL® 408 Resins *(continued)*

| Property | ASTM Method | Units | ZYTEL 408 L NC-10<br>ZYTEL 408 L NC-10 FC<br>ZYTEL 408 HS L NC-10 DAM | 50% RH |
|---|---|---|---|---|
| **Stiffness & Creep** | | | | |
| Deformation under Load<br>2,000 psi 122°F<br>13.8 MPa 50°C | D621 | % | 1.4 | — |
| Heat Deflection Temp.<br>264 psi<br>1.8 × 10⁶ Pa<br>66 psi<br>0.5 × 10⁶ Pa | D648 | °F<br>°C<br>°F<br>°C | 167<br>75<br>446<br>230 | —<br>—<br>—<br>— |
| **Toughness** | | | | |
| Brittleness Temp. | D746 | °F<br>°C | −155<br>−104 | −120<br>−84 |
| Izod Impact Strength<br>−40°F<br>−40°C<br>73°F<br>23°C | D256 | ft lb/in<br>J/m<br>ft lb/in<br>J/m | 1.3<br>69<br>4.3<br>230 | 1.2<br>64<br>4.5<br>240 |
| Gardner Impact<br>−30°F<br>−34°C<br>73°F<br>23°C | D3029<br>Method G,<br>Geometry<br>GB | ft lbs.<br>J<br>ft lbs<br>J | >320<br>>434<br>>320<br>>434 | —<br>—<br>—<br>— |
| Tensile Impact Strength<br>Long Specimen 73°F<br>23°C<br>Short Specimen 73°F<br>23°C<br>Repeated Impact 73°F<br>23°C<br>"Dynatup" 73°F<br>23°C | D1822<br><br><br><br>(4)<br><br>(5) | ft lb/in²<br>KJ/m²<br>ft lb/in²<br>KJ/m²<br>ft lbs<br>J<br>ft lbs<br>J | 262<br>550<br>90<br>189<br>1.0<br>1.4<br>63<br>85 | 800<br>1,680<br>126<br>265<br>—<br>—<br>—<br>— |

*(continued)*

## Typical Properties—Family of ZYTEL® 408 Resins

| Property | ASTM Method | Units | ZYTEL 408 L NC-10<br>ZYTEL 408 L NC-10 FC<br>ZYTEL 408 HS L NC-10<br>DAM | 50% RH |
|---|---|---|---|---|
| **Thermal** | | | | |
| Melting Point (Fisher-Johns) | D789 | °F<br>°C | 491<br>255 | 491<br>255 |
| Melting Point (Crystalline) | D2117 | °F<br>°C | 518<br>270 | —<br>— |
| Coefficient of Linear Thermal Expansion | | | | |
| −40°F | D696 | in/in/°F | $3.4 \times 10^{-5}$ | |
| −40°C | | m/m/°C | $6.1 \times 6^{-5}$ | |
| 32°F | | in/in/°F | $3.6 \times 10^{-5}$ | |
| 0°C | | m/m/°C | $6.5 \times 10^{-5}$ | |
| 73°F | | in/in/°F | $4.0 \times 10^{-5}$ | |
| 23°C | | m/m/°C | $7.2 \times 10^{-5}$ | |
| 170°F | | in/in/°F | $5.0 \times 10^{-5}$ | |
| 77°C | | m/m/°C | $9.0 \times 10^{-5}$ | |
| **Electrical** | | | | |
| Volume Resistivity | D257 | ohm-cm | $10^{15}$ | $10^{13}$ |
| Dielectric Constant | | | | |
| 100 Hz (cps) | D150 | | 3.1 | 5.9 |
| $10^3$ Hz (cps) | D150 | | 3.1 | 4.8 |
| $10^6$ Hz (cps) | D150 | | 2.9 | 3.3 |
| Dissipation Factor | | | | |
| 100 Hz (cps) | | | 0.02 | 0.1 |
| $10^3$ Hz (cps) | | | 0.02 | 0.11 |
| $10^6$ Hz (cps) | | | 0.02 | 0.10 |
| **Flammability** | | | | |
| UL Flammability (Class) | UL 94 | | 94 HB | 94 HB |
| Oxygen Index | D2863 | % | 19 | 20 |
| **Miscellaneous** | | | | |
| Specific Gravity | D792 | | 1.09 | 1.09 |
| Water Absorption 24 hr immersion 73°F (23°C) | D570 | % | 1.2 | — |
| Water Absorption Saturation 73°F (23°C) | D570 | % | 7.0 | — |
| Hardness Rockwell M | D785 | | M71 | M50 |

*(continued)*

### Typical Properties—Family of ZYTEL® 408 Resins *(continued)*

| Property | ASTM Method | Units | ZYTEL 408 L NC-10<br>ZYTEL 408 L NC-10 FC<br>ZYTEL 408 HS L NC-10 | |
|---|---|---|---|---|
| | | | DAM | 50% RH |
| **Miscellaneous** | | | | |
| Hardness Rockwell R | D785 | | R115 | R102 |
| Durometer Hardness (D Scale) | D676 | | 83 | 76 |
| Mold Shrinkage for 1/8″ thick (approx.) | | % | 1.5 | — |
| **Chemical** | | | | |
| Acid Resistance | Limited; Attacked by strong acids; General order of resistance 612 > 66 > copolymers or 6 | | | |
| Base Resistance | Excellent at room temperature; Attacked by strong bases at elevated temperatures | | | |
| Solvent Resistance | Generally excellent; Some absorption of such polar solvents as water, alcohols, and certain halogenated hydrocarbons causing plasticization and dimension changes | | | |

# E. I. du Pont de Nemours & Co. Inc.

## Contact Address:

E. I. du Pont de Nemours & Co. Inc.
Polymer Products Department
Wilmington, DE 19898

Phone: 1 (302) 999-4592

## ZYTEL ST 801

### General Information:

ZYTEL is Du Pont's registered trademark for nylon based polymer alloys.

### Product Information:

ZYTEL ST 801 HS is a super tough resin that has been heat stabilized to resist embrittlement at elevated temperatures while maintaining extremely high impact strength and toughness at low moisture or low temperature. The resin has outstanding resistance to crack propagation and excellent retention of properties following exposure to high temperatures.

ZYTEL ST 801 HS is suitable for use in applications involving exposure to elevated temperature and requiring outstanding impact resistance. These include automotive components, power tool housings, machinery parts, and electrical component parts.

### Typical Properties:

**ZYTEL® ST 801 HS Super Tough Nylon Resin[1]**

| Property | ASTM Method | English Units[2] | SI Units |
|---|---|---|---|
| Tensile Strength | D638 | | |
| Dry, 23°C (73°F) | | 7,500 psi | 51.7 MPa |
| 50% RH, 23°C (73°F) | | 6,000 psi | 4.14 MPa |
| Elongation | D638 | | |
| Dry, 23°C (73°F) | | 60% | |
| 50% RH, 23°C (73°F) | | 210% | |
| Flexural Modulus | D790 | | |
| Dry, 23°C (73°F) | | 245 K psi | 1689 MPa |
| 50% RH, 23°C (73°F) | | 125 K psi | 862 MPa |
| Rockwell Hardness | D785 | | |
| Dry, 23°C (73°F) | | 112 R Scale | |

[1]These values are for natural color (NC-10) resins only. Addition of colorants of any kind may alter some or all of these properties. The data listed here fall within the normal range of product properties; they are not intended to represent minimum values nor should they be used alone as a basis of design, or to establish specification limits.
[2]Based on Du Pont test results.

*(continued)*

NOTE—Information provided in this book has been condensed. The publisher recommends that the product manufacturer be contacted, at the address or phone number provided, for more complete information on product safety, processing parameters, and applications not provided in this information sheet.

**ZYTEL® ST 801 HS Super Tough Nylon Resin[1] *(continued)***

| Property | ASTM Method | English Units[2] | SI Units |
|---|---|---|---|
| Izod Impact Strength | D256 | | |
| Dry, 23°C (73°F) | (10 mil notch radius) | 17.0 ft lb/in | 908 J/m |
| 50% RH, 23°C (73°F) | | 20.0[3] ft lb/in | 1068 J/m |
| Dry, 23°C (73°F) | Special Test (2 mil notch radius) | 17.0 ft lb/in | 908 J/m |
| Heat Deflection Temperature (annealed) | D648 | | |
| 0.5 MPa (66 psi) | | 421°F | 216°C |
| 1.8 MPa (264 psi) | | 160°F | 71°C |
| Melting Point | D789 | 491°F | 255°C |
| Specific Gravity | D792 | 1.08 | |
| Mold Shrinkage | | .015 in/in | |
| Combustibility | UL 94 | 94 HB | |
| Water Absorption (saturation) 100% RH | D570 | 6.7% | |

[3]Samples do not break into two parts, but break part way through the sample and then bend out of way of hammer.

# E. I. du Pont de Nemours & Co. Inc.

## Contact Address:

E. I. du Pont de Nemours & Co. Inc.
Polymer Products Department
Wilmington, DE 19898

Phone: 1 (302) 999-4592

## ZYTEL ST 811 HS

### General Information:

ZYTEL is Du Pont's registered trademark for nylon based polymer alloys.

### Product Information:

ZYTEL ST 811 HS is the companion heat stabilized extrusion resin.

### Product Data:

**ZYTEL® ST-811 HS Super-Tough, Heat-Stabilized Nylon Extrusion Resin (High Flexibility)**

| | Preliminary Data | | |
|---|---|---|---|
| **Property**[1] | **ASTM Method** | **Units** | **Dry As Molded**[2] |
| Tensile Strength | D638 | psi<br>MPa | 7,000<br>48 |
| Elongation | D638 | % | >500 |
| Flex Modulus | D790 | psi<br>MPa | 85,000<br>550 |
| Izod | | | |
| 73°F | D256 | ft lb/in | 40[3] |
| 23°C | | J/m | 2,140 |
| 32°F | | ft lb/in | 22[3] |
| 0°C | | J/m | 1,170 |
| −4°F | | ft lb/in | 15[3] |
| −20°C | | J/m | 800 |
| −22°F | | ft lb/in | 10[3] |
| −30°C | | J/m | 530 |
| −31°F | | ft lb/in | 9[3] |
| −35°C | | J/m | 480 |

[1]Measured at 73°F (23°C) unless noted otherwise.
[2]These values may change as additional data are obtained.
[3]No breaks.

*(continued)*

NOTE – Information provided in this book has been condensed. The publisher recommends that the product manufacturer be contacted, at the address or phone number provided, for more complete information on product safety, processing parameters, and applications not provided in this information sheet.

**ZYTEL® ST-811 HS Super-Tough, Heat-Stabilized Nylon Extrusion Resin (High Flexibility)** ***(continued)***

| Property[1] | Preliminary Data ASTM Method | Units | Dry As Molded[2] |
|---|---|---|---|
| HDT @ 66 psi<br>0.5 MPa<br>@ 264 psi<br>1.8 MPa | D648 | °F<br>°C<br>°F<br>°C | 340<br>170<br>126<br>52 |
| Specific Gravity | D792 | | 1.04 |
| Melt Point | | °F<br>°C | 410<br>210 |
| Mold Shrinkage | | % | 0.8–1.4 |

# E. I. du Pont de Nemours & Co. Inc.

## Contact Address:

E. I. du Pont de Nemours & Co. Inc.
Polymer Products Department
Wilmington, DE 19898

Phone: 1 (302) 999-4592

# ZYTEL ST 901 L

## General Information:

ZYTEL is Du Pont's registered trademark for nylon based polymer alloys.

## Product Information:

ZYTEL ST 901 L is a super-tough amorphous nylon resin designed for injection molding, but it is also suitable for extruded profiles, blow molding, and the manufacture of sheet for thermoforming. The resin produces products with a unique combination of excellent impact resistance, high stiffness, good chemical resistance, property retention over a wide range of temperatures, low mold shrinkage, and good dimensional stability.

## Typical Properties:

**ZYTEL® ST-901 L Super-Tough Nylon Resin**

| Property[1] | ASTM Method | Units | Dry As Molded | 50% RH Values |
|---|---|---|---|---|
| Specific Gravity | D792 | — | 1.11 | — |
| Tensile Strength | D638 | psi<br>MPa | 9000<br>62 | 7500<br>52 |
| Elongation at Break | D638 | % | 120 | 150 |
| Shear Strength | D732 | psi<br>MPa | 8600<br>59 | — |
| Flexural Strength | D790 | psi<br>MPa | 12,500<br>86 | — |
| Flexural Modulus | D790 | psi<br>MPa | 290,000<br>2000 | 300,000<br>2100 |

[1]Measured at 73°F (23°C).

*(continued)*

NOTE – Information provided in this book has been condensed. The publisher recommends that the product manufacturer be contacted, at the address or phone number provided, for more complete information on product safety, processing parameters, and applications not provided in this information sheet.

## ZYTEL® ST-901 L Super-Tough Nylon Resin *(continued)*

| Property[1] | ASTM Method | Units | Dry As Molded | 50% RH Values |
|---|---|---|---|---|
| Izod Impact (1/4″ bar) | D256 | | | |
| @ −4°F | | ft lb/in | 5 | 5 |
| @ −20°C | | J/m | 270 | 270 |
| @ 32°F | | ft lb/in | 15 | 17 |
| @ 0°F | | J/m | 800 | 910 |
| @ 73°F | | ft lb/in | 19 | 24 |
| @ 23°C | | J/m | 1026 | 1280 |
| Tensile Impact | D785 | ft lb/in$^2$ | 614 | — |
| | | KJ/m$^2$ | 1290 | — |
| Rockwell Hardness, R Scale | | | 108 | — |
| M Scale | | | 58 | — |
| Heat Deflection Temperature | D648 | | | |
| @ 264 psi (1.8 MPa) | | °F | 239 | — |
| | | °C | 115 | — |
| @ 66 psi (0.5 MPa) | | °F | 248 | |
| | | °C | 120 | — |
| Coefficient of Linear Thermal Expansion | E831 | in/in/°F | $4 \times 10^{-5}$ | — |
| | | m/m/°C | $7.3 \times 10^{-5}$ | — |
| Mold Shrinkage, 1/4″ (0.35m) Flow/Transverse | | % | 0.6/0.6 | |
| Flammability | UL 94 | — | HB at 0.032 inch (0.80 mm) | |
| 1/16 in (1.59mm) | FMVSS 302 | in/min max/min | Did not ignite | — |
| Limiting Oxygen Index | | % | 20 | — |

[1]Measured at 73°F (23°C).

# E. I. du Pont de Nemours & Co. Inc.

## Contact Address:

E. I. du Pont de Nemours & Co. Inc.
Polymer Products Department
Wilmington, DE 19898

Phone: 1 (302) 999-4592

## ZYTEL 80 G 33 HS1-L

### General Information:

ZYTEL is Du Pont's registered trademark for nylon based polymer alloys.

### Product Information:

ZYTEL 80 G 33 HS1-L is a super tough nylon resin reinforced with 33% short glass fibers. It contains a heat stabilizer for better performance at elevated temperatures and a lubricant to improve machine feeding characteristics. It was formulated for applications requiring a high degree of impact strength and stiffness, having twice the impact strength of comparable glass reinforced 66 nylon.

ZYTEL 80 G 33 HS1-L is a logical candidate for bearing retainers, power tool housings, ski bindings, rifle stocks, and valve bodies. It can be readily processed in all types of thermoplastic injection molding machines.

### Product Data:

**ZYTEL 80 G 33 HS1-L**

| Property[1] | ASTM Method | English Units | SI Units |
|---|---|---|---|
| **Physical** | | | |
| Tensile Strength | D638 | | |
| DAM[2] | | 21,000 psi | 145 MPa |
| 50% RH[3] | | 16,000 psi | 110 MPa |
| Elongation | D638 | | |
| DAM | | 4% | |
| 50% RH | | 5% | |

[1]These values are for natural color (NC-10) resins only. Colorants or additives of any kind may alter some or all of these properties. The data listed here fall within the normal range of product properties but they should not be used to establish specification limits nor used alone as the basis for design.
[2]Dry-as-molded. Normally contains approximately 0.2% moisture.
[3]Equilibrated to 50% relative humidity.

*(continued)*

NOTE – Information provided in this book has been condensed. The publisher recommends that the product manufacturer be contacted, at the address or phone number provided, for more complete information on product safety, processing parameters, and applications not provided in this information sheet.

## ZYTEL 80 G 33 HS1-L *(continued)*

| Property[1] | ASTM Method | English Units | SI Units |
|---|---|---|---|
| **Physical** | | | |
| Flexural Modulus<br>DAM<br>50% RH | D790 | <br>1,000,000<br>735,000 | <br>6895 MPa<br>5068 MPa |
| Flexural Strength, DAM | D790 | 29,800 | 206 MPa |
| Compressive Strength, DAM | D790 | 24,000 | 165 MPa |
| Izod Impact Strength<br>DAM<br>50% RH | D256 | <br>4.1 ft lb/in<br>4.4 ft lb/in | <br>219 J/m<br>235 J/m |
| Melting Point | D789 | 491°F | 255°C |
| Heat Deflection Temperature[4]<br>264 psi (1.8 MPa) | D648 | <br>482°F | <br>250°C |
| Flammability (class) | UL 94 | 94 HB | |
| Specific Gravity | D792 | 1.34 | |
| Glass Content | | 33% | |
| Mold Shrinkage[5]<br>Flow direction<br>Transverse direction | | <br>0.3%<br>1.1% | |

[4]Annealed 30 minutes in oil at 90°F (50°C) below the melting point of resin.
[5]Measured on 3″ × 5″ × ⅛″ (7.62 × 127 × 3.2 mm) plaques.

# E. I. du Pont de Nemours & Co. Inc.

**Contact Address:**

E. I. du Pont de Nemours & Co. Inc.
Polymer Products Department
Wilmington, DE 19898

Phone: 1 (302) 999-4592

## ZYTEL 82 G 33L

**General Information:**

ZYTEL is Du Pont's registered trademark for nylon based polymer alloys.

**Product Information:**

ZYTEL 82 G 33L is a toughened version of 72G33 which is based on a nylon copolymer. It contains a lubricant to improve processing characteristics, and has been formulated to have almost twice the Izod impact strength of comparable glass reinforced copolymer nylon. It is also a candidate for ski bindings, power tools, and valve bodies, and can be injection molded in all types of thermoplastic injection molding machines.

**Product Data:**

ZYTEL 82 G 33L

| Property | ASTM Method | Units | Value |
|---|---|---|---|
| **Physical** | | | |
| Tensile Strength | D638 | | |
| DAM | | psi | 22,200 |
| 50% RH | | psi | 16,000 |
| Elongation | D638 | | |
| DAM | | | 4% |
| 50% RH | | | 8% |
| Flexural Modulus | D790 | | |
| DAM | | psi | 1,100,000 |
| 50% RH | | | 650,000 |
| Flexural Strength, DAM | D790 | psi | 33,500 |
| Izod Impact Strength | D256 | | |
| DAM | | ft lb/in | 4.2 |
| 50% RH | | ft lb/in | 5.4 |

*(continued)*

NOTE – Information provided in this book has been condensed. The publisher recommends that the product manufacturer be contacted, at the address or phone number provided, for more complete information on product safety, processing parameters, and applications not provided in this information sheet.

ZYTEL 82 G 33L *(continued)*

| Property | ASTM Method | Units | Value |
|---|---|---|---|
| **Physical** | | | |
| Melting Point | D789 | °C | 233 |
| Heat Deflection Temperature<br>264 psi (1.8 MPa) | D648 | °C | 220 |
| Specific Gravity | D792 | | 1.34 |
| Glass Content | | % | 33 |
| Mold Shrinkage | | % | 0.2 |

# EMS-Emser Industries

## Contact Address:

EMS-Emser Industries
P.O. Box 1717
Industrial Park and Corporate Way
Sumter, SC 29151-1717

Phone: 1 (803) 481-3172

## GRILON BT40

### Product Information

Grilon BT40 is a unique thermoplastic polyamide with an unusual combination of properties. It is suitable for both injection molding and extrusion applications.

The water absorption of Grilon BT40 is lower than nylon 6.6, making it more dimensionally stable. The impact resistance of Grilon BT40 is substantially higher even dry-as-molded, than nylon 6.6, POM and PBTP. In the conditioned state, however, Grilon BT40 is stiffer than nylon 6.6.

Grilon BT40 has a low specific gravity, high resistance to stress cracking, good insulation properties and a reduced mold shrinkage.

Typical applications include drive pulleys, fan blades, levers, clamps, housings for electrical appliances, electronic parts and helmets.

### Processing:

Grilon BT40 is supplied ready for processing in free-flowing pellet form, predried to 0.1% moisture and packaged in multiwall, foil-lined, moisture-proof bags containing 25 kg net.

For optimum processing and parts quality, the resin should be protected from exposure to a humid environment prior to processing. If hopper residence times are likely to exceed two hours, use of a hopper dryer operating at 80°C (176°F) is recommended.

If the material is otherwise exposed to a humid environment for more than two hours, drying is recommended to ensure that the moisture content does not exceed 0.2% for processing. Use of a vacuum dryer operating at 110°C (230°F) for 3–4 hours is strongly recommended.

A dessicant dryer may also be used, but to avoid oxidation of the material and consequent discoloration, temperatures should be held below 80°C (176°F), with drying times between six and ten hours. For longer periods, overnight for example, dessicant drying is acceptable at temperatures not exceeding 70°C (158°F).

*(continued)*

NOTE – Information provided in this book has been condensed. The publisher recommends that the product manufacturer be contacted, at the address or phone number provided, for more complete information on product safety, processing parameters, and applications not provided in this information sheet.

## Preliminary Property Data:

### Grilon BT-40

| Properties | Test Method | Units | Values | |
|---|---|---|---|---|
| **General** | | | | |
| Melting point | DSC max. | °C/°F | 220/430 | |
| Specific gravity | ASTM D792 | — | 1.06 | |
| Water absorption, 24h | ASTM D570 | % | 0.62 | |
| Linear mold shrinkage | EMS | % | 0.7–1.0 | |
| Heat deflection temperature | ASTM D648 | °C/°F | | |
| at 66 psi | | | 130/266 | |
| at 264 psi | | | 113/235 | |
| **Mechanical** | | | **Dry-as-molded** | **Conditioned** |
| Tensile strength at yield | ASTM D638 | psi (MPa) | 9,000(62) | 7,800(54) |
| Tensile strength at break | ASTM D638 | psi (MPa) | 7,900(54) | 6,900(46) |
| Elongation at yield | ASTM D790 | % | 10 | 10 |
| Elongation at break | ASTM D638 | % | 110 | 190 |
| Flexural modulus | ASTM D790 | psi (MPa) | 350,000(2413) | 221,000(152) |
| Flexural strength | ASTM D790 | psi (MPa) | 14,000(97) | 8,500(59) |
| Izod impact strength (notched) | ASTM D256 | ft-lb/in(J/m) | 4.6(246) | 19.9(1062) |
| Hardness | | | | |
| Shore D | ASTM D2240 | — | 80 | 78 |
| Rockwell R | ASTM D785 | — | 80 | 78 |
| **Electrical** | | | | |
| Volume resistivity | DIN 53482 | ohm-cm | $10^{14}$ | $10^{14}$ |
| Surface resistivity | DIN 53482 | ohm | $10^{12}$ | $10^{12}$ |
| Dielectric constant, $10^5$Hz | DIN 53481 | — | 3.0 | 3.2 |
| Dielectric strength | DIN 53481 | kV/mm | 41 | 38 |
| Comparative tracking index | ASTM D3638 | V | 600+ | 600+ |

**Processing Guidelines**

| Injection Molding | Extrusion |
|---|---|
| Barrel temperatures (°C): 250-265-270<br>(°F): 482-509-518<br>Nozzel temperature (°C/°F): 265/509<br>Melt temperature (°C/°F): 285/545<br>Mold temperature (°C/°F): 80/176 | Barrel temperatures (°C): 235-240-245<br>(°F): 455-464-473<br>Die temperature (°C/°F): 240/464<br>Melt temperature (°C/°F): 285/545 |

All data generated on natural color materials at 23°C unless otherwise noted. Inclusion of color pigments or other additives may change some or all of these properties. Specimens dry-as-molded contain less than 0.1% moisture; conditioned specimens were stored at 23°C/50% r.h. to achieve an equilibrium moisture level before testing. This data is typical of normal production material, and should not be used to establish specification limits or alone as the basis of design.

# Ferro Corporation

**Contact Address:**

Ferro Corporation
Thermoplastics Division
3 Railroad Avenue
Stryker, OH 43557

Phone: 1 (419) 682-3311

## FERROFLEX FF-100

**General Information:**

The Ferroflex products are modified polyolefins (ethylene propylene diene monomer or ethylene propylene blended with polyethylene or polypropylene) which are available in injection molding and extrusion grades. The entire line comes in black, natural or can be custom color matched. If the desired properties fall between two products, a custom material match can be made.

**Product Information:**

A low flexural modulus material with good impact at low temperatures, FF 100 can be used for interior and/or exterior applications. Processing temperatures are approximately 350°F (175°C) and no drying is required.

**Product Data:**

**FERROFLEX FF-100**

| Physical Test | ASTM Test Method | Value |
|---|---|---|
| Specific Gravity | D-792 | 0.89 |
| Hardness | D-2240 | A-68 |
| Tensile Strength | D-638 | 750 psi |
| Elongation | D-638 | 1300% |
| Tear Strength | D-624 | 150 lb./in |
| Flexural Modulus | D-790 | 5000 psi |
| Gardner Impact (73°F) | 4 lb wgt. | >160 in/lb |
| Gardner Impact (−20°F) | 4 lb wgt. | >160 in/lb |
| MFR Condition | D-1238 | 0.5 g/10 min |
| Mold Shrinkage | | 20 mils/in |

NOTE – Information provided in this book has been condensed. The publisher recommends that the product manufacturer be contacted, at the address or phone number provided, for more complete information on product safety, processing parameters, and applications not provided in this information sheet.

# Ferro Corporation

Contact Address:

Ferro Corporation
Thermoplastics Division
3 Railroad Avenue
Stryker, OH 43557

Phone: 1 (419) 682-3311

## FERROFLEX FF-200

General Information:

The Ferroflex products are modified polyolefins (ethylene propylene diene monomer or ethylene propylene blended with polyethylene or polypropylene) which are available in injection molding and extrusion grades. The entire line comes in black, natural or can be custom color matched. If the desired properties fall between two products, a custom material match can be made.

Product Information:

This material is slightly stiffer than FF-100 but still retains the good impact properties. FF-200 was developed for both automotive and non-automotive applications. Processing should be done around 350°F (175°C).

Product Data:

**FERROFLEX FF-200**

| Physical Test | ASTM Test Method | Value |
|---|---|---|
| Specific Gravity | D-792 | 0.89 |
| Hardness | D-2240 | A-92 |
| Tensile Strength | D-638 | 1600 psi |
| Elongation | D-638 | 1500% |
| Tear Strength | D-624 | 300 lb./in |
| Flexural Modulus | D-790 | 20,000 psi |
| Gardner Impact (73°F) | 4 lb wgt. | >160 in/lb |
| Gardner Impact (−20°F) | 4 lb wgt. | >160 in/lb |
| MFR Condition | D-1238 | 0.9 g/10 min |
| Mold Shrinkage | | 18–20 mils/in |

NOTE—Information provided in this book has been condensed. The publisher recommends that the product manufacturer be contacted, at the address or phone number provided, for more complete information on product safety, processing parameters, and applications not provided in this information sheet.

# Ferro Corporation

## Contact Address:

Ferro Corporation
Thermoplastics Division
3 Railroad Avenue
Stryker, OH 43557

Phone: 1 (419) 682-3311

## FERROFLEX FF-300

### General Information:

The Ferroflex products are modified polyolefins (ethylene propylene diene monomer or ethylene propylene blended with polyethylene or polypropylene) which are available in injection molding and extrusion grades. The entire line comes in black, natural or can be custom color matched. If the desired properties fall between two products, a custom material match can be made.

### Product Information:

The Ferroflex with lowest Shore D hardness is FF-300. It is harder and less flexible than FF-200. The temperatures required for processing are approximately 375°F (190°C).

### Product Data:

**FERROFLEX FF-300**

| Physical Test | ASTM Test Method | Value |
|---|---|---|
| Specific Gravity | D-792 | 0.89 |
| Hardness | D-2240 | D-45 |
| Tensile Strength | D-638 | 2500 psi |
| Elongation | D-638 | 1500% |
| Tear Strength | D-624 | 450 lb./in |
| Flexural Modulus | D-790 | 30,000 psi |
| Gardner Impact (73°F) | 4 lb wgt. | >160 in/lb |
| Gardner Impact (−20°F) | 4 lb wgt. | >160 in/lb |
| MFR Condition | D-1238 | 2.0 g/10 min |
| Mold Shrinkage | | 15–17 mils/in |

NOTE – Information provided in this book has been condensed. The publisher recommends that the product manufacturer be contacted, at the address or phone number provided, for more complete information on product safety, processing parameters, and applications not provided in this information sheet.

# Ferro Corporation

## Contact Address:

Ferro Corporation
Thermoplastics Division
3 Railroad Avenue
Stryker, OH 43557

Phone: 1 (419) 682-3311

## FERROFLEX FF-400

### General Information:

The Ferroflex products are modified polyolefins (ethylene propylene diene monomer or ethylene propylene blended with polyethylene or polypropylene) which are available in injection molding and extrusion grades. The entire line comes in black, natural or can be custom color matched. If the desired properties fall between two products, a custom material match can be made.

### Product Information:

This material was developed with the automotive industry in mind. It has the impact strength and rigidity required for such parts as rubstrips and end caps. Processing can be done around 375°F (190°C).

### Product Data:

**FERROFLEX FF-400**

| Physical Test | ASTM Test Method | Value |
|---|---|---|
| Specific Gravity | D-792 | 0.89 |
| Hardness | D-2240 | D-50 |
| Tensile Strength | D-638 | 2550 psi |
| Elongation | D-638 | 1450% |
| Tear Strength | D-624 | 500 lb./in |
| Flexural Modulus | D-790 | 40,000 psi |
| Gardner Impact (73°F) | 4 lb wgt. | >160 in/lb |
| Gardner Impact (−20°F) | 4 lb wgt. | >160 in/lb |
| MFR Condition | D-1238 | 2.0 g/10 min |
| Mold Shrinkage | | 15–17 mils/in |

NOTE—Information provided in this book has been condensed. The publisher recommends that the product manufacturer be contacted, at the address or phone number provided, for more complete information on product safety, processing parameters, and applications not provided in this information sheet.

# Ferro Corporation

## Contact Address:

Ferro Corporation
Thermoplastics Division
3 Railroad Avenue
Stryker, OH 43557

Phone: 1 (419) 682-3311

## FERROFLEX FF-600

### General Information:

The Ferroflex products are modified polyolefins (ethylene propylene diene monomer or ethylene propylene blended with polyethylene or polypropylene) which are available in injection molding and extrusion grades. The entire line comes in black, natural or can be custom color matched. If the desired properties fall between two products, a custom material match can be made.

### Product Information:

FF-600 is another material developed for the automotive industry and can fill a need for a material with good impact strength and higher stiffness. The best processing can be achieved around 390°F (195°C).

### Product Data:

**FERROFLEX FF-600**

| Physical Test | ASTM Test Method | Value |
|---|---|---|
| Specific Gravity | D-792 | 0.90 |
| Hardness | D-2240 | D-55 |
| Tensile Strength | D-638 | 2900 psi |
| Elongation | D-638 | 1200% |
| Tear Strength | D-624 | 550 lb./in |
| Flexural Modulus | D-790 | 60,000 psi |
| Gardner Impact (73°F) | 4 lb wgt. | >140 in/lb |
| Gardner Impact (−20°F) | 4 lb wgt. | >140 in/lb |
| MFR Condition | D-1238 | 4.0 g/10 min |
| Mold Shrinkage | | 8–10 mils/in |

NOTE – Information provided in this book has been condensed. The publisher recommends that the product manufacturer be contacted, at the address or phone number provided, for more complete information on product safety, processing parameters, and applications not provided in this information sheet.

# Ferro Corporation

## Contact Address:

Ferro Corporation
Thermoplastics Division
3 Railroad Avenue
Stryker, OH 43557

Phone: 1 (419) 682-3311

---

# FERROFLEX FF-800

## General Information:

The Ferroflex products are modified polyolefins (ethylene propylene diene monomer or ethylene propylene blended with polyethylene or polypropylene) which are available in injection molding and extrusion grades. The entire line comes in black, natural or can be custom color matched. If the desired properties fall between two products, a custom material match can be made.

## Product Information:

Originally designed for injection molding, FF-800 can give good impact strength while still being economical. Processing temperatures of approximately 390°F (195°C) are required.

## Product Data:

**FERROFLEX FF-800**

| Physical Test | ASTM Test Method | Value |
|---|---|---|
| Specific Gravity | D-792 | 0.91 |
| Hardness | D-2240 | D-60 |
| Tensile Strength | D-638 | 3000 psi |
| Elongation | D-638 | 800% |
| Tear Strength | D-624 | 650 lb./in |
| Flexural Modulus | D-790 | 80,000 psi |
| Gardner Impact (73°F) | 4 lb wgt. | >160 in/lb |
| Gardner Impact (−20°F) | 4 lb wgt. | >160 in/lb |
| MFR Condition | D-1238 | 6.0 g/10 min |
| Mold Shrinkage | | 9–11 mils/in |

NOTE—Information provided in this book has been condensed. The publisher recommends that the product manufacturer be contacted, at the address or phone number provided, for more complete information on product safety, processing parameters, and applications not provided in this information sheet.

# Ferro Corporation

## Contact Address:

Ferro Corporation
Thermoplastics Division
3 Railroad Avenue
Stryker, OH 43557

Phone: 1 (419) 682-3311

## FERROFLEX FF-1000

### General Information:

The Ferroflex products are modified polyolefins (ethylene propylene diene monomer or ethylene propylene blended with polyethylene or polypropylene) which are available in injection molding and extrusion grades. The entire line comes in black, natural or can be custom color matched. If the desired properties fall between two products, a custom material match can be made.

### Product Information:

FF-1000 has a higher flexural modulus than lower grades of Ferroflex. Processing should be done around 400°F (200°C).

### Product Data:

**FERROFLEX FF-1000**

| Physical Test | ASTM Test Method | Value |
|---|---|---|
| Specific Gravity | D-792 | 0.91 |
| Hardness | D-2240 | D-65 |
| Tensile Strength | D-638 | 3200 psi |
| Elongation | D-638 | 800% |
| Tear Strength | D-624 | 700 lb./in |
| Flexural Modulus | D-790 | 100,000 psi |
| Gardner Impact (73°F) | 4 lb wgt. | 160 in/lb |
| Gardner Impact (−20°F) | 4 lb wgt. | 100 in/lb |
| MFR Condition | D-1238 | 7.0 g/10 min |
| Mold Shrinkage | | 9-11 mils/in |

NOTE—Information provided in this book has been condensed. The publisher recommends that the product manufacturer be contacted, at the address or phone number provided, for more complete information on product safety, processing parameters, and applications not provided in this information sheet.

# Ferro Corporation

## Contact Address:

Ferro Corporation
Thermoplastics Division
3 Railroad Avenue
Stryker, OH 43557

Phone: 1 (419) 682-3311

# FERROFLEX FF-1200

## General Information:

The Ferroflex products are modified polyolefins (ethylene propylene diene monomer or ethylene propylene blended with polyethylene or polypropylene) which are available in injection molding and extrusion grades. The entire line comes in black, natural or can be custom color matched. If the desired properties fall between two products, a custom material match can be made.

## Product Information:

Based on its properties and economics, Ferroflex FF-1200 is a prime candidate for many applications. FF-1200 retains its properties over a wide range of conditions. Temperatures required for processing are approximately 425°F (220°C).

## Product Data:

**FERROFLEX FF-1200**

| Physical Test | ASTM Test Method | Value |
|---|---|---|
| Specific Gravity | D-792 | 0.91 |
| Hardness | D-2240 | D-68 |
| Tensile Strength | D-638 | 3200 psi |
| Elongation | D-638 | 750% |
| Tear Strength | D-624 | 800 lb./in |
| Flexural Modulus | D-790 | 120,000 psi |
| Gardner Impact (73°F) | 4 lb wgt. | 100 in/lb |
| Gardner Impact (−20°F) | 4 lb wgt. | 40 in/lb |
| MFR Condition | D-1238 | 10 g/10 min |
| Mold Shrinkage | | 9–11 mils/in |

NOTE – Information provided in this book has been condensed. The publisher recommends that the product manufacturer be contacted, at the address or phone number provided, for more complete information on product safety, processing parameters, and applications not provided in this information sheet.

# Ferro Corporation

## Contact Address:

Ferro Corporation
Thermoplastics Division
3 Railroad Avenue
Stryker, OH 43557

Phone: 1 (419) 682-3311

## FERROFLEX FF-1500

### General Information:

The Ferroflex products are modified polyolefins (ethylene propylene diene monomer or ethylene propylene blended with polyethylene or polypropylene) which are available in injection molding and extrusion grades. The entire line comes in black, natural or can be custom color matched. If the desired properties fall between two products, a custom material match can be made.

### Product Information:

Because of its basic nature, the properties of FF-1500 are maintained over a broad distribution of conditions. The best processing can be done around 425°F (220°C).

### Product Data:

**FERROFLEX FF-1500**

| Physical Test | ASTM Test Method | Value |
|---|---|---|
| Specific Gravity | D-792 | 0.91 |
| Hardness | D-2240 | D-70 |
| Tensile Strength | D-638 | 3300 psi |
| Elongation | D-638 | 400% |
| Tear Strength | D-624 | 800 lb./in |
| Flexural Modulus | D-790 | 150,000 psi |
| Gardner Impact (73°F) | 4 lb wgt. | 65 in/lb |
| Gardner Impact (−20°F) | 4 lb wgt. | 20 in/lb |
| MFR Condition | D-1238 | 12 g/10 min |
| Mold Shrinkage | | 8–10 mils/in |

NOTE – Information provided in this book has been condensed. The publisher recommends that the product manufacturer be contacted, at the address or phone number provided, for more complete information on product safety, processing parameters, and applications not provided in this information sheet.

# Ferro Corporation

## Contact Address:

Ferro Corporation
Thermoplastics Division
3 Railroad Avenue
Stryker, OH 43557

Phone: 1 (419) 682-3311

## FERROFLEX FF-1900

### General Information:

The Ferroflex products are modified polyolefins (ethylene propylene diene monomer or ethylene propylene blended with polyethylene or polypropylene) which are available in injection molding and extrusion grades. The entire line comes in black, natural or can be custom color matched. If the desired properties fall between two products, a custom material match can be made.

### Product Information:

Another material developed for the automotive industry, FF-1900 has the processibility of copolymer polypropylene, but some of its properties exceed that of polypropylene. Processing temperatures of approximately 450°F (230°C) should be employed.

### Product Data:

**FERROFLEX FF-1900**

| Physical Test | ASTM Test Method | Value |
|---|---|---|
| Specific Gravity | D-792 | 0.92 |
| Hardness | D-2240 | D-75 |
| Tensile Strength | D-638 | 3500 psi |
| Elongation | D-638 | 300% |
| Tear Strength | D-624 | 850 lb./in |
| Flexural Modulus | D-790 | 190,000 psi |
| Gardner Impact (73°F) | 4 lb wgt. | 40 in/lb |
| Gardner Impact (−20°F) | 4 lb wgt. | 10 in/lb |
| MFR Condition | D-1238 | 12 g/10 min |
| Mold Shrinkage | | 8–10 mils/in |

NOTE—Information provided in this book has been condensed. The publisher recommends that the product manufacturer be contacted, at the address or phone number provided, for more complete information on product safety, processing parameters, and applications not provided in this information sheet.

# General Electric Company

## Contact Address:

General Electric Company
Plastics Group
One Plastics Avenue
Pittsfield, MA 01201

Phone: 1 (800) 845-0600

---

## GELOY

### Product Information:

GELOY® resins are a family of outdoor-weatherable polymers based on an advanced terpolymer of acrylic-styrene-acrylonitrile (ASA). These resins offer a superior combination of color and property retention as well as ease of processing on conventional equipment.

Available in sheet extrusion, profile extrusion, blow molding and injection molding grades, GELOY resins have demonstrated excellent property and color retention in both light and dark colors after prolonged outdoor exposure.

NOTE – Information provided in this book has been condensed. The publisher recommends that the product manufacturer be contacted, at the address or phone number provided, for more complete information on product safety, processing parameters, and applications not provided in this information sheet.

# General Electric Company

## Contact Address:

General Electric Company
Plastics Group
One Plastics Avenue
Pittsfield, MA 01201

Phone: 1 (800) 845-0600

## GELOY 1020

### Product Information:

GELOY 1020 resin is an ASA granular powder perfect for blending with PVC. Geloy 1020 resin is easily mixed with PVC powder, providing a blend with improved heat distortion temperature, impact, and dark-color weatherability while maintaining excellent processability using equipment and tooling designed for PVC. Products and colors can thus be formulated to fit the manufacturer's exact material needs for extrusion applications such as window profiles, mobile home skirting and siding. The blends can also be easily coextruded or laminated over PVC, providing a cost-effective weatherable capstock.

GELOY 1120 resin is a pelletized form of GELOY 1020 grade. It is used for applications such as truck tops, RV parts, spas, and pool steps where high heat, gloss, and mechanical properties are needed in addition to excellent outdoor-weatherability. GELOY 1120 resin can be used in profile extrusion, sheet extrusion (monolithic or coextruded over ABS), and in extrusion blow molding (door panels, shutters) where color and property retention is desired, eliminating the need for painting.

### Product Data:

**GELOY 1020—Typical Property Values**

| Property | ASTM Test Method | English Units | SI Units |
|---|---|---|---|
| **Thermal** | | | |
| Deflection Temperature | D648 | | |
| @ 66 psi, 1/4″ (0.46 MPa, 6.4 mm) | | 205°F | 96°C |
| @ 264 psi, 1/4″ (1.82 MPa, 6.4 mm) | | 190°F | 88°C |
| Coefficient of Thermal Expansion | D696 | $5.9 \times 10^{-5}$ in/in/°F | $10.6 \times 10^{-5}$ m/m/°C |

*(continued)*

NOTE—Information provided in this book has been condensed. The publisher recommends that the product manufacturer be contacted, at the address or phone number provided, for more complete information on product safety, processing parameters, and applications not provided in this information sheet.

GELOY 1020—Typical Property Values *(continued)*

| Property | ASTM Test Method | English Units | SI Units |
|---|---|---|---|
| **Mechanical** | | | |
| Izod Impact Strength, Notched<br>@ 73°F (23°C)<br>@ −20°F (−29°C) | D256 | <br>10.0 ft-lbs/in<br>1.5 ft-lbs/in | <br>544 J/m<br>80 J/m |
| Gardner Impact Strength<br>@ 73°F (23°C) | D3029 | <br>200 in-lbs | <br>23 J |
| Instrumented Impact Strength<br>Total Energy @ 73°F<br>J/m @ 23°C<br>Total Energy @ −20°F<br>J/m @ −29°C | D3763 | <br>340 in-lbs<br><br>270 in-lbs<br> | <br><br>37 J<br><br>30 J |
| Tensile Strength | D638 | 5600 psi | 38 MPa |
| Tensile Elongation @ Break | D638 | 25% | |
| Flexural Strength, 1/4″ | D790 | 8000 psi | 55 MPa |
| Flexural Modulus, 1/4″ | D790 | 240,000 psi | 1654 MPa |
| Rockwell Hardness | D785 | R85 | |
| **Electrical** | | | |
| Dielectric Strength | D149 | | |
| Volume Resistivity<br>@ 73°F, dry (23°C) | D257 | | |
| **General** | | | |
| Specific Gravity | D792 | 1.06 | |
| Mold Shrinkage | D955 | | |
| Water Absorption<br>24 hrs @ 73°F (23°C) | D570 | <br>0.25% | |
| Flammability | UL94 | HB | |
| 60° Gloss | D523 | 85–95 | |
| Melt Viscosity: Poise @ °F<br>@ 100 $sec^{-1}$<br>@ 1500 $sec^{-1}$ | | @ 446°F<br>14,000<br>1800 | |

# General Electric Company

**Contact Address:**

General Electric Company
Plastics Group
One Plastics Avenue
Pittsfield, MA 01201

Phone: 1 (800) 845-0600

## GELOY 1220

**Product Information:**

GELOY 1220 resin is an alloy of ASA and PVC provided in pellet form. It is perfect for profile extruded parts that need higher heat and/or outdoor-weatherability in either dark or light colors. Typical applications include window and door profiles, gutter systems, fencing, vending machine trim, and automotive exterior trim.

**Product Data:**

**GELOY 1220—Typical Property Values**

| Property | ASTM Test Method | English Units | SI Units |
|---|---|---|---|
| **Thermal** | | | |
| Deflection Temperature<br>@ 66 psi, 1/4″ (0.46 MPa, 6.4 mm)<br>@ 264 psi, 1/4″ (1.82 MPa, 6.4 mm) | D648 | <br>185°F<br>170°F | <br>85°C<br>76°C |
| Coefficient of Thermal Expansion | D696 | $4.9 \times 10^{-5}$ in/in/°F | $8.8 \times 10^{-5}$ m/m/°C |
| **Mechanical** | | | |
| Izod Impact Strength, Notched<br>@ 73°F (23°C)<br>@ −20°F (−29°C) | D256 | <br>20.0 ft-lbs/in<br>2.0 ft-lbs/in | <br>1068 J/m<br>107 J/m |
| Gardner Impact Strength<br>@ 73°F (23°C) | D3029 | <br>300 in-lbs | <br>34 J |
| Instrumented Impact Strength<br>Total Energy @ 73°F<br>J/m @ 23°C<br>Total Energy @ −20°F<br>J/m @ −29°C | D3763 | | |

*(continued)*

NOTE—Information provided in this book has been condensed. The publisher recommends that the product manufacturer be contacted, at the address or phone number provided, for more complete information on product safety, processing parameters, and applications not provided in this information sheet.

GELOY 1220—Typical Property Values *(continued)*

| Property | ASTM Test Method | English Units | SI Units |
|---|---|---|---|
| **Mechanical** | | | |
| Tensile Strength | D638 | 6600 psi | 45 MPa |
| Tensile Elongation @ Break | D638 | 35% | |
| Flexural Strength, 1/4″ | D790 | 9200 psi | 63 MPa |
| Flexural Modulus, 1/4″ | D790 | 310,000 psi | 2140 MPa |
| Rockwell Hardness | D785 | R96 | |
| **Electrical** | | | |
| Dielectric Strength | D149 | | |
| Volume Resistivity @ 73°F, dry (23°C) | D257 | | |
| **General** | | | |
| Specific Gravity | D792 | 1.21 | |
| Mold Shrinkage | D955 | | |
| Water Absorption 24 hrs @ 73°F (23°C) | D570 | 0.10% | |
| Flammability | UL94 | V-1 | |
| 60° Gloss | D523 | 80–90 | |
| Melt Viscosity: Poise @ °F | | @ 400°F | |
| @ 100 $sec^{-1}$ | | 26,000 | |
| @ 1500 $sec^{-1}$ | | 3700 | |

# General Electric Company

Contact Address:

General Electric Company
Plastics Group
One Plastics Avenue
Pittsfield, MA 01201

Phone: 1 (800) 845-0600

## GELOY XP 1001 Injection Molding Grades

Product Information:

GELOY XP 1001 resin is an injection molding ASA resin offering excellent weatherability. Typical applications are interior or exterior automotive/truck trim and parts, providing color stability without painting.

Product Data:

**GELOY XP 1001—Preliminary Property Values**

| Property | ASTM Test Method | English Units | SI Units |
|---|---|---|---|
| **Thermal** | | | |
| Deflection Temperature | D648 | | |
| @ 66 psi, 1/4″ (0.46 MPa, 6.4 mm) | | 205°F | 96°C |
| @ 264 psi, 1/4″ (1.82 MPa, 6.4 mm) | | 185°F | 85°C |
| Coefficient of Thermal Expansion | D696 | | |
| **Mechanical** | | | |
| Izod Impact Strength, Notched | D256 | | |
| @ 73°F (23°C) | | 9.0 ft-lbs/in | 480 J/m |
| @ −20°F (−29°C) | | 1.2 ft-lbs/in | 65 J/m |
| Gardner Impact Strength | D3029 | | |
| @ 73°F (23°C) | | 200 in-lbs | 23 J |
| @ −20°F (−29°C) | | 100 in-lbs | 11 J |
| Instrumented Impact Strength | D3763 | | |
| Total Energy @ 73°F | | 340 in-lbs | |
| J/m @ 23°C | | | 37 J |
| Total Energy @ −20°F | | 270 in-lbs | |
| J/m @ −29°C | | | 30 J |

*(continued)*

NOTE – Information provided in this book has been condensed. The publisher recommends that the product manufacturer be contacted, at the address or phone number provided, for more complete information on product safety, processing parameters, and applications not provided in this information sheet.

### GELOY XP 1001—Preliminary Property Values *(continued)*

| Property | ASTM Test Method | English Units | SI Units |
|---|---|---|---|
| **Mechanical** | | | |
| Tensile Strength | D638 | 5600 psi | 38 MPa |
| Tensile Elongation @ Break | D638 | 30% | |
| Flexural Strength<br>@ 1/8″<br>@ 1/4″ | D790 | <br>6300 psi<br>7000 psi | <br>43 MPa<br>48 MPa |
| Flexural Modulus<br>@ 1/8″<br>@ 1/4″ | D790 | <br>270,000 psi<br>240,000 psi | <br>1856 MPa<br>1654 MPa |
| Rockwell Hardness | D785 | R84 | |
| **Electrical** | | | |
| Dielectric Strength | D149 | | |
| Volume Resistivity<br>@ 73°F, dry (23°C) | D257 | | |
| **General** | | | |
| Specific Gravity | D792 | 1.06 | |
| Mold Shrinkage | D955 | 5–7 $10^{-3}$ in/in | |
| Water Absorption<br>24 hrs @ 73°F (23°C) | D570 | | |
| Flammability | UL94 | HB | |
| 60° Gloss | D523 | | |
| Melt Viscosity: Poise @ °F<br>@ 100 $sec^{-1}$<br>@ 1500 $sec^{-1}$ | | @ 446°F<br>14,000<br>1800 | |

# General Electric Company

## Contact Address:

General Electric Company
Plastics Group
One Plastics Avenue
Pittsfield, MA 01201

Phone: 1 (800) 845-0600

# GELOY XP 2001

## Product Information:

GELOY XP 2001 resin is an ASA/PVC alloy offering flame retardance along with good low-temperature impact. It is used for outdoor electrical boxes, NIDs, and appliances.

## Product Data:

**GELOY XP 2001—Preliminary Property Values**

| Property | ASTM Test Method | English Units | SI Units |
|---|---|---|---|
| **Thermal** | | | |
| Deflection Temperature | D648 | | |
| @ 66 psi, 1/4″ (0.46 MPa, 6.4 mm) | | 175°F | 79°C |
| @ 264 psi, 1/4″ (1.82 MPa, 6.4 mm) | | 160°F | 70°C |
| Coefficient of Thermal Expansion | D696 | | |
| **Mechanical** | | | |
| Izod Impact Strength, Notched | D256 | | |
| @ 73°F (23°C) | | 17.0 ft-lbs/in | 910 J/m |
| @ −20°F (−29°C) | | 1.2 ft-lbs/in | 65 J/m |
| Gardner Impact Strength | D3029 | | |
| @ 73°F (23°C) | | 220 in-lbs | 25 J |
| @ −20°F (−29°C) | | 100 in-lbs | 11 J |
| Instrumented Impact Strength | D3763 | | |
| Total Energy @ 73°F | | | |
| J/m @ 23°C | | | |
| Total Energy @ −20°F | | | |
| J/m @ −29°C | | | |

*(continued)*

NOTE—Information provided in this book has been condensed. The publisher recommends that the product manufacturer be contacted, at the address or phone number provided, for more complete information on product safety, processing parameters, and applications not provided in this information sheet.

**GELOY XP 2001—Preliminary Property Values *(continued)***

| Property | ASTM Test Method | English Units | SI Units |
|---|---|---|---|
| **Mechanical** | | | |
| Tensile Strength | D638 | 6000 psi | 41 MPa |
| Tensile Elongation @ Break | D638 | 40% | |
| Flexural Strength<br>@ 1/8″<br>@ 1/4″ | D790 | <br><br>9000 psi | <br><br>62 MPa |
| Flexural Modulus<br>@ 1/8″<br>@ 1/4″ | D790 | <br><br>280,000 psi | <br><br>1930 MPa |
| Rockwell Hardness | D785 | R96 | |
| **Electrical** | | | |
| Dielectric Strength | D149 | 862 volts/mil | 33.9 kV/mm |
| Volume Resistivity<br>@ 73°F, dry (23°C) | D257 | <br>$10^{14}$ ohm-cm | |
| **General** | | | |
| Specific Gravity | D792 | 1.21 | |
| Mold Shrinkage | D955 | 5–7 $10^{-3}$ in/in | |
| Water Absorption<br>24 hrs @ 73°F (23°C) | D570 | <br>0.16% | |
| Flammability | UL94 | V-1 @ 0.060″<br>5V @ 0.125″ | |
| 60° Gloss | D523 | | |
| Melt Viscosity: Poise @ °F<br>@ 100 $sec^{-1}$<br>@ 1500 $sec^{-1}$ | | @ 390°F<br>22,000<br>3900 | |

# General Electric Company

## Contact Address:

General Electric Company
Plastics Group
One Plastics Avenue
Pittsfield, MA 01201

Phone: 1 (800) 845-0600

## GELOY XP 2003

### Product Information:

GELOY XP 2003 resin is an ASA/PVC alloy with excellent weatherability and chemical resistance. RV parts and accessories, marine parts, and automotive injection trim are typical applications.

### Product Data:

**GELOY XP 2003—Preliminary Property Values**

| Property | ASTM Test Method | English Units | SI Units |
|---|---|---|---|
| **Thermal** | | | |
| Deflection Temperature | D648 | | |
| @ 66 psi, 1/4″ (0.46 MPa, 6.4 mm) | | 180°F | 82°C |
| @ 264 psi, 1/4″ (1.82 MPa, 6.4 mm) | | 165°F | 73°C |
| Coefficient of Thermal Expansion | D696 | | |
| **Mechanical** | | | |
| Izod Impact Strength, Notched | D256 | | |
| @ 73°F (23°C) | | 16.0 ft-lbs/in | 855 J/m |
| @ −20°F (−29°C) | | 1.0 ft-lbs/in | 54 J/m |
| Gardner Impact Strength | D3029 | | |
| @ 73°F (23°C) | | 220 in-lbs | 25 J |
| @ −20°F (−29°C) | | 50 in-lbs | 6 J |
| Instrumented Impact Strength | D3763 | | |
| Total Energy @ 73°F | | | |
| J/m @ 23°C | | | |
| Total Energy @ −20°F | | | |
| J/m @ −29°C | | | |

*(continued)*

NOTE—Information provided in this book has been condensed. The publisher recommends that the product manufacturer be contacted, at the address or phone number provided, for more complete information on product safety, processing parameters, and applications not provided in this information sheet.

GELOY XP 2003—Preliminary Property Values *(continued)*

| Property | ASTM Test Method | English Units | SI Units |
|---|---|---|---|
| **Mechanical** | | | |
| Tensile Strength | D638 | 6500 psi | 45 MPa |
| Tensile Elongation @ Break | D638 | 50% | |
| Flexural Strength<br>@ 1/8″<br>@ 1/4″ | D790 | <br><br>9900 psi | <br><br>68 MPa |
| Flexural Modulus<br>@ 1/8″<br>@ 1/4″ | D790 | <br><br>300,000 psi | <br><br>2100 MPa |
| Rockwell Hardness | D785 | R98 | |
| **Electrical** | | | |
| Dielectric Strength | D149 | | |
| Volume Resistivity<br>@ 73°F, dry (23°C) | D257 | | |
| **General** | | | |
| Specific Gravity | D792 | 1.20 | |
| Mold Shrinkage | D955 | 5–7 $10^{-3}$ in/in | |
| Water Absorption<br>24 hrs @ 73°F (23°C) | D570 | | |
| Flammability | UL94 | V-2 @ 0.125″ | |
| 60° Gloss | D523 | | |
| Melt Viscosity: Poise @ °F<br>@ 100 $sec^{-1}$<br>@ 1500 $sec^{-1}$ | | @ 390°F<br>21,800<br>3700 | |

# General Electric Company

## Contact Address:

General Electric Company
Plastics Group
One Plastics Avenue
Pittsfield, MA 01201

Phone: 1 (800) 845-0600

# GELOY XP 4001 HG

## Product Information:

GELOY XP 4001 resin is an ASA alloy with polycarbonate. It offers higher heat resistance and impact strength than GELOY XP 1001 resin, along with good weatherability. Unpainted interior and exterior automotive parts, either injection molded or blow molded, are good applications for GELOY XP 4001 resin. Both low gloss and high gloss versions are available.

## Product Data:

**GELOY XP 4001 HG—Preliminary Property Values**

| Property | ASTM Test Method | English Units | SI Units |
|---|---|---|---|
| **Thermal** | | | |
| Deflection Temperature | D648 | | |
| @ 66 psi, 1/4″ (0.46 MPa, 6.4 mm) | | 230°F | 110°C |
| @ 264 psi, 1/4″ (1.82 MPa, 6.4 mm) | | 210°F | 99°C |
| Coefficient of Thermal Expansion | D696 | | |
| **Mechanical** | | | |
| Izod Impact Strength, Notched | D256 | | |
| @ 73°F (23°C) | | 13.0 ft-lbs/in | 700 J/m |
| @ −20°F (−29°C) | | 3.5 ft-lbs/in | 187 J/m |
| Gardner Impact Strength | D3029 | | |
| @ 73°F (23°C) | | 250 in-lbs | 28 J |
| @ −20°F (−29°C) | | 220 in-lbs | 25 J |
| Instrumented Impact Strength | D3763 | | |
| Total Energy @ 73°F | | 450 in-lbs | |
| J/m @ 23°C | | | 51 J |
| Total Energy @ −20°F | | 420 in-lbs | |
| J/m @ −29°C | | | 47 J |

*(continued)*

NOTE—Information provided in this book has been condensed. The publisher recommends that the product manufacturer be contacted, at the address or phone number provided, for more complete information on product safety, processing parameters, and applications not provided in this information sheet.

## GELOY XP 4001 HG—Preliminary Property Values *(continued)*

| Property | ASTM Test Method | English Units | SI Units |
|---|---|---|---|
| **Mechanical** | | | |
| Tensile Strength | D638 | 7000 psi | 48 MPa |
| Tensile Elongation @ Break | D638 | 120% | |
| Flexural Strength<br>@ 1/8″<br>@ 1/4″ | D790 | <br>7300 psi<br>9700 psi | <br>50 MPa<br>66 MPa |
| Flexural Modulus<br>@ 1/8″<br>@ 1/4″ | D790 | <br>295,000 psi<br>277,000 psi | <br>2035 MPa<br>1911 MPa |
| Rockwell Hardness | D785 | R97 | |
| **Electrical** | | | |
| Dielectric Strength | D149 | | |
| Volume Resistivity<br>@ 73°F, dry (23°C) | D257 | | |
| **General** | | | |
| Specific Gravity | D792 | 1.11 | |
| Mold Shrinkage | D955 | | |
| Water Absorption<br>24 hrs @ 73°F (23°C) | D570 | | |
| Flammability | UL94 | HB | |
| 60° Gloss | D523 | 90–100 | |
| Melt Viscosity: Poise @ °F<br>@ 100 $sec^{-1}$<br>@ 1500 $sec^{-1}$ | | @ 446°F<br>13,150<br>2090 | |

# General Electric Company

## Contact Address:

General Electric Company
Plastics Group
One Plastics Avenue
Pittsfield, MA 01201

Phone: 1 (800) 845-0600

## GELOY XP 4001 LG

### Product Information:

GELOY XP 4001 resin is an ASA alloy with polycarbonate. It offers higher heat resistance and impact strength than GELOY XP 1001 resin, along with good weatherability. Unpainted interior and exterior automotive parts, either injection molded or blow molded, are good applications for GELOY XP 4001 resin. Both low gloss and high gloss versions are available.

### Product Data:

**GELOY XP 4001 LG—Preliminary Property Values**

| Property | ASTM Test Method | English Units | SI Units |
|---|---|---|---|
| **Thermal** | | | |
| Deflection Temperature<br>@ 66 psi, 1/4″ (0.46 MPa, 6.4 mm)<br>@ 264 psi, 1/4″ (1.82 MPa, 6.4 mm) | D648 | <br>225°F<br>205°F | <br>107°C<br>96°C |
| Coefficient of Thermal Expansion | D696 | | |
| **Mechanical** | | | |
| Izod Impact Strength, Notched<br>@ 73°F (23°C)<br>@ −20°F (−29°C) | D256 | <br>10.0 ft-lbs/in<br>1.5 ft-lbs/in | <br>544 J/m<br>80 J/m |
| Gardner Impact Strength<br>@ 73°F (23°C)<br>@ −20°F (−29°C) | D3029 | <br>190 in-lbs<br>130 in-lbs | <br>21 J<br>15 J |
| Instrumented Impact Strength<br>Total Energy @ 73°F<br>J/m @ 23°C<br>Total Energy @ −20°F<br>J/m @ −29°C | D3763 | <br>346 in-lbs<br><br>220 in-lbs<br> | <br><br>39 J<br><br>25 J |

*(continued)*

NOTE–Information provided in this book has been condensed. The publisher recommends that the product manufacturer be contacted, at the address or phone number provided, for more complete information on product safety, processing parameters, and applications not provided in this information sheet.

GELOY XP 4001 LG—Preliminary Property Values *(continued)*

| Property | ASTM Test Method | English Units | SI Units |
|---|---|---|---|
| **Mechanical** | | | |
| Tensile Strength | D638 | 6200 psi | 42 MPa |
| Tensile Elongation @ Break | D638 | 120% | |
| Flexural Strength<br>@ 1/8″<br>@ 1/4″ | D790 | <br>6900 psi<br>8600 psi | <br>48 MPa<br>58 MPa |
| Flexural Modulus<br>@ 1/8″<br>@ 1/4″ | D790 | <br>295,000 psi<br>265,000 psi | <br>2035 MPa<br>1828 MPa |
| Rockwell Hardness | D785 | R97 | |
| **Electrical** | | | |
| Dielectric Strength | D149 | | |
| Volume Resistivity<br>@ 73°F, dry (23°C) | D257 | | |
| **General** | | | |
| Specific Gravity | D792 | 1.11 | |
| Mold Shrinkage | D955 | | |
| Water Absorption<br>24 hrs @ 73°F (23°C) | D570 | | |
| Flammability | UL94 | HB | |
| 60° Gloss | D523 | | |
| Melt Viscosity: Poise @ °F<br>@ 100 $sec^{-1}$<br>@ 1500 $sec^{-1}$ | | @ 446°F<br>12,500<br>2110 | |

# General Electric Company

Contact Address:

General Electric Company
Plastics Group
One Plastics Avenue
Pittsfield, MA 01201

Phone: 1 (800) 845-0600

## NORYL GTX

Product Information:

NORYL GTX resins are polyphenylene oxide based polyamide alloys. These molding resins are either unreinforced or glass reinforced injection moldable pellets. Noryl GTX resins have excellent chemical resistance, high temperature dimensional stability, low temperature impact resistance, good melt flow providing a wide processing temperature range, low water absorption and low creep under stress.

Noryl GTX 625 is an unreinforced blow molding resin.

In addition, the new Noryl SPN and RFN superior productivity grades offer even faster cycles and higher yields with improved across-the-board performance: high strength, wear resistance and superior out-of-mold surface quality. Noryl SPN grades are unreinforced resins for improved flow and lower mold pressure requirements. Noryl RFN grades are reinforced resins for higher strength and modulus with wide processing parameters and thin-wall performance.

## NORYL GTX 625

Product Data:

**NORYL GTX 625—Blow Molding Grade**

| Property | ASTM | English Units | SI Units |
|---|---|---|---|
| **Mechanical** | | | |
| Tensile Strength @ Yield | D638 | 8,000 psi | 56 N/mm$^2$ |
| Tensile Elongation @ Break | D638 | 50% | 50% |
| Flexural Strength @ Yield | D790 | 12,000 psi | 83 N/mm$^2$ |
| Flexural Modulus | D790 | 290,000 psi | 2000 N/mm$^2$ |
| Izod Impact Strength, Notched<br>@ 73°F (23°C), 1/8″ (3.2mm)<br>@ −40°F (−40°C), 1/8″ (3.2mm) | D256 | <br>9.0 ft-lb/in<br>3.0 ft-lb/in | <br>477 J/m<br>156 J/m |
| Dynatup Impact Strength (5 mph)<br>@ 73°F (23°C), Energy at Peak | D3763 | <br>400 in-lb | |

*(continued)*

NOTE—Information provided in this book has been condensed. The publisher recommends that the product manufacturer be contacted, at the address or phone number provided, for more complete information on product safety, processing parameters, and applications not provided in this information sheet.

**NORYL GTX 625—Blow Molding Grade*(continued)***

| Property | ASTM | English Units | SI Units |
|---|---|---|---|
| **Thermal** | | | |
| Heat Deflection Temperature | D648 | | |
| @ 264 psi (1.82 N/mm²) 1/4″ (6.4mm) | | 270°F | 132°C |
| @ 66 psi (0.45 N/mm²) 1/4″ (6.4mm) | | 350°F | 177°C |
| **Physical** | | | |
| Specific Gravity (natural color) | D792 | 1.10 | 1.10 |
| Density | | 0.040 lb/in³ | 1.10 g/cc |
| Water Absorption (50% RH) | | | |
| @ 24 hrs, 73°F (23°C) | | .3% | .3% |
| @ Equilibrium, 73°F (23°C) | | 1.0% | 1.0% |
| Mold Shrinkage | | 0.013–0.017 | 1.3–1.7% |

# General Electric Company

## Contact Address:

General Electric Company
Plastics Group
One Plastics Avenue
Pittsfield, MA 01201

Phone: 1 (800) 845-0600

## NORYL GTX and NORYL PLUS

### Product Information:

NORYL GTX and NORYL PLUS are polyphenylene oxide based/polyamide alloys. The molding resins are either unreinforced or glass reinforced injection moldable pellets. These NORYL resins have excellent chemical resistance, high temperature dimensional stability, low temperature impact resistance, good melt flow providing a wide processing temperature range, low water absorbtion, and low creep under stress. The molded product provides a class "A" finish.

### Product Data:

**NORYL GTX 810**

| Property | ASTM | English Units | SI Units |
|---|---|---|---|
| **Mechanical** | | | |
| Tensile Strength @ Yield<br>@ Break | D638 | 13,500 psi<br>12,900 psi | 93 N/mm²<br>89 N/mm² |
| Tensile Elongation @ Break | D638 | 10% | 10% |
| Flexural Strength @ Yield | D790 | 19,000 psi | 131 N/mm² |
| Flexural Modulus | D790 | 560,000 psi | 3860 N/mm² |
| Shear Strength | D732 | 10,000 psi | 69 N/mm² |
| Compressive Strength | D695 | 17,000 psi | 117 N/mm² |
| Izod Impact Strength, Notched<br>@ 73°F (23°C), 1/8″ (3.2mm)<br>@ −20°F (−29°C), 1/8″ (3.2mm) | D256 | <br>1.5 ft-lb/in<br>1.0 ft-lb/in | <br>80 J/m<br>55 J/m |

*(continued)*

NOTE – Information provided in this book has been condensed. The publisher recommends that the product manufacturer be contacted, at the address or phone number provided, for more complete information on product safety, processing parameters, and applications not provided in this information sheet.

NORYL GTX 810 *(continued)*

| Property | ASTM | English Units | SI Units |
|---|---|---|---|
| **Thermal** | | | |
| Heat Deflection Temperature | D648 | | |
| @ 264 psi (1.8 $N/mm^2$), 1/4″ (6.4mm) | | 370°F | 188°C |
| @ 66 psi (0.45 $N/mm^2$), 1/4″ (6.4mm) | | 485°F | 252°C |
| Vicat Softening Point (Method B) | D1525 | 456°F | 241°C |
| Coefficient of Thermal Expansion, −20 to 275°F (−29 to 135°C) | D696 | 2.2–2.8 × $10^{-5}$ in/in-°F | 4.0–5.0 × $10^{-5}$ m/m-°C |
| **Physical** | | | |
| Specific Gravity (black color) | D792 | 1.16 | 1.16 |
| Density | | 0.042 $lb/in^3$ | 1.16 g/cc |
| Rockwell Hardness | D785 | R119 | R119 |
| Water Absorption (50% RH) | D570 | | |
| @ 24 hrs, 73°F (23°C) | | 0.5% | 0.5% |
| @ Equilibrium, 73°F (23°C) | | 1.0% | 1.0% |
| Mold Shrinkage | D955 | | |
| Unannealed | | .005–.010 in/in | 0.5–1.0% |
| Annealed, 325°F (163°C), 1 hr | | .006–.012 in/in | 0.6–1.2% |

# General Electric Company

Contact Address:

General Electric Company
Plastics Group
One Plastics Avenue
Pittsfield, MA 01201

Phone: 1 (800) 845-0600

## NORYL GTX 820

Product Data:

**NORYL GTX 820**

| Property | ASTM | English Units | SI Units |
|---|---|---|---|
| **Mechanical** | | | |
| Tensile Strength @ Yield<br>@ Break | D638 | 18,000 psi<br>18,000 psi | 124 N/mm²<br>124 N/mm² |
| Tensile Elongation @ Break | D638 | 10% | 10% |
| Flexural Strength @ Yield | D790 | 25,000 psi | 172 N/mm² |
| Flexural Modulus | D790 | 800,000 psi | 5515 N/mm² |
| Shear Strength | D732 | 10,000 psi | 69 N/mm² |
| Compressive Strength | D695 | 23,000 psi | 159 N/mm² |
| Izod Impact Strength, Notched<br>@ 73°F (23°C), 1/8″ (3.2mm)<br>@ −20°F (−29°C), 1/8″ (3.2mm) | D256 | <br>1.5 ft-lb/in<br>1.0 ft-lb/in | <br>80 J/m<br>55 J/m |
| **Thermal** | | | |
| Heat Deflection Temperature<br>@ 264 psi (1.8 N/mm²), 1/4″ (6.4mm)<br>@ 66 psi (0.45 N/mm²), 1/4″ (6.4mm) | D648 | <br>430°F<br>485°F | <br>221°C<br>252°C |
| Vicat Softening Point (Method B) | D1525 | 480°F | 249°C |
| Coefficient of Thermal Expansion,<br>−20 to 275°F (−29 to 135°C) | D696 | $1.7\text{–}2.2 \times 10^{-5}$<br>in/in-°F | $3.0\text{–}4.0 \times 10^{-5}$<br>m/m-°C |

*(continued)*

NOTE—Information provided in this book has been condensed. The publisher recommends that the product manufacturer be contacted, at the address or phone number provided, for more complete information on product safety, processing parameters, and applications not provided in this information sheet.

**NORYL GTX 820 *(continued)***

| Property | ASTM | English Units | SI Units |
|---|---|---|---|
| **Physical** | | | |
| Specific Gravity (black color) | D792 | 1.24 | 1.24 |
| Density | | 0.045 lb/in$^3$ | 1.24 g/cc |
| Rockwell Hardness | D785 | R119 | R119 |
| Water Absorption (50% RH) | D570 | | |
| @ 24 hrs, 73°F (23°C) | | 0.5% | 0.5% |
| @ Equilibrium, 73°F (23°C) | | 1.0% | 1.0% |
| Mold Shrinkage | D955 | | |
| Unannealed | | .004–.006 in/in | 0.4–0.6% |
| Annealed, 325°F (163°C), 1 hr | | .006–.007 in/in | 0.6–0.7% |

# General Electric Company

## Contact Address:

General Electric Company
Plastics Group
One Plastics Avenue
Pittsfield, MA 01201

Phone: 1 (800) 845-0600

# NORYL GTX 830

## Product Data:

**NORYL GTX 830**

| Property | ASTM | English Units | SI Units |
|---|---|---|---|
| **Mechanical** | | | |
| Tensile Strength @ Yield<br>@ Break | D638 | 23,000 psi<br>23,000 psi | 159 N/mm²<br>159 N/mm² |
| Tensile Elongation @ Break | D638 | 8% | 8% |
| Flexural Strength @ Yield | D790 | 32,500 psi | 224 N/mm² |
| Flexural Modulus | D790 | 1,110,000 psi | 7,655 N/mm² |
| Shear Strength | D732 | 11,000 psi | 76 N/mm² |
| Compressive Strength | D695 | 26,000 psi | 179 N/mm² |
| Izod Impact Strength, Notched<br>@ 73°F (23°C), 1/8″ (3.2mm)<br>@ −20°F (−29°C), 1/8″ (3.2mm) | D256 | <br>2.0 ft-lb/in<br>1.5 ft-lb/in | <br>105 J/m<br>80 J/m |
| **Thermal** | | | |
| Heat Deflection Temperature<br>@ 264 psi (1.8 N/mm²), 1/4″ (6.4mm)<br>@ 66 psi (0.45 N/mm²), 1/4″ (6.4mm) | D648 | <br>455°F<br>485°F | <br>235°C<br>252°C |
| Vicat Softening Point (Method B) | D1525 | 480°F | 249°C |
| Coefficient of Thermal Expansion,<br>−20 to 275°F (−29 to 135°C) | D696 | 1.1–1.7 × $10^{-5}$<br>in/in-°F | 2.0–3.0 × $10^{-5}$<br>m/m-°C |

*(continued)*

NOTE – Information provided in this book has been condensed. The publisher recommends that the product manufacturer be contacted, at the address or phone number provided, for more complete information on product safety, processing parameters, and applications not provided in this information sheet.

## NORYL GTX 830 *(continued)*

| Property | ASTM | English Units | SI Units |
|---|---|---|---|
| **Physical** | | | |
| Specific Gravity (natural color) | D792 | 1.33 | 1.33 |
| Density | | 0.048 lb/in$^3$ | 1.33 g/cc |
| Rockwell Hardness | D785 | R120 | R120 |
| Water Absorption (50% RH)<br>@ 24 hrs, 73°F (23°C)<br>@ Equilibrium, 73°F (23°C) | D570 | <br>0.5%<br>1.0% | <br>0.5%<br>1.0% |
| Mold Shrinkage<br>Unannealed<br>Annealed, 325°F (163°C), 1 hr | D955 | <br>.004–.006 in/in<br>.006–.007 in/in | <br>0.4–0.6%<br>0.6–0.7% |

# General Electric Company

Contact Address:

General Electric Company
Plastics Group
One Plastics Avenue
Pittsfield, MA 01201

Phone: 1 (800) 845-0600

## NORYL GTX 910

Product Data:

**NORYL GTX 910**

| Property | ASTM | English Units | SI Units |
|---|---|---|---|
| **Mechanical** | | | |
| Tensile Strength @ Yield<br>@ Break | D638 | 7,850 psi<br>8,100 psi | 54 $N/mm^2$<br>56 $N/mm^2$ |
| Tensile Elongation @ Break | D638 | 50% | 50% |
| Flexural Strength @ Yield | D790 | 11,000 psi | 76 $N/mm^2$ |
| Flexural Modulus | D790 | 290,000 psi | 2000 $N/mm^2$ |
| Izod Impact Strength, Notched<br>@ 73°F (23°C), 1/8″ (3.2mm)<br>@ −20°F (−29°C), 1/8″ (3.2mm) | D256 | <br>3.3 ft-lb/in<br>2.4 ft-lb/in | <br>175 J/m<br>125 J/m |
| Dynatup Impact Strength (5 mph)<br>@ 73°F (23°C), Energy Peak | D3763 | <br>340 in-lb | |
| **Thermal** | | | |
| Heat Deflection Temperature<br>@ 264 psi (1.82 $N/mm^2$)<br>1/4″ (6.4mm)<br>@ 66 psi (0.45 $N/mm^2$)<br>1/4″ (6.4mm) | D648 | <br>250°F<br><br>320°F | <br>121°C<br><br>160°C |
| Vicat Softening Point (Method B) | D1525 | 420°F | 215°C |
| Coefficient of Thermal Expansion,<br>−20 to 275°F (−29 to 135°C) | D696 | 5.0–7.0 × $10^{-5}$<br>in/in-°F | 9.0–12.6 × $10^{-5}$<br>m/m-°C |

*(continued)*

NOTE—Information provided in this book has been condensed. The publisher recommends that the product manufacturer be contacted, at the address or phone number provided, for more complete information on product safety, processing parameters, and applications not provided in this information sheet.

NORYL GTX 910 *(continued)*

| Property | ASTM | English Units | SI Units |
|---|---|---|---|
| **Physical** | | | |
| Specific Gravity (natural color) | D792 | 1.10 | 1.10 |
| Density | | 0.040 lb/in$^3$ | 1.10 g/cc |
| Rockwell Hardness | D785 | R116 | R116 |
| Water Absorption (50% RH)<br>@ 24 hrs, 73°F (23°C)<br>@ Equilibrium, 73°F (23°C) | | <br>0.3%<br>1.0% | <br>0.3%<br>1.0% |
| Mold Shrinkage<br>Unannealed | D955 | <br>0.012–0.016 in/in | <br>1.2–1.6% |
| Annealed, 325°F (163°C), 1 hr | | 0.015–0.017 in/in | 1.5–1.7% |

# General Electric Company

## Contact Address:

General Electric Company
Plastics Group
One Plastics Avenue
Pittsfield, MA 01201

Phone: 1 (800) 845-0600

# NORYL PLUS CRN 410

## Product Data:

- Chemical Resistance
- Heat Resistance
- Impact Performance
- Consistent/Predictable Shrinkage

## Technical Data:

**NORYL PLUS CRN 410—Preliminary Property Profile**

| Property | ASTM Method | System | |
|---|---|---|---|
| | | English Units | SI Units |
| **Physical** | | | |
| Specific Gravity | D792 | 1.10 | 1.10 |
| Specific Volume | D892 | | |
| Water Absorption—24 hours | D570 | 0.5% | 0.5% |
| Mold Shrinkage Flow Direction and Cross Flow Direction (125 mil) (3.175mm) | D955-51 | 12–16 in/in × $10^{-3}$ | 1.2–1.6% |

*(continued)*

NOTE—Information provided in this book has been condensed. The publisher recommends that the product manufacturer be contacted, at the address or phone number provided, for more complete information on product safety, processing parameters, and applications not provided in this information sheet.

## NORYL PLUS CRN 410—Preliminary Property Profile *(continued)*

| Property | ASTM Method | System | |
|---|---|---|---|
| | | English Units | SI Units |
| **Mechanical** | | | |
| Tensile Strength at Yield | D638 | 8,600 psi | 59.3 MPa |
| Elongation at Break | D638 | 60% | 60% |
| Flexural Strength | D790 | 11,000 psi | 75.8 MPa |
| Flexural Modulus | D790 | 310,000 psi | 2136 MPa |
| Compressive Strength | D695 | 11,500 psi | .79 MPa |
| Shear Strength | D732 | | |
| Izod Impact Strength<br>Notched, 1/8″ thick (3.2mm)<br>Notched @ −40°C<br>Unnotched, 1/8″ thick (3.2mm) | D256 | <br>4.0 ft lb/in | <br>214 J/m |
| Gardner Impact, 1/8″ thick (3.2mm) | | | |
| Instrumented Impact Test | D3763-85 | 450 in-lbs | 50.9 N-m |
| Rockwell Hardness<br>R-scale | D785 | R116 | R116 |
| **Thermal** | | | |
| Heat Deflection Temp.<br>66 psi (0.46 MPa)<br>264 psi (1.82 MPa) | D648 | <br>380°F<br>290°F | <br>193°C<br>143°C |
| Vicat Softening Point (Method B) | D1525 | 450°F | 232°C |
| Coefficient of Thermal Expansion<br>(linear) | E831 | | 9.0–12.6 × $10^{-5}$<br>m/m/°C |
| Mold Direction<br>Range: −40 to 100°F (−40 to 40°C)<br>Range: 240 to 280°F (80 to 140°C) | | <br>5.0–7.0 × $10^{-5}$<br>in/in/°F | |
| **UL Properties** | | | |
| Flammability | UL94 | HB* | HB* |

*Electrical properties.

# General Electric Company

Contact Address:

General Electric Company
Plastics Group
One Plastics Avenue
Pittsfield, MA 01201

Phone: 1 (800) 845-0600

## NORYL PLUS CRN 630

Product Data:

- Chemical Resistance
- Heat Resistance
- Dimensional Stability
- Excellent Rigidity

Technical Data:

**NORYL PLUS CRN 630—Preliminary Property Profile**

| Property | ASTM Method | System | |
|---|---|---|---|
| | | English Units | SI Units |
| **Physical** | | | |
| Specific Gravity | D792 | 1.33 | 1.33 |
| Specific Volume | D892 | | |
| Water Absorption—24 hours | D570 | 0.5% | 0.5% |
| Mold Shrinkage<br>Flow Direction and<br>Cross Flow Direction<br>(125 mil) (3.175mm) | D955-51 | 4–6 in/in × $10^{-3}$ | 0.4–0.6% |

*(continued)*

NOTE—Information provided in this book has been condensed. The publisher recommends that the product manufacturer be contacted, at the address or phone number provided, for more complete information on product safety, processing parameters, and applications not provided in this information sheet.

NORYL PLUS CRN 630—Preliminary Property Profile *(continued)*

| Property | ASTM Method | System | |
|---|---|---|---|
| | | English Units | SI Units |
| **Mechanical** | | | |
| Tensile Strength at Yield | D638 | 23,000 psi | 159 MPa |
| Elongation at Break | D638 | 8% | 8% |
| Flexural Strength | D790 | 32,500 psi | 224 MPa |
| Flexural Modulus | D790 | 1,110,000 psi | 7,648 MPa |
| Compressive Strength | D695 | 25,000 psi | 179 MPa |
| Shear Strength | D732 | 11,000 | 76 MPa |
| Izod Impact Strength | D256 | | |
| Notched, 1/8″ thick (3.2mm) | | 2.0 ft lb/in | 107 J/m |
| Notched @ −40°C | | 1.5 ft lb/in | 80 J/m |
| Unnotched, 1/8″ thick (3.2mm) | | | |
| Gardner Impact, 1/8″ thick (3.2mm) | | | |
| Instrumented Impact Test | D3763-85 | | |
| Rockwell Hardness R-scale | D785 | R120 | R120 |
| **Thermal** | | | |
| Heat Deflection Temp. | D648 | | |
| 66 psi (0.46 MPa) | | 485°F | 252°C |
| 264 psi (1.82 MPa) | | 450°F | 235°C |
| Vicat Softening Point (Method B) | D1525 | 480°F | 249°C |
| Coefficient of Thermal Expansion (linear) | E831 | | |
| Mold Direction | | | |
| Range: −40 to 100°F (−40 to 40°C) | | $1.1\text{–}1.7 \times 10^{-5}$ in/in/°F | $1.0\text{–}3.0 \times 10^{-5}$ m/m/°C |
| Range: 240 to 280°F (80 to 140°C) | | | |
| **UL Properties** | | | |
| Flammability | UL94 | HB* | HB* |

*Electrical properties.

# General Electric Company

## Contact Address:

General Electric Company
Plastics Group
One Plastics Avenue
Pittsfield, MA 01201

Phone: 1 (800) 845-0600

# NORYL PLUS RFN 420

## Product Data:

- Rigidity/Dimensional Stability
- Heat Resistance
- Excellent Processability

## Technical Data:

**NORYL PLUS RFN 420—Preliminary Property Profile**

| Property | ASTM Method | System | |
|---|---|---|---|
| | | English Units | SI Units |
| **Physical** | | | |
| Specific Gravity | D792 | 1.21 | 1.21 |
| Specific Volume | D892 | | |
| Water Absorption—24 hours | D570 | .06% | .06% |
| Mold Shrinkage<br>Flow Direction and<br>Cross Flow Direction<br>(125 mil) (3.175mm) | D955-51 | 2–4 in/in × $10^{-3}$ | 0.2–0.4% |

*(continued)*

NOTE—Information provided in this book has been condensed. The publisher recommends that the product manufacturer be contacted, at the address or phone number provided, for more complete information on product safety, processing parameters, and applications not provided in this information sheet.

**NORYL PLUS RFN 420—Preliminary Property Profile *(continued)***

| Property | ASTM Method | System | |
|---|---|---|---|
| | | English Units | SI Units |
| **Mechanical** | | | |
| Tensile Strength at Yield | D638 | 12,500 psi | 86 MPa |
| Elongation at Break | D638 | 6% | 6% |
| Flexural Strength | D790 | 17,000 psi | 117 MPa |
| Flexural Modulus | D790 | 750,000 psi | 5,168 MPa |
| Compressive Strength | D695 | | |
| Shear Strength | D732 | | |
| Izod Impact Strength | D256 | | |
| Notched, 1/8″ thick (3.2mm) | | 1.5 ft lb/in | 30 J/m |
| Notched @ −40°C | | | |
| Unnotched, 1/8″ thick (3.2mm) | | | |
| Gardner Impact, 1/8″ thick (3.2mm) | | | |
| Instrumented Impact Test | D3763-85 | | |
| Rockwell Hardness | D785 | | |
| R-scale | | | |
| **Thermal** | | | |
| Heat Deflection Temp. | D648 | | |
| 66 psi (0.46 MPa) | | | |
| 264 psi (1.82 MPa) | | 270°F | 132°C |
| Coefficient of Thermal Expansion (linear) | E831 | | |
| Mold Direction | | | |
| Range: −40 to 100°F (−40 to 40°C) | | | |
| Range: 240 to 280°F (80 to 140°C) | | | |
| **UL Properties** | | | |
| Flammability | UL94 | HB* | HB* |

*Electrical properties.

# General Electric Company

## Contact Address:

General Electric Company
Plastics Group
One Plastics Avenue
Pittsfield, MA 01201

Phone: 1 (800) 845-0600

# NORYL PLUS RFN 430

## Product Data:

- Flame Retardant
- Toughness
- Superior Processability

## Technical Data:

**NORYL PLUS RFN 430—Preliminary Property Profile**

| Property | ASTM Method | System | |
|---|---|---|---|
| | | English Units | SI Units |
| **Physical** | | | |
| Specific Gravity | D792 | 1.15 | 1.15 |
| Specific Volume | D892 | | |
| Water Absorption—24 hours | D570 | 0.06% | 0.06% |
| Mold Shrinkage<br>Flow Direction and<br>Cross Flow Direction<br>(125 mil) (3.175mm) | D955-51 | 5–7 in/in × $10^{-3}$ | 0.5–0.7% |

*(continued)*

NOTE – Information provided in this book has been condensed. The publisher recommends that the product manufacturer be contacted, at the address or phone number provided, for more complete information on product safety, processing parameters, and applications not provided in this information sheet.

**NORYL PLUS RFN 430—Preliminary Property Profile *(continued)***

| Property | ASTM Method | System: English Units | System: SI Units |
|---|---|---|---|
| **Mechanical** | | | |
| Tensile Strength at Yield | D638 | 6,100 psi | 42 MPa |
| Elongation at Break | D638 | 60% | 60% |
| Flexural Strength | D790 | 8,700 psi | 59.9 MPa |
| Flexural Modulus | D790 | 330,000 psi | 2,274 MPa |
| Compressive Strength | D695 | | |
| Shear Strength | D732 | | |
| Izod Impact Strength | D256 | | |
| Notched, 1/8″ thick (3.2mm) | | 6.0 ft lb/in | 320 J/m |
| Notched @ −40°C | | 1.5 ft lb/in | 80 J/m |
| Unnotched, 1/8″ thick (3.2mm) | | | |
| Gardner Impact, 1/8″ thick (3.2mm) | | 180 in-lbs | 20.3 N-m |
| Instrumented Impact Test | D3763-85 | | |
| Rockwell Hardness<br>R-scale | D785 | | |
| **Thermal** | | | |
| Heat Deflection Temp. | D648 | | |
| 66 psi (0.46 MPa) | | | |
| 264 psi (1.82 MPa) | | 200°F | 93°C |
| Coefficient of Thermal Expansion (linear) | E831 | | |
| Mold Direction | | | |
| Range: −40 to 100°F (−40 to 40°C) | | | |
| Range: 240 to 280°F (80 to 140°C) | | | |
| **UL Properties** | | | |
| Flammability | UL94 | VO@0.063″ | VO@1.6mm |

# General Electric Company

Contact Address:

General Electric Company
Plastics Group
One Plastics Avenue
Pittsfield, MA 01201

Phone: 1 (800) 845-0600

## NORYL PLUS SPN 410

Product Data:

- Heat Resistance
- Impact Resistance
- Excellent Processability

Technical Data:

**NORYL PLUS SPN 410—Preliminary Property Profile**

| Property | ASTM Method | System | |
|---|---|---|---|
| | | English Units | SI Units |
| **Physical** | | | |
| Specific Gravity | D792 | 1.06 | 1.06 |
| Specific Volume | D892 | | |
| Water Absorption—24 hours | D570 | .07% | .07% |
| Mold Shrinkage<br>Flow Direction and<br>Cross Flow Direction<br>(125 mil) (3.175mm) | D955-51 | 5–7 in/in × $10^{-3}$ | 0.5–0.7% |

*(continued)*

NOTE – Information provided in this book has been condensed. The publisher recommends that the product manufacturer be contacted, at the address or phone number provided, for more complete information on product safety, processing parameters, and applications not provided in this information sheet.

NORYL PLUS SPN 410—Preliminary Property Profile *(continued)*

| Property | ASTM Method | System | |
|---|---|---|---|
| | | English Units | SI Units |
| **Mechanical** | | | |
| Tensile Strength at Yield | D638 | 5,200 psi | 36 MPa |
| Elongation at Break | D638 | 82% | 82% |
| Flexural Strength | D790 | 7,300 psi | 50 MPa |
| Flexural Modulus | D790 | 265,000 psi | 1,826 MPa |
| Compressive Strength | D695 | | |
| Shear Strength | D732 | | |
| Izod Impact Strength | D256 | | |
| Notched, 1/8″ thick (3.2mm) | | 7.0 ft lb/in | 374 J/m |
| Notched @ −40°C | | 3.5 ft lb/in | 187 J/m |
| Unnotched, 1/8″ thick (3.2mm) | | | |
| Gardner Impact, 1/87″ thick (3.2mm) | | 200 in-lbs | 23 N-m |
| Instrumented Impact Test | D3763-85 | | |
| Rockwell Hardness R-scale | D785 | | |
| **Thermal** | | | |
| Heat Deflection Temp. | D648 | | |
| 66 psi (0.46 MPa) | | | |
| 264 psi (1.82 MPa) | | 210°F | 96°C |
| Coefficient of Thermal Expansion (linear) | E831 | | |
| Mold Direction | | | |
| Range: −40 to 100°F (−40 to 40°C) | | | |
| Range: 240 to 280°F (80 to 140°C) | | | |
| **UL Properties** | | | |
| Flammability | UL94 | HB* | HB* |

*Electrical properties.

# General Electric Company

## Contact Address:

General Electric Company
Plastics Group
One Plastics Avenue
Pittsfield, MA 01201

Phone: 1 (800) 845-0600

## NORYL PLUS SPN 420

### Product Data:

- Heat Resistance
- Impact Performance
- Excellent Processability

### Technical Data:

**NORYL PLUS SPN 420—Preliminary Property Profile**

| Property | ASTM Method | System | |
|---|---|---|---|
| | | English Units | SI Units |
| **Physical** | | | |
| Specific Gravity | D792 | 1.07 | 1.07 |
| Specific Volume | D892 | | |
| Water Absorption—24 hours | D570 | .07% | .07% |
| Mold Shrinkage<br>Flow Direction and<br>Cross Flow Direction<br>(125 mil) (3.175mm) | D955-51 | 5–7 in/in × $10^{-3}$ | 0.5–0.7% |

*(continued)*

NOTE – Information provided in this book has been condensed. The publisher recommends that the product manufacturer be contacted, at the address or phone number provided, for more complete information on product safety, processing parameters, and applications not provided in this information sheet.

NORYL PLUS SPN 420—Preliminary Property Profile *(continued)*

| Property | ASTM Method | System | |
|---|---|---|---|
| | | English Units | SI Units |
| **Mechanical** | | | |
| Tensile Strength at Yield | D638 | 6,400 psi | 44 MPa |
| Elongation at Break | D638 | 60% | 60% |
| Flexural Strength | D790 | 8,600 psi | 59 MPa |
| Flexural Modulus | D790 | 310,000 psi | 2,136 MPa |
| Compressive Strength | D695 | | |
| Shear Strength | D732 | | |
| Izod Impact Strength | D256 | | |
| Notched, 1/8″ thick (3.2mm) | | 5.5 ft lb/in | 294 J/m |
| Notched @ −40°C | | 2.5 ft lb/in | 133 J/m |
| Unnotched, 1/8″ thick (3.2mm) | | | |
| Gardner Impact, 1/8″ thick (3.2mm) | | 240 in-lbs | 27 N-m |
| Instrumented Impact Test | D3763-85 | | |
| Rockwell Hardness<br>R-scale | D785 | | |
| **Thermal** | | | |
| Heat Deflection Temp. | D648 | | |
| 66 psi (0.46 MPa) | | | 110°C |
| 264 psi (1.82 MPa) | | 230°F | |
| Coefficient of Thermal Expansion (linear) | E831 | | |
| Mold Direction | | | |
| Range: −40 to 100°F (−40 to 40°C) | | | |
| Range: 240 to 280°F (80 to 140°C) | D955 | | |
| **UL Properties** | | | |
| Flammability | UL94 | HB* | HB* |

*Electrical properties.

# General Electric Company

## Contact Address:

General Electric Company
Plastics Group
One Plastics Avenue
Pittsfield, MA 01201

Phone: 1 (800) 845-0600

## NORYL PLUS SPN 422L

### Product Data:

- Wear Resistance
- Heat Resistance
- Excellent Processability

### Technical Data:

**NORYL PLUS SPN 422L—Preliminary Property Profile**

| Property | ASTM Method | System | |
|---|---|---|---|
| | | English Units | SI Units |
| **Physical** | | | |
| Specific Gravity | D792 | 1.08 | 1.08 |
| Specific Volume | D892 | | |
| Water Absorption—24 hours | D570 | | |
| Mold Shrinkage<br>Flow Direction and<br>Cross Flow Direction<br>(125 mil) (3.175mm) | D955-51 | 5–7 in/in × $10^{-3}$ | 0.5–0.7% |

*(continued)*

NOTE—Information provided in this book has been condensed. The publisher recommends that the product manufacturer be contacted, at the address or phone number provided, for more complete information on product safety, processing parameters, and applications not provided in this information sheet.

**NORYL PLUS SPN 422L—Preliminary Property Profile *(continued)***

| Property | ASTM Method | System | |
|---|---|---|---|
| | | English Units | SI Units |
| **Mechanical** | | | |
| Tensile Strength at Yield | D638 | 7,100 psi | 48.9 MPa |
| Elongation at Break | D638 | 35% | 35% |
| Flexural Strength | D790 | 10,000 psi | 68.9 MPa |
| Flexural Modulus | D790 | 350,000 psi | 24,110 MPa |
| Compressive Strength | D695 | | |
| Shear Strength | D732 | | |
| Izod Impact Strength | D256 | | |
| Notched, 1/8″ thick (3.2mm) | | 5.0 ft lb/in | 267 J/m |
| Notched @ −40°C | | 1.8 ft lb/in | 96 J/m |
| Unnotched, 1/8″ thick (3.2mm) | | | |
| Gardner Impact, 1/8″ thick (3.2mm) | | 125 in-lbs | 14 N-m |
| Instrumented Impact Test | D3763-85 | | |
| Rockwell Hardness<br>R-scale | D785 | | |
| **Thermal** | | | |
| Heat Deflection Temp. | D648 | | |
| 66 psi (0.46 MPa) | | | |
| 264 psi (1.82 MPa) | | 235°F | 113°C |
| Coefficient of Thermal Expansion (linear) | E831 | | |
| Mold Direction<br>Range: −40 to 100°F (−40 to 40°C)<br>Range: 240 to 280°F (80 to 140°C) | | | |
| **UL Properties** | | | |
| Flammability | UL94 | HB* | HB* |

*Electrical properties.

# General Electric Company

## Contact Address:

General Electric Company
Plastics Group
One Plastics Avenue
Pittsfield, MA 01201

Phone: 1 (800) 845-0600

# NORYL PLUS SPN 580

## Product Data:

- Rigidity/Dimensional Stability
- Heat Resistance
- Excellent Processability

## Technical Data:

**NORYL PLUS SPN 580—Preliminary Property Profile**

| Property | ASTM Method | System | |
|---|---|---|---|
| | | English Units | SI Units |
| **Physical** | | | |
| Specific Gravity | D792 | 1.27 | 1.27 |
| Specific Volume | D892 | | |
| Water Absorption—24 hours | D570 | .06% | .06% |
| Mold Shrinkage<br>Flow Direction and<br>Cross Flow Direction<br>(125 mil) (3.175mm) | D955-51 | 1–3 in/in × $10^{-3}$ | 0.1–0.3% |

*(continued)*

NOTE—Information provided in this book has been condensed. The publisher recommends that the product manufacturer be contacted, at the address or phone number provided, for more complete information on product safety, processing parameters, and applications not provided in this information sheet.

**NORYL PLUS SPN 580—Preliminary Property Profile *(continued)***

| Property | ASTM Method | System | |
|---|---|---|---|
| | | English Units | SI Units |
| **Mechanical** | | | |
| Tensile Strength at Yield | D638 | 14,500 psi | 100 MPa |
| Elongation at Break | D638 | 5% | 5% |
| Flexural Strength | D790 | 19,000 psi | 131 MPa |
| Flexural Modulus | D790 | 1,000,000 psi | 6,890 MPa |
| Compressive Strength | D695 | | |
| Shear Strength | D732 | | |
| Izod Impact Strength | D256 | | |
| Notched, 1/8″ thick (3.2mm) | | 1.5 ft lb/in | 80 J/m |
| Notched @ −40°C | | | |
| Unnotched, 1/8″ thick (3.2mm) | | | |
| Gardner Impact, 1/8″ thick (3.2mm) | | | |
| Instrumented Impact Test | D3763-85 | | |
| Rockwell Hardness<br>R-scale | D785 | | |
| **Thermal** | | | |
| Heat Deflection Temp. | D648 | | |
| 66 psi (0.46 MPa) | | | |
| 264 psi (1.82 MPa) | | 275°F | 135°C |
| Coefficient of Thermal Expansion (linear) | E831 | | |
| Mold Direction | | | |
| Range: −40 to 100°F (−40 to 40°C) | | | |
| Range: 240 to 280°F (80 to 140°C) | | | |
| **UL Properties** | | | |
| Flammability | UL94 | HB* | HB* |

*Electrical properties.

# General Electric Company

**Contact Address:**

General Electric Company
Plastics Group
One Plastics Avenue
Pittsfield, MA 01201

Phone: 1 (800) 845-0600

## NORYL PLUS SPN 585H

**Product Data:**

- High Gloss
- Flame Retardant
- Impact Resistance
- Excellent Processability

**Technical Data:**

**NORYL PLUS SPN 585H—Preliminary Property Profile**

| Property | ASTM Method | System | |
|---|---|---|---|
| | | English Units | SI Units |
| **Physical** | | | |
| Specific Gravity | D792 | 1.18 | 1.18 |
| Specific Volume | D892 | | |
| Water Absorption—24 hours | D570 | | |
| Mold Shrinkage<br>Flow Direction and<br>Cross Flow Direction<br>(125 mil) (3.175mm) | D955-51 | 5–7 in/in × $10^{-3}$ | 0.5–0.7% |

*(continued)*

NOTE – Information provided in this book has been condensed. The publisher recommends that the product manufacturer be contacted, at the address or phone number provided, for more complete information on product safety, processing parameters, and applications not provided in this information sheet.

**NORYL PLUS SPN 585H—Preliminary Property Profile *(continued)***

| Property | ASTM Method | System | |
|---|---|---|---|
| | | English Units | SI Units |
| **Mechanical** | | | |
| Tensile Strength at Yield | D638 | 6,800 psi | 47 MPa |
| Elongation at Break | D638 | 40% | 40% |
| Flexural Strength | D790 | 9,900 psi | 68 MPa |
| Flexural Modulus | D790 | 328,000 psi | 2,260 MPa |
| Compressive Strength | D695 | | |
| Shear Strength | D732 | | |
| Izod Impact Strength | D256 | | |
| Notched, 1/8″ thick (3.2mm) | | 5.8 ft lb/in | 310 J/m |
| Notched @ −40°C | | | |
| Unnotched, 1/8″ thick (3.2mm) | | | |
| Gardner Impact, 1/8″ thick (3.2mm) | | 200 in-lbs | 23 N-m |
| Instrumented Impact Test | D3763-85 | | |
| Rockwell Hardness R-scale | D785 | | |
| **Thermal** | | | |
| Heat Deflection Temp. | D648 | | |
| 66 psi (0.46 MPa) | | | |
| 264 psi (1.82 MPa) | | 205°F | 96°C |
| Coefficient of Thermal Expansion (linear) | E831 | | |
| Mold Direction | | | |
| Range: −40 to 100°F (−40 to 40°C) | | | |
| Range: 240 to 280°F (80 to 140°C) | | | |
| **UL Properties** | | | |
| Flammability | UL94 | VO@0.063″ | VO@1.6mm |

# General Electric Company

Contact Address:

General Electric Company
Plastics Group
One Plastics Avenue
Pittsfield, MA 01201

Phone: 1 (800) 845-0600

## NORYL PLUS UNV 580

Product Data:

- Enhanced Color Stability
- Flame Retardant
- Excellent Processability

Technical Data:

**NORYL PLUS UNV 580—Preliminary Property Profile**

| Property | ASTM Method | System | |
|---|---|---|---|
| | | English Units | SI Units |
| **Physical** | | | |
| Specific Gravity | D792 | 1.14 | 1.14 |
| Specific Volume | D892 | | |
| Water Absorption—24 hours | D570 | 0.08% | 0.08% |
| Mold Shrinkage<br>Flow Direction and<br>Cross Flow Direction<br>(125 mil) (3.175mm) | D955-51 | | |

*(continued)*

NOTE—Information provided in this book has been condensed. The publisher recommends that the product manufacturer be contacted, at the address or phone number provided, for more complete information on product safety, processing parameters, and applications not provided in this information sheet.

NORYL PLUS UNV 580—Preliminary Property Profile *(continued)*

| Property | ASTM Method | System | |
|---|---|---|---|
| | | English Units | SI Units |
| **Mechanical** | | | |
| Tensile Strength at Yield | D638 | 5,800 psi | 40 MPa |
| Elongation at Break | D638 | 25% | 25% |
| Flexural Strength | D790 | 10,900 psi | 75 MPa |
| Flexural Modulus | D790 | 406,000 psi | 2,800 MPa |
| Compressive Strength | D695 | | |
| Shear Strength | D732 | | |
| Izod Impact Strength<br>Notched, 1/8″ thick (3.2mm)<br>Notched @ −40°C<br>Unnotched, 1/8″ thick (3.2mm) | D256 | | |
| Gardner Impact, 1/8″ thick (3.2mm) | | | |
| Instrumented Impact Test | D3763-85 | | |
| Rockwell Hardness<br>R-scale | D785 | L-93 | L-93 |
| **Thermal** | | | |
| Heat Deflection Temp.<br>66 psi (0.46 MPa)<br>264 psi (1.82 MPa) | D648 | <br><br>185°F | <br><br>86°C |
| Coefficient of Thermal Expansion<br>(linear) | E831 | | |
| Mold Direction<br>Range: −40 to 100°F (−40 to 40°C)<br>Range: 240 to 280°F (80 to 140°C) | | $3.9 \times 10^{-5}$ in/in/°F | $7 \times 10^{-5}$ m/m/°C |
| **UL Properties** | | | |
| Flammability | UL94 | | |

# General Electric Company

## Contact Address:

General Electric Company
Plastics Group
One Plastics Avenue
Pittsfield, MA 01201

Phone: 1 (800) 845-0600

## VALOX 815

### Product Information:

VALOX is a PET/PBT thermoplastic blend available unreinforced or with glass/mineral reinforcement. It has been modified to improve impact resistance. It can be foamed. Flame retardant grades are available, and the resin can be purchased in a wide range of colors. It can be extruded or injection molded. VALOX has excellent chemical resistance, good electrical properties, and a natural gloss. The inherent lubricity of VALOX makes it a candidate for mechanical component applications.

### Product Data:

**VALOX 815—Typical Property Values**

| Property | Units (SI Units) | ASTM Method | 15% Glass Reinforced PBT Alloy | |
|---|---|---|---|---|
| | | | English Units | SI Units |
| **Physical** | | | | |
| Specific Gravity | | D792 | 1.43 | |
| Specific Volume | in³/lb(cm³/kg) | D892 | 19.5 | 701 |
| Water Absorption 24 hours | % | | .06 | |
| Mold Shrinkage | in/in × $10^{-3}$ (mm/mm × $10^{-3}$) | | | |
| Flow Direction | | | | |
| 30–90 mil (0.76–2.3mm) | | | 5–7 | |
| 90–180 mil (2.3–4.6mm) | | | 7–9 | |
| Cross Flow Direction | | | | |
| 30–90 mil (0.76–2.3mm) | | | 7–8 | |
| 90–180 mil (2.3–4.6mm) | | | 8–11 | |
| **Mechanical** | | | | |
| Tensile Strength | psi (MPa) | D638 | 12,000 | 83 |
| Elongation at Break | % | D638* | 3 | |
| Flexural Strength | psi (MPa) | D790 | 20,000 | 140 |
| Flexural Modulus | psi (MPa) | D790 | 650,000 | 4,500 |

NOTE – Information provided in this book has been condensed. The publisher recommends that the product manufacturer be contacted, at the address or phone number provided, for more complete information on product safety, processing parameters, and applications not provided in this information sheet.

**VALOX 815—Typical Property Values *(continued)***

| Property | Units (SI Units) | ASTM Method | 15% Glass Reinforced PBT Alloy | |
|---|---|---|---|---|
| | | | English Units | SI Units |
| **Mechanical** | | | | |
| Compressive Strength | psi (MPa) | D695 | 21,800 | 150 |
| Shear Strength | psi (MPa) | D732 | 8,900 | 61 |
| Izod Impact Strength | | D256 | | |
| Notched, 1/8″ thick (3.2mm) | ft-lb/in (J/m) | | 0.7 | 37 |
| Unnotched, 1/8″ thick (3.2mm) | ft-lb/in (J/m) | | 5 | 270 |
| Rockwell Hardness R-scale | | D785 | 119 | |
| **Thermal** | | | | |
| Heat Deflection Temp. | | D648 | | |
| @ 66 psi (0.46 MPa) | °F(°C) | | 410 | 210 |
| @ 264 psi (1.82 MPa) | °F(°C) | | 320 | 160 |
| Coeff. of Thermal Expansion | in/in/°F | D696 | | |
| Mold Direction × $10^{-5}$ | (m/m/°C) | | | |
| Range: −40–100°F (−40–40°C) | | | 2.5 | 4.5 |
| Range: 140–280°F (60–140°C) | | | 3.0 | 5.4 |
| **Electrical** | | | | |
| Dielectric Strength | | | | |
| 1/16″ (1.6mm) | V/mil(kV/mm) | D149 | 600 | 24 |
| 1/8″ (3.2mm) | V/mil(kV/mm) | | 560 | 22 |
| Dielectric Constant | | | | |
| 100 Hz | | D150 | 3.6 | |
| $10^6$ Hz | | | 3.5 | |
| Dissipation Factor | | | | |
| 100 Hz | | D150 | .002 | |
| 10 $Hz^6$ | ohm-cm × $10^{16}$ | | .02 | |
| Volume Resistivity | (ohm-m × $10^{14}$) | D257 | 5.6 | |
| **UL** | | | | |
| Flammability** | | UL 94 | HB/.063″ | |
| Arc Resistance | sec | D495 | 136/.125″ | |
| High Voltage Arc Tracking Rate | in/min | UL 746A | .8/.125″ | |
| High Ampere Arc Ignition | arcs | UL 746A | 44/.125″ | |
| Hot Wire Ignition | sec | UL 746A | 50/.125″ | |
| Comparative Track Index (CTI) | volts | UL 746A | 275/.125″ | |
| UL Temp. Index | | UL 746B | | |
| Elec. Properties | °C | | 125/.063″ | |
| Mech. Properties with impact | °C | | 110/.063″ | |
| Mech. Properties without impact | °C | | 125/.063″ | |

*ASTM D638 type V @ 0.5°/min.
**This rating is not intended to reflect hazards presented by this or any other material under actual fire conditions.

# General Electric Company

Contact Address:

General Electric Company
Plastics Group
One Plastics Avenue
Pittsfield, MA 01201

Phone: 1 (800) 845-0600

## VALOX 830

Product Information:

VALOX is a PET/PBT thermoplastic blend available unreinforced or with glass/mineral reinforcement. It has been modified to improve impact resistance. It can be foamed. Flame retardant grades are available, and the resin can be purchased in a wide range of colors. It can be extruded or injection molded. VALOX has excellent chemical resistance, good electrical properties, and a natural gloss. The inherent lubricity of VALOX makes it a candidate for mechanical component applications.

Product Data:

**VALOX 830—Typical Property Values**

| Property | Units (SI Units) | ASTM Method | 30% Glass Reinforced PBT Alloy | |
|---|---|---|---|---|
| | | | English Units | SI Units |
| **Physical** | | | | |
| Specific Gravity | | D792 | 1.54 | |
| Specific Volume | $in^3/lb(cm^3/kg)$ | D892 | 18.6 | 669 |
| Water Absorption 24 hours | % | | .06 | |
| Mold Shrinkage | in/in × $10^{-3}$ (mm/mm × $10^{-3}$) | | | |
| Flow Direction | | | | |
| 30–90 mil (0.76–2.3mm) | | | 3–5 | |
| 90–180 mil (2.3–4.6mm) | | | 5–7 | |
| Cross Flow Direction | | | | |
| 30–90 mil (0.76–2.3mm) | | | 6–7 | |
| 90–180 mil (2.3–4.6mm) | | | 7–9 | |
| **Mechanical** | | | | |
| Tensile Strength | psi (MPa) | D638 | 15,000 | 100 |
| Elongation at Break | % | D638* | 3 | |
| Flexural Strength | psi (MPa) | D790 | 25,000 | 170 |
| Flexural Modulus | psi (MPa) | D790 | 1,000,000 | 6,900 |

NOTE—Information provided in this book has been condensed. The publisher recommends that the product manufacturer be contacted, at the address or phone number provided, for more complete information on product safety, processing parameters, and applications not provided in this information sheet.

VALOX 830—Typical Property Values *(continued)*

| Property | Units (SI Units) | ASTM Method | 30% Glass Reinforced PBT Alloy | |
|---|---|---|---|---|
| | | | English Units | SI Units |
| **Mechanical** | | | | |
| Compressive Strength | psi (MPa) | D695 | 21,800 | 150 |
| Shear Strength | psi (MPa) | D732 | 8,900 | 61 |
| Izod Impact Strength | | D256 | | |
| Notched, 1/8″ thick (3.2mm) | ft-lb/in (J/m) | | 1.5 | 80 |
| Unnotched, 1/8″ thick (3.2mm) | ft-lb/in (J/m) | | 12 | 640 |
| Rockwell Hardness R-scale | | D785 | 119 | |
| **Thermal** | | | | |
| Heat Deflection Temp. | | D648 | | |
| @ 66 psi (0.46 MPa) | °F(°C) | | 430 | 221 |
| @ 264 psi (1.82 MPa) | °F(°C) | | 380 | 193 |
| Coeff. of Thermal Expansion | in/in/°F | D696 | | |
| Mold Direction × $10^{-5}$ | (m/m/°C) | | | |
| Range: −40–100°F (−40–40°C) | | | 1.4 | 2.5 |
| Range: 140–280°F (60–140°C) | | | 1.4 | 2.5 |
| **Electrical** | | | | |
| Dielectric Strength | | | | |
| 1/16″ (1.6mm) | V/mil(kV/mm) | D149 | 630 | 25 |
| 1/8″ (3.2mm) | V/mil(kV/mm) | | 530 | 21 |
| Dielectric Constant | | | | |
| 100 Hz | | D150 | 3.6 | |
| $10^6$ Hz | | | 3.5 | |
| Dissipation Factor | | | | |
| 100 Hz | | D150 | .002 | |
| 10 $Hz^6$ | ohm-cm × $10^{16}$ | | .02 | |
| Volume Resistivity | (ohm-m × $10^{14}$) | D257 | 4.0 | |
| **UL** | | | | |
| Flammability** | | UL 94 | HB/.063″ | |
| Arc Resistance | sec | D495 | 84/.125″ | |
| High Voltage Arc Tracking Rate | in/min | UL 746A | 1.5/.125″ | |
| High Ampere Arc Ignition | arcs | UL 746A | 18/.125″ | |
| Hot Wire Ignition | sec | UL 746A | 139/.125″ | |
| Comparative Track Index (CTI) | volts | UL 746A | 260/.125″ | |
| UL Temp. Index | | UL 746B | | |
| Elec. Properties | °C | | 120/.063″ | |
| Mech. Properties with impact | °C | | 75/.063″ | |
| Mech. Properties without impact | °C | | 120/.063″ | |

*ASTM D638 type V @ 0.5°/min.
**This rating is not intended to reflect hazards presented by this or any other material under actual fire conditions.

# General Electric Company

## Contact Address:

General Electric Company
Plastics Group
One Plastics Avenue
Pittsfield, MA 01201

Phone: 1 (800) 845-0600

# VALOX 850

## Product Information:

VALOX is a PET/PBT thermoplastic blend available unreinforced or with glass/mineral reinforcement. It has been modified to improve impact resistance. It can be foamed. Flame retardant grades are available, and the resin can be purchased in a wide range of colors. It can be extruded or injection molded. VALOX has excellent chemical resistance, good electrical properties, and a natural gloss. The inherent lubricity of VALOX makes it a candidate for mechanical component applications.

## Product Data:

**VALOX 850—Typical Property Values**

| Property | Units (SI Units) | ASTM Method | 15% Glass Reinforced PBT Alloy | |
|---|---|---|---|---|
| | | | English Units | SI Units |
| **Physical** | | | | |
| Specific Gravity | | D792 | 1.51 | |
| Specific Volume | in³/lb(cm³/kg) | D892 | 18.3 | 659 |
| Water Absorption 24 hours | % | | .05 | |
| Mold Shrinkage | in/in × $10^{-3}$ (mm/mm × $10^{-3}$) | | | |
| Flow Direction | | | | |
| 30–90 mil (0.76–2.3mm) | | | 5–7 | |
| 90–180 mil (2.3–4.6mm) | | | 7–9 | |
| Cross Flow Direction | | | | |
| 30–90 mil (0.76–2.3mm) | | | 7–8 | |
| 90–180 mil (2.3–4.6mm) | | | 8–11 | |
| **Mechanical** | | | | |
| Tensile Strength | psi (MPa) | D638 | 13,000 | 90 |
| Elongation at Break | % | D638* | 5 | |
| Flexural Strength | psi (MPa) | D790 | 20,000 | 140 |
| Flexural Modulus | psi (MPa) | D790 | 750,000 | 5,200 |

NOTE – Information provided in this book has been condensed. The publisher recommends that the product manufacturer be contacted, at the address or phone number provided, for more complete information on product safety, processing parameters, and applications not provided in this information sheet.

## VALOX 850—Typical Property Values *(continued)*

| Property | Units (SI Units) | ASTM Method | 15% Glass Reinforced PBT Alloy | |
|---|---|---|---|---|
| | | | English Units | SI Units |
| **Physical** | | | | |
| Compressive Strength | psi (MPa) | D695 | 14,970 | 100 |
| Shear Strength | psi (MPa) | D732 | 7,510 | 52 |
| Izod Impact Strength | | D256 | | |
| Notched, 1/8″ thick (3.2mm) | ft-lb/in (J/m) | | 0.8 | 43 |
| Unnotched, 1/8″ thick (3.2mm) | ft-lb/in (J/m) | | 5 | 270 |
| Rockwell Hardness R-scale | | D785 | 121 | |
| **Thermal** | | | | |
| Heat Deflection Temp. | | D648 | | |
| @ 66 psi (0.46 MPa) | °F(°C) | | 400 | 204 |
| @ 264 psi (1.82 MPa) | °F(°C) | | 370 | 187 |
| Coeff. of Thermal Expansion | in/in/°F (m/m/°C) | D696 | | |
| Mold Direction × $10^{-5}$ | | | | |
| Range: −40–100°F (−40–40°C) | | | 2.3 | 4.1 |
| Range: 140–280°F (60–140°C) | | | 2.3 | 4.1 |
| **Electrical** | | | | |
| Dielectric Strength | | | | |
| 1/16″ (1.6mm) | V/mil(kV/mm) | D149 | 580 | 23 |
| 1/8″ (3.2mm) | V/mil(kV/mm) | | 520 | 20 |
| Dielectric Constant | | | | |
| 100 Hz | | D150 | 3.5 | |
| $10^6$ Hz | | | 3.4 | |
| Dissipation Factor | | | | |
| 100 Hz | | D150 | .001 | |
| $10^6$ Hz | ohm-cm × $10^{16}$ | | .01 | |
| Volume Resistivity | (ohm-m × $10^{14}$) | D257 | 3.9 | |
| **UL** | | | | |
| Flammability** | | UL 94 | V-2/.062″ | |
| Arc Resistance | sec | D495 | 81/.125″ | |
| High Voltage Arc Tracking Rate | in/min | UL 746A | 6.9/.125″ | |
| High Ampere Arc Ignition | arcs | UL 746A | 14/.125″ | |
| Hot Wire Ignition | sec | UL 746A | 56/.125″ | |
| Comparative Track Index (CTI) | volts | UL 746A | 225/.125″ | |
| UL Temp. Index | | UL 746B | | |
| Elec. Properties | °C | | 125/.062″ | |
| Mech. Properties with impact | °C | | 110/.062″ | |
| Mech. Properties without impact | °C | | 125/.062″ | |

*ASTM D638 type V @ 0.5°/min.
**This rating is not intended to reflect hazards presented by this or any other material under actual fire conditions.

# General Electric Company

## Contact Address:

General Electric Company
Plastics Group
One Plastics Avenue
Pittsfield, MA 01201

Phone: 1 (800) 845-0600

# VALOX 855

## Product Information:

VALOX is a PET/PBT thermoplastic blend available unreinforced or with glass/mineral reinforcement. It has been modified to improve impact resistance. It can be foamed. Flame retardant grades are available, and the resin can be purchased in a wide range of colors. It can be extruded or injection molded. VALOX has excellent chemical resistance, good electrical properties, and a natural gloss. The inherent lubricity of VALOX makes it a candidate for mechanical component applications.

## Product Data:

**VALOX 855—Typical Property Values**

| Property | Units (SI Units) | ASTM Method | 15% Glass Reinforced PBT Alloy English Units | 15% Glass Reinforced PBT Alloy SI Units |
|---|---|---|---|---|
| **Physical** | | | | |
| Specific Gravity | | D792 | 1.54 | |
| Specific Volume | $in^3/lb(cm^3/kg)$ | D892 | 18.0 | 649 |
| Water Absorption 24 hours | % | | .06 | |
| Mold Shrinkage | $in/in \times 10^{-3}$ ($mm/mm \times 10^{-3}$) | | | |
| Flow Direction | | | | |
| 30–90 mil (0.76–2.3mm) | | | 5–7 | |
| 90–180 mil (2.3–4.6mm) | | | 7–9 | |
| Cross Flow Direction | | | | |
| 30–90 mil (0.76–2.3mm) | | | 7–8 | |
| 90–180 mil (2.3–4.6mm) | | | 8–11 | |
| **Mechanical** | | | | |
| Tensile Strength | psi (MPa) | D638 | 14,000 | 96 |
| Elongation at Break | % | D638* | 5 | |
| Flexural Strength | psi (MPa) | D790 | 22,000 | 150 |
| Flexural Modulus | psi (MPa) | D790 | 750,000 | 5,200 |

*ASTM D638 type V @ 0.5°/min.

NOTE – Information provided in this book has been condensed. The publisher recommends that the product manufacturer be contacted, at the address or phone number provided, for more complete information on product safety, processing parameters, and applications not provided in this information sheet.

**VALOX 855—Typical Property Values *(continued)***

| Property | Units (SI Units) | ASTM Method | 15% Glass Reinforced PBT Alloy | |
|---|---|---|---|---|
| | | | English Units | SI Units |
| **Mechanical** | | | | |
| Compressive Strength | psi (MPa) | D695 | 15,970 | 110 |
| Shear Strength | psi (MPa) | D732 | 7,390 | 51 |
| Izod Impact Strength | | D256 | | |
| Notched, 1/8″ thick (3.2mm) | ft-lb/in (J/m) | | 1.0 | 53 |
| Unnotched, 1/8″ thick (3.2mm) | ft-lb/in (J/m) | | 6 | 320 |
| Rockwell Hardness R-scale | | D785 | 119 | |
| **Thermal** | | | | |
| Heat Deflection Temp. | | D648 | | |
| @ 66 psi (0.46 MPa) | °F(°C) | | 400 | 204 |
| @ 264 psi (1.82 MPa) | °F(°C) | | 370 | 187 |
| Coeff. of Thermal Expansion | in/in/°F | D696 | | |
| Mold Direction × $10^{-5}$ | (m/m/°C) | | | |
| Range: −40–100°F (−40–40°C) | | | 2.5 | 4.5 |
| Range: 140–280°F (60–140°C) | | | 3.4 | 6.1 |
| **Electrical** | | | | |
| Dielectric Strength | | | | |
| 1/16″ (1.6mm) | V/mil(kV/mm) | D149 | 570 | 22 |
| 1/8″ (3.2mm) | V/mil(kV/mm) | | 510 | 20 |
| Dielectric Constant | | | | |
| 100 Hz | | D150 | 3.5 | |
| 10 Hz$^6$ Hz | | | 3.4 | |
| Dissipation Factor | | | | |
| 100 Hz | | D150 | .001 | |
| $10^6$ | ohm-cm × $10^{16}$ | | .01 | |
| Volume Resistivity | (ohm-m × $10^{14}$) | D257 | 4.5 | |
| **UL** | | | | |
| Flammability** | | UL 94 | V-0/.062″ | |
| Arc Resistance | sec | D495 | 68/.127″ | |
| High Voltage Arc Tracking Rate | in/min | UL 746A | 13.4/.127″ | |
| High Ampere Arc Ignition | arcs | UL 746A | 13/.127″ | |
| Hot Wire Ignition | sec | UL 746A | 33/.062″ | |
| Comparative Track Index (CTI) | volts | UL 746A | 205/.127″ | |
| UL Temp. Index | | UL 746B | | |
| Elec. Properties | °C | | 125/.062″ | |
| Mech. Properties with impact | °C | | 110/.062″ | |
| Mech. Properties without impact | °C | | 125/.062″ | |

**This rating is not intended to reflect hazards presented by this or any other material under actual fire conditions.

# General Electric Company

## Contact Address:

General Electric Company
Plastics Group
One Plastics Avenue
Pittsfield, MA 01201

Phone: 1 (800) 845-0600

# VALOX 865

## Product Information:

VALOX is a PET/PBT thermoplastic blend available unreinforced or with glass/mineral reinforcement. It has been modified to improve impact resistance. It can be foamed. Flame retardant grades are available, and the resin can be purchased in a wide range of colors. It can be extruded or injection molded. VALOX has excellent chemical resistance, good electrical properties, and a natural gloss. The inherent lubricity of VALOX makes it a candidate for mechanical component applications.

## Product Data:

**VALOX 865—Typical Property Values**

| Property | Units (SI Units) | ASTM Method | 30% Glass Reinforced PBT Alloy | |
|---|---|---|---|---|
| | | | English Units | SI Units |
| **Physical** | | | | |
| Specific Gravity | | D792 | 1.66 | |
| Specific Volume | $in^3/lb(cm^3/kg)$ | D892 | 16.8 | 607 |
| Water Absorption 24 hours | % | | .03 | |
| Mold Shrinkage | in/in × $10^{-3}$ (mm/mm × $10^{-3}$) | | | |
| Flow Direction | | | | |
| 30–90 mil (0.76–2.3mm) | | | 3–5 | |
| 90–180 mil (2.3–4.6mm) | | | 5–7 | |
| Cross Flow Direction | | | | |
| 30–90 mil (0.76–2.3mm) | | | 6–7 | |
| 90–180 mil (2.3–4.6mm) | | | 7–9 | |

*(continued)*

NOTE – Information provided in this book has been condensed. The publisher recommends that the product manufacturer be contacted, at the address or phone number provided, for more complete information on product safety, processing parameters, and applications not provided in this information sheet.

**VALOX 865—Typical Property Values *(continued)***

| Property | Units (SI Units) | ASTM Method | 30% Glass Reinforced PBT Alloy | |
|---|---|---|---|---|
| | | | English Units | SI Units |
| **Mechanical** | | | | |
| Tensile Strength | psi (MPa) | D638 | 17,000 | 117 |
| Elongation at Break | % | D638* | — | |
| Flexural Strength | psi (MPa) | D790 | 26,000 | 180 |
| Flexural Modulus | psi (MPa) | D790 | 1,100,000 | 7,600 |
| Compressive Strength | psi (MPa) | D695 | 19,600 | 135 |
| Shear Strength | psi (MPa) | D732 | 9,000 | 60 |
| Izod Impact Strength<br>Notched, 1/8″ thick (3.2mm)<br>Unnotched, 1/8″ thick (3.2mm) | <br>ft-lb/in (J/m)<br>ft-lb/in (J/m) | D256 | <br>1.4<br>12 | <br>75<br>640 |
| Rockwell Hardness R-scale | | D785 | 119 | |
| **Thermal** | | | | |
| Heat Deflection Temp.<br>@ 66 psi (0.46 MPa)<br>@ 264 psi (1.82 MPa) | <br>°F(°C)<br>°F(°C) | D648 | <br>415<br>380 | <br>212<br>192 |
| Coeff. of Thermal Expansion<br>Mold Direction $\times 10^{-5}$<br>Range: −40–100°F (−40–40°C)<br>Range: 140–280°F (60–140°C) | in/in/°F<br>(m/m/°C) | D696 | <br><br>1.2<br>1.2 | <br><br>2.2<br>2.2 |
| **Electrical** | | | | |
| Dielectric Strength<br>1/16″ (1.6mm)<br>1/8″ (3.2mm) | <br>V/mil(kV/mm)<br>V/mil(kV/mm) | D149 | <br>590<br>480 | <br>23<br>20 |
| Dielectric Constant<br>100 Hz<br>$10^6$ Hz | | D150 | <br>3.8<br>3.7 | |
| Dissipation Factor<br>100 Hz<br>$10^6$ Hz | <br><br>ohm-cm $\times 10^{16}$ | D150 | <br>.002<br>.01 | |
| Volume Resistivity | (ohm-m $\times 10^{14}$) | D257 | 1.8 | |

*ASTM D638 type V @ 0.5°/min.

*(continued)*

**VALOX 865—Typical Property Values *(continued)***

| Property | Units (SI Units) | ASTM Method | 30% Glass Reinforced PBT Alloy | |
|---|---|---|---|---|
| | | | English Units | SI Units |
| **UL** | | | | |
| Flammability** | | UL 94 | 5V/.090″ | |
| Arc Resistance | sec | D495 | 86/.124″ | |
| High Voltage Arc Tracking Rate | in/min | UL 746A | 11.4/.124″ | |
| High Ampere Arc Ignition | arcs | UL 746A | 16/.124″ | |
| Hot Wire Ignition | sec | UL 746A | 24/.062″ | |
| Comparative Track Index (CTI) | volts | UL 746A | 225/.124″ | |
| UL Temp. Index | | UL 746B | | |
| Elec. Properties | °C | | 110/.062″ | |
| Mech. Properties with impact | °C | | 105/.062″ | |
| Mech. Properties without impact | °C | | 110/.062″ | |

**This rating is not intended to reflect hazards presented by this or any other material under actual fire conditions.

# General Electric Company

Contact Address:

General Electric Company
Plastics Group
One Plastics Avenue
Pittsfield, MA 01201

Phone: 1 (800) 845-0600

## XENLOY 1100

Product Information:

XENLOY is a polycarbonate based composite polymer alloy, also available unreinforced or with glass reinforcement, formulated to improve impact resistance. XENLOY resins can be injection molded, extruded, blow molded, or thermoformed. XENLOY has excellent physical strength, and is resistant to impact. Its excellent melt flow properties allows it to be processed over a wide range of temperature and pressure. XENLOY has excellent chemical resistance, UV-stability, dimensional stability, low mold shrinkage, and low water absorbtion.

Product Data:

**XENLOY 1100—Typical Property Values**

| Property | ASTM Method | English Units | SI Units |
|---|---|---|---|
| **Mechanical** | | | |
| Izod Impact Strength | D256 | | |
| Notched | | | |
| @ 73°F ft-lbs/in (23°C J/m) | | 14.8 | 790 |
| @ 32°F ft-lbs/in (0°C J/m) | | 12.2 | 700 |
| @ −4°F ft-lbs/in (−20°C J/m) | | 4.7 | 250 |
| @ −40°F ft-lbs/in (−40°C J/m) | | 3.0 | 160 |
| Tensile Strength | D638 | | |
| Yield psi (MPa) | | 7400 | 51 |
| Break psi (MPa) | | 8100 | 56 |
| Elongation @ Break % (%) | | 145 | 145 |
| Flexural Strength psi (MPa) | D790 | 12,300 | 85 |
| Flexural Modulus psi (MPa) | | 297,000 | 2050 |
| **Thermal** | | | |
| Deflection Temperature under Load | D648 | | |
| @ 66 psi °F (°C) | | 239 | 115 |
| @ 264 psi °F (°C) | | 203 | 95 |

*(continued)*

NOTE—Information provided in this book has been condensed. The publisher recommends that the product manufacturer be contacted, at the address or phone number provided, for more complete information on product safety, processing parameters, and applications not provided in this information sheet.

**XENLOY 1100—Typical Property Values *(continued)***

| Property | ASTM Method | English Units | SI Units |
|---|---|---|---|
| **Other** | | | |
| Coefficient of Thermal Expansion in/in°F (mm/mm/°C) | D696 | $5.3 \times 10^{-3}$ | $9.5 \times 10^{-3}$ |
| Mold Shrinkage % (%) | D955 | 0.9 | 0.9 |
| Specific Gravity | D792 | 1.22 | |

# General Electric Company

Contact Address:

General Electric Company
Plastics Group
One Plastics Avenue
Pittsfield, MA 01201

Phone: 1 (800) 845-0600

## XENLOY 1101

Product Information:

XENLOY is a polycarbonate based composite polymer alloy, also available unreinforced or with glass reinforcement, formulated to improve impact resistance. XENLOY resins can be injection molded, extruded, blow molded, or thermoformed. XENLOY has excellent physical strength, and is resistant to impact. Its excellent melt flow properties allows it to be processed over a wide range of temperature and pressure. XENLOY has excellent chemical resistance, UV-stability, dimensional stability, low mold shrinkage, and low water absorbtion.

Product Data:

**XENLOY 1101—Typical Property Values**

| Property | ASTM Method | English Units | SI Units |
|---|---|---|---|
| **Mechanical** | | | |
| Izod Impact Strength | D256 | | |
| Notched | | | |
| @ 73°F ft-lbs/in (23°C J/m) | | 13.3 | 710 |
| @ 32°F ft-lbs/in (0°C J/m) | | 12.7 | 680 |
| @ −4°F ft-lbs/in (−20°C J/m) | | 11.4 | 610 |
| @ −40°F ft-lbs/in (−40°C J/m) | | 5.6 | 300 |
| Tensile Strength | D638 | | |
| Yield psi (MPa) | | 7680 | 53 |
| Break psi (MPa) | | 8500 | 59 |
| Elongation @ Break % (%) | | 120 | 120 |
| Flexural Strength psi (MPa) | D790 | 12,300 | 85 |
| Flexural Modulus psi (MPa) | | 296,000 | 2040 |
| **Thermal** | | | |
| Deflection Temperature under Load | D648 | | |
| @ 66 psi °F (°C) | | 223 | 106 |
| @ 264 psi °F (°C) | | 210 | 99 |

*(continued)*

NOTE—Information provided in this book has been condensed. The publisher recommends that the product manufacturer be contacted, at the address or phone number provided, for more complete information on product safety, processing parameters, and applications not provided in this information sheet.

XENLOY 1101—Typical Property Values *(continued)*

| Property | ASTM Method | English Units | SI Units |
|---|---|---|---|
| **Other** | | | |
| Coefficient of Thermal Expansion in/in°F (mm/mm/°C) | D696 | $5.0 \times 10^{-3}$ | $9.0 \times 10^{-3}$ |
| Mold Shrinkage % (%) | D955 | 0.9 | 0.9 |
| Specific Gravity | D792 | 1.21 | |

# General Electric Company

## Contact Address:

General Electric Company
Plastics Group
One Plastics Avenue
Pittsfield, MA 01201

Phone: 1 (800) 845-0600

---

## XENLOY 1200

### Product Information:

XENLOY is a polycarbonate based composite polymer alloy, also available unreinforced or with glass reinforcement, formulated to improve impact resistance. XENLOY resins can be injection molded, extruded, blow molded, or thermoformed. XENLOY has excellent physical strength, and is resistant to impact. Its excellent melt flow properties allows it to be processed over a wide range of temperature and pressure. XENLOY has excellent chemical resistance, UV-stability, dimensional stability, low mold shrinkage, and low water absorbtion.

### Product Data:

**XENLOY 1200—Typical Property Values**

| Property | ASTM Method | English Units | SI Units |
|---|---|---|---|
| **Mechanical** | | | |
| Izod Impact Strength | D256 | | |
| Notched | | | |
| @ 73°F ft-lbs/in (23°C J/m) | | 16.0 | 855 |
| @ 32°F ft-lbs/in (0°C J/m) | | 15.0 | 800 |
| @ −4°F ft-lbs/in (−20°C J/m) | | 15.0 | 800 |
| @ −40°F ft-lbs/in (−40°C J/m) | | 12.0 | 640 |
| Tensile Strength | D638 | | |
| Yield psi (MPa) | | 5000 | 34 |
| Break psi (MPa) | | 4500 | 31 |
| Elongation @ Break % (%) | | 100 | 100 |
| Flexural Strength psi (MPa) | D790 | 8500 | 58 |
| Flexural Modulus psi (MPa) | | 240,000 | 1654 |
| **Thermal** | | | |
| Deflection Temperature under Load | D648 | | |
| @ 66 psi °F (°C) | | 210 | 99 |
| @ 264 psi °F (°C) | | 140 | 60 |

*(continued)*

NOTE – Information provided in this book has been condensed. The publisher recommends that the product manufacturer be contacted, at the address or phone number provided, for more complete information on product safety, processing parameters, and applications not provided in this information sheet.

**XENLOY 1200—Typical Property Values *(continued)***

| Property | ASTM Method | English Units | SI Units |
|---|---|---|---|
| **Other** | | | |
| Coefficient of Thermal Expansion in/in°F (mm/mm/°C) | D696 | $5.5 \times 10^{-3}$ | $9.9 \times 10^{-3}$ |
| Mold Shrinkage % (%) | D955 | 1.7 | 1.7 |
| Specific Gravity | D792 | 1.20 | |

# General Electric Company

## Contact Address:

General Electric Company
Plastics Group
One Plastics Avenue
Pittsfield, MA 01201

Phone: 1 (800) 845-0600

## XENLOY 2300

### Product Information:

XENLOY is a polycarbonate based composite polymer alloy, also available unreinforced or with glass reinforcement, formulated to improve impact resistance. XENLOY resins can be injection molded, extruded, blow molded, or thermoformed. XENLOY has excellent physical strength, and is resistant to impact. Its excellent melt flow properties allows it to be processed over a wide range of temperature and pressure. XENLOY has excellent chemical resistance, UV-stability, dimensional stability, low mold shrinkage, and low water absorbtion.

### Product Data:

**XENLOY 2300—Typical Property Values**

| Property | ASTM Method | English Units | SI Units |
|---|---|---|---|
| **Physical** | | | |
| Specific Gravity | D792 | 1.22 | 1.22 |
| Water Absorption | D570 | | |
| 24 hrs @ 73°F % (%) | | 0.14 | 0.14 |
| Equilibrium @ 73°F % (%) | | 0.60 | 0.60 |
| Mold Shrinkage/Flow | D955 | | |
| (0.125 mil) mil/in (%) | | 5–7 | 5–7 |
| **Optical** | | | |
| Transmittance % (%) | D1003 | | |
| Haze % (%) | D1003 | | |
| **Mechanical** | | | |
| Tensile Strength | D638 | | |
| Yield psi (MPa) | | 8000 | 55 |
| Break psi (MPa) | | 8200 | 56 |
| Tensile Elongation | D638 | | |
| Break % (%) | | 120 | 120 |
| Flexural Strength | D790 | | |
| Yield psi (MPa) | | 12,500 | 86 |

NOTE – Information provided in this book has been condensed. The publisher recommends that the product manufacturer be contacted, at the address or phone number provided, for more complete information on product safety, processing parameters, and applications not provided in this information sheet.

**XENLOY 2300—Typical Property Values *(continued)***

| Property | ASTM Method | English Units | SI Units |
|---|---|---|---|
| **Mechanical** | | | |
| Flexural Modulus psi (MPa) | D790 | 325,000 | 2240 |
| Izod Impact Strength 1/8″<br>Notched<br>.010″ rad. @ 73°F ft-lb/in (J/m)<br>.020″ rad. @ 73°F ft-lb/in (J/m) | D256 | <br><br>15.0<br>20.0 | <br><br>800<br>1068 |
| Gardner Impact<br>73°F ft-lb (J) | D3029 | <br>>28 | <br>36 |
| Rockwell Hardness R Scale | D785 | 115 | 115 |
| **Thermal** | | | |
| Deflection Temperature under Load<br>@ 66 psi °F (°C)<br>@ 264 psi °F (°C) | D648 | <br>265<br>250 | <br>129<br>121 |
| Coefficient of Linear<br>Thermal Expansion<br>−22°F to 86°F in/in—°F × E5<br>(m/m—°C × E5) | E831 | <br><br>4.4 | <br><br>8.0 |
| **Electrical** | | | |
| Dielectric Strength<br>In oil @ 0.062 in V/mil (kV/mm) | D149 | <br>500 | <br>20 |
| Dielectric Constant<br>100 Hz | D150 | <br>2.96 | <br>2.96 |
| Dissipation Factor<br>100 Hz | D150 | <br>0.0017 | <br>0.0017 |
| Volume Resistivity<br>ohm-cm × E16 (ohm-cm × E16) | D257 | <br>1.40 | <br>1.40 |
| **UL** | | | |
| Flammability | UL 94 | | |
| Arc Resistance<br>Stainless stl. elec. sec. | D495 | | |
| High Voltage Arc<br>Tracking Rate in/min | UL 746A | | |
| High Ampere Arc Ign arcs | UL 746A | | |
| Hot Wire Ignition sec | UL 746A | | |
| Comparative Trk Index volts | UL 746A | | |
| UL Temperature Index<br>Elec Prop °C<br>Mech Prop w/impt °C<br>Mech Prop wo/impt °C | UL 746B | | |
| Oxygen Index % | D2863 | | |

# General Electric Company

## Contact Address:

General Electric Company
Plastics Group
One Plastics Avenue
Pittsfield, MA 01201

Phone: 1 (800) 845-0600

## XENLOY 2310

### Product Information:

XENLOY is a polycarbonate based composite polymer alloy, also available unreinforced or with glass reinforcement, formulated to improve impact resistance. XENLOY resins can be injection molded, extruded, blow molded, or thermoformed. XENLOY has excellent physical strength, and is resistant to impact. Its excellent melt flow properties allows it to be processed over a wide range of temperature and pressure. XENLOY has excellent chemical resistance, UV-stability, dimensional stability, low mold shrinkage, and low water absorbtion.

### Product Data:

**XENLOY 2310—Typical Property Values**

| Property | ASTM Method | English Units | SI Units |
|---|---|---|---|
| **Physical** | | | |
| Specific Gravity | D792 | 1.33 | 1.33 |
| Water Absorption | D570 | | |
| 24 hrs @ 73°F % (%) | | 0.14 | 0.14 |
| Equilibrium @ 73°F % (%) | | 0.60 | 0.60 |
| Mold Shrinkage/Flow | D955 | | |
| (0.125 mil) mil/in (%) | | 5–7 | 5–7 |
| **Optical** | | | |
| Transmittance % (%) | D1003 | | |
| Haze % (%) | D1003 | | |
| **Mechanical** | | | |
| Tensile Strength | D638 | | |
| Yield psi (MPa) | | 8000 | 55 |
| Break psi (MPa) | | 8200 | 56 |
| Tensile Elongation | D638 | | |
| Break % (%) | | 120 | 120 |
| Flexural Strength | D790 | | |
| Yield psi (MPa) | | 12,500 | 86 |

NOTE – Information provided in this book has been condensed. The publisher recommends that the product manufacturer be contacted, at the address or phone number provided, for more complete information on product safety, processing parameters, and applications not provided in this information sheet.

**XENLOY 2310—Typical Property Values *(continued)***

| Property | ASTM Method | English Units | SI Units |
|---|---|---|---|
| **Mechanical** | | | |
| Flexural Modulus psi (MPa) | D790 | 325,000 | 2240 |
| Izod Impact Strength 1/8″<br>Notched<br>.010″ rad. @ 73°F ft-lb/in (J/m)<br>.020″ rad. @ 73°F ft-lb/in (J/m) | D256 | <br><br>15.0<br>20.0 | <br><br>800<br>1068 |
| Gardner Impact<br>73°F ft-lb (J) | D3029 | <br>>28 | <br>36 |
| Rockwell Hardness R Scale | D785 | 115 | 115 |
| **Thermal** | | | |
| Deflection Temperature under Load<br>@ 66 psi °F (°C)<br>@ 264 psi °F (°C) | D648 | <br>265<br>250 | <br>129<br>121 |
| Coefficient of Linear<br>Thermal Expansion<br>−22°F to 86°F in/in—°F × E5<br>(m/m—°C × E5) | E831 | <br><br>4.4 | <br><br>8.0 |
| **Electrical** | | | |
| Dielectric Strength<br>In oil @ 0.062 in V/mil (kV/mm) | D149 | | |
| Dielectric Constant<br>100 Hz | D150 | | |
| Dissipation Factor<br>100 Hz | D150 | | |
| Volume Resistivity<br>ohm-cm × E16 (ohm-cm × E16) | D257 | | |
| **UL** | | | |
| Flammability | UL 94 | V-0 @ .062″ | |
| Arc Resistance<br>Stainless stl. elec. sec. | D495 | <br>75 @ .125″ | |
| High Voltage Arc<br>Tracking Rate in/min | UL 746A | <br>3.10 @ .125″ | |
| High Ampere Arc Ign arcs | UL 746A | 14 @ .062″ | |
| Hot Wire Ignition sec | UL 746A | 22 @ .125″ | |
| Comparative Trk Index volts | UL 746A | 165 @ .125″ | |
| UL Temperature Index<br>Elec Prop °C<br>Mech Prop w/impt °C<br>Mech Prop wo/impt °C | UL 746B | <br>65 @ .062″<br>65 @ .062″<br>65 @ .062″ | |
| Oxygen Index % | D2863 | | |

# General Electric Company

## Contact Address:

General Electric Company
Plastics Group
One Plastics Avenue
Pittsfield, MA 01201

Phone: 1 (800) 845-0600

## XENLOY 3500

### Product Information:

XENLOY is a polycarbonate based composite polymer alloy, also available unreinforced or with glass reinforcement, formulated to improve impact resistance. XENLOY resins can be injection molded, extruded, blow molded, or thermoformed. XENLOY has excellent physical strength, and is resistant to impact. Its excellent melt flow properties allows it to be processed over a wide range of temperature and pressure. XENLOY has excellent chemical resistance, UV-stability, dimensional stability, low mold shrinkage, and low water absorbtion.

### Product Data:

**XENLOY 3500—Typical Property Values**

| Property | ASTM Method | English Units | SI Units |
|---|---|---|---|
| **Physical** | | | |
| Specific Gravity | D792 | 1.20 | 1.20 |
| Water Absorption | D570 | | |
| 24 hrs @ 73°F % (%) | | 0.14 | 0.14 |
| Equilibrium @ 73°F % (%) | | 0.35 | 0.35 |
| Mold Shrinkage/Flow | D955 | | |
| (0.125 mil) mil/in (%) | | 5–7 | 5–7 |
| **Optical** | | | |
| Transmittance % (%) | D1003 | 4–5 | 4–5 |
| Haze % (%) | D1003 | 87–88 | 87–88 |
| **Mechanical** | | | |
| Tensile Strength | D638 | | |
| Yield psi (MPa) | | 8300 | 57 |
| Break psi (MPa) | | 8000 | 55 |
| Tensile Elongation | D638 | | |
| Break % (%) | | 125 | 125 |
| Flexural Strength | D790 | | |
| Yield psi (MPa) | | 12,000 | 83 |

NOTE – Information provided in this book has been condensed. The publisher recommends that the product manufacturer be contacted, at the address or phone number provided, for more complete information on product safety, processing parameters, and applications not provided in this information sheet.

**XENLOY 3500—Typical Property Values *(continued)***

| Property | ASTM Method | English Units | SI Units |
|---|---|---|---|
| **Mechanical** | | | |
| Flexural Modulus psi (MPa) | D790 | 280,000 | 1930 |
| Izod Impact Strength 1/8″<br>Notched<br>.010″ rad. @ 73°F ft-lb/in (J/m)<br>.020″ rad. @ 73°F ft-lb/in (J/m) | D256 | <br><br>2.0<br>20.0 | <br><br>107<br>1068 |
| Gardner Impact<br>73°F ft-lb (J) | D3029 | <br>>40 | <br>54 |
| Rockwell Hardness R Scale | D785 | 122 | 122 |
| **Thermal** | | | |
| Deflection Temperature under Load<br>@ 66 psi °F (°C)<br>@ 264 psi °F (°C) | D648 | <br>225<br>210 | <br>107<br>99 |
| Coefficient of Linear<br>Thermal Expansion<br>−22°F to 86°F in/in—°F × E5<br>(m/m—°C × E5) | E831 | <br><br>3.9 | <br><br>7.1 |
| **Electrical** | | | |
| Dielectric Strength<br>In oil @ 0.062 in V/mil (kV/mm) | D149 | <br>440 | <br>17 |
| Dielectric Constant<br>100 Hz | D150 | <br>3.02 | <br>3.02 |
| Dissipation Factor<br>100 Hz | D150 | <br>0.0015 | <br>0.0015 |
| Volume Resistivity<br>ohm-cm × E16 (ohm-cm × E16) | D257 | <br>4.20 | <br>4.20 |
| **UL** | | | |
| Flammability | UL 94 | V-2 @ .062″ | |
| Arc Resistance<br>Stainless stl. elec. sec. | D495 | <br>124 @ .125″ | |
| High Voltage Arc<br>Tracking Rate in/min | UL 746A | <br>0.42 @ .062″ | |
| High Ampere Arc Ign arcs | UL 746A | 200 @ .062″ | |
| Hot Wire Ignition sec | UL 746A | 11 @ .062″ | |
| Comparative Trk Index volts | UL 746A | 600 @ .125″ | |
| UL Temperature Index<br>Elec Prop °C<br>Mech Prop w/impt °C<br>Mech Prop wo/impt °C | UL 746B | <br>75 @ .062″<br>75 @ .062″<br>75 @ .062″ | |
| Oxygen Index % | D2863 | | |

# General Electric Company

## Contact Address:

General Electric Company
Plastics Group
One Plastics Avenue
Pittsfield, MA 01201

Phone: 1 (800) 845-0600

## XENLOY 3505

### Product Information:

XENLOY is a polycarbonate based composite polymer alloy, also available unreinforced or with glass reinforcement, formulated to improve impact resistance. XENLOY resins can be injection molded, extruded, blow molded, or thermoformed. XENLOY has excellent physical strength, and is resistant to impact. Its excellent melt flow properties allows it to be processed over a wide range of temperature and pressure. XENLOY has excellent chemical resistance, UV-stability, dimensional stability, low mold shrinkage, and low water absorbtion.

### Product Data:

**XENLOY 3505—Typical Property Values**

| Property | ASTM Method | English Units | SI Units |
|---|---|---|---|
| **Physical** | | | |
| Specific Gravity | D792 | 1.26 | 1.26 |
| Water Absorption | D570 | | |
| 24 hrs @ 73°F % (%) | | 0.14 | 0.14 |
| Equilibrium @ 73°F % (%) | | 0.35 | 0.35 |
| Mold Shrinkage/Flow | D955 | | |
| (0.125 mil) mil/in (%) | | 5–7 | 5–7 |
| **Optical** | | | |
| Transmittance % (%) | D1003 | 4–5 | 4–5 |
| Haze % (%) | D1003 | 87–88 | 87–88 |
| **Mechanical** | | | |
| Tensile Strength | D638 | | |
| Yield psi (MPa) | | 8300 | 57 |
| Break psi (MPa) | | 8000 | 55 |
| Tensile Elongation | D638 | | |
| Break % (%) | | 125 | 125 |
| Flexural Strength | D790 | | |
| Yield psi (MPa) | | 12,000 | 83 |

NOTE – Information provided in this book has been condensed. The publisher recommends that the product manufacturer be contacted, at the address or phone number provided, for more complete information on product safety, processing parameters, and applications not provided in this information sheet.

**XENLOY 3505—Typical Property Values *(continued)***

| Property | ASTM Method | English Units | SI Units |
|---|---|---|---|
| **Mechanical** | | | |
| Flexural Modulus psi (MPa) | D790 | 280,000 | 1930 |
| Izod Impact Strength 1/8″<br>Notched<br>.010″ rad. @ 73°F ft-lb/in (J/m)<br>.020″ rad. @ 73°F ft-lb/in (J/m) | D256 | <br><br>2.0<br>20.0 | <br><br>107<br>1068 |
| Gardner Impact<br>73°F ft-lb (J) | D3029 | <br>>40 | <br>54 |
| Rockwell Hardness R Scale | D785 | 122 | 122 |
| **Thermal** | | | |
| Deflection Temperature under Load<br>@ 66 psi °F (°C)<br>@ 264 psi °F (°C) | D648 | <br>225<br>210 | <br>107<br>99 |
| Coefficient of Linear<br>Thermal Expansion<br>−22°F to 86°F in/in—°F × E5<br>(m/m—°C × E5) | E831 | <br><br>3.9 | <br><br>7.1 |
| **Electrical** | | | |
| Dielectric Strength<br>In oil @ 0.062 in V/mil (kV/mm) | D149 | | |
| Dielectric Constant<br>100 Hz | D150 | | |
| Dissipation Factor<br>100 Hz | D150 | | |
| Volume Resistivity<br>ohm-cm × E16 (ohm-cm × E16) | D257 | | |
| **UL** | | | |
| Flammability | UL 94 | V-2 @ .062″ | |
| Arc Resistance<br>Stainless stl. elec. sec. | D495 | <br>130 @ .125″ | |
| High Voltage Arc<br>Tracking Rate in/min | UL 746A | <br>2.50 @ .125″ | |
| High Ampere Arc Ign arcs | UL 746A | 200 @ .062″ | |
| Hot Wire Ignition sec | UL 746A | 19 @ .062″ | |
| Comparative Trk Index volts | UL 746A | 420 @ .125″ | |
| UL Temperature Index<br>Elec Prop °C<br>Mech Prop w/impt °C<br>Mech Prop wo/impt °C | UL 746B | <br>75 @ .062″<br>75 @ .062″<br>75 @ .062″ | |
| Oxygen Index % | D2863 | 28 | |

# General Electric Company

Contact Address:

General Electric Company
Plastics Group
One Plastics Avenue
Pittsfield, MA 01201

Phone: 1 (800) 845-0600

## XENLOY 3700

Product Information:

XENLOY is a polycarbonate based composite polymer alloy, also available unreinforced or with glass reinforcement, formulated to improve impact resistance. XENLOY resins can be injection molded, extruded, blow molded, or thermoformed. XENLOY has excellent physical strength, and is resistant to impact. Its excellent melt flow properties allows it to be processed over a wide range of temperature and pressure. XENLOY has excellent chemical resistance, UV-stability, dimensional stability, low mold shrinkage, and low water absorbtion.

Product Data:

**XENLOY 3700—Typical Property Values**

| Property | ASTM Method | English Units | SI Units |
|---|---|---|---|
| **Physical** | | | |
| Specific Gravity | D792 | 1.20 | 1.20 |
| Water Absorption | D570 | | |
| 24 hrs @ 73°F % (%) | | 0.14 | 0.14 |
| Equilibrium @ 73°F % (%) | | 0.35 | 0.35 |
| Mold Shrinkage/Flow | D955 | | |
| (0.125 mil) mil/in (%) | | 5–7 | 5–7 |
| **Optical** | | | |
| Transmittance % (%) | D1003 | 4–5 | 4–5 |
| Haze % (%) | D1003 | 87–88 | 87–88 |
| **Mechanical** | | | |
| Tensile Strength | D638 | | |
| Yield psi (MPa) | | 8300 | 57 |
| Break psi (MPa) | | 8000 | 55 |
| Tensile Elongation | D638 | | |
| Break % (%) | | 125 | 125 |
| Flexural Strength | D790 | | |
| Yield psi (MPa) | | 12,000 | 83 |

NOTE – Information provided in this book has been condensed. The publisher recommends that the product manufacturer be contacted, at the address or phone number provided, for more complete information on product safety, processing parameters, and applications not provided in this information sheet.

**XENLOY 3700—Typical Property Values *(continued)***

| Property | ASTM Method | English Units | SI Units |
|---|---|---|---|
| **Mechanical** | | | |
| Flexural Modulus psi (MPa) | D790 | 280,000 | 1930 |
| Izod Impact Strength 1/8″<br>Notched<br>.010″ rad. @ 73°F ft-lb/in (J/m)<br>.020″ rad. @ 73°F ft-lb/in (J/m) | D256 | <br><br>2.0<br>20.0 | <br><br>107<br>1068 |
| Gardner Impact<br>73°F ft-lb (J) | D3029 | <br>>40 | <br>54 |
| Rockwell Hardness R Scale | D785 | 122 | 122 |
| **Thermal** | | | |
| Deflection Temperature under Load<br>@ 66 psi °F (°C)<br>@ 264 psi °F (°C) | D648 | <br>225<br>210 | <br>107<br>99 |
| Coefficient of Linear<br>Thermal Expansion<br>−22°F to 86°F in/in—°F × E5<br>(m/m—°C × E5) | E831 | <br><br>3.9 | <br><br>7.1 |
| **Electrical** | | | |
| Dielectric Strength<br>In oil @ 0.062 in V/mil (kV/mm) | D149 | <br>440 | <br>17 |
| Dielectric Constant<br>100 Hz | D150 | <br>3.02 | <br>3.02 |
| Dissipation Factor<br>100 Hz | D150 | <br>0.0015 | <br>0.0015 |
| Volume Resistivity<br>ohm-cm × E16 (ohm-cm × E16) | D257 | <br>4.20 | <br>4.20 |
| **UL** | | | |
| Flammability | UL 94 | V-2 @ .062″ | |
| Arc Resistance<br>Stainless stl. elec. sec. | D495 | <br>124 @ .125″ | |
| High Voltage Arc<br>Tracking Rate in/min | UL 746A | <br>0.42 @ .062″ | |
| High Ampere Arc Ign arcs | UL 746A | 200 @ .062″ | |
| Hot Wire Ignition sec | UL 746A | 11 @ .062″ | |
| Comparative Trk Index volts | UL 746A | 600 @ .125″ | |
| UL Temperature Index<br>Elec Prop °C<br>Mech Prop w/impt °C<br>Mech Prop wo/impt °C | UL 746B | <br>75 @ .062″<br>75 @ .062″<br>75 @ .062″ | |
| Oxygen Index % | D2863 | | |

# General Electric Company

## Contact Address:

General Electric Company
Plastics Group
One Plastics Avenue
Pittsfield, MA 01201

Phone: 1 (800) 845-0600

## XENOY 1101

### Product Information:

XENOY is a polycarbonate based composite polymer alloy, also available unreinforced or with glass reinforcement, formulated to improve impact resistance. XENOY resins can be injection molded, extruded, blow molded, or thermoformed. XENOY has excellent physical strength, and is resistant to impact. Its excellent melt flow properties allows it to be processed over a wide range of temperature and pressure. XENOY has excellent chemical resistance, UV-stability, dimensional stability, low mold shrinkage, and low water absorbtion.

### Product Data:

An unreinforced polycarbonate-polyester blend. It offers superior chemical resistance and low temperature impact.

**XENOY 1101**

| Property | ASTM Method | English Units | SI Units |
|---|---|---|---|
| **Physical** | | | |
| Specific Gravity | D792 | 1.21 | 1.21 |
| Specific Volume | D892 | 22.9 in$^3$/lb | 829 cm$^3$/kg |
| Water Absorption, 24 hours | D570 | | |
| Mold Shrinkage<br>Flow Direction and<br>Cross Flow Direction<br>(125 mil) (3.175 mm) | | 8–10 in/in $\times$ $10^{-3}$ | 8–10 mm/mm $\times$ $10^{-3}$ |
| **Mechanical** | | | |
| Tensile Strength | D638 | 7680 psi | 53 MPa |
| Elongation at Break | D638 | 120% | 120% |
| Flexural Strength | D790 | 12,300 psi | 85 MPa |
| Flexural Modulus | D790 | 296,000 psi | 2040 MPa |

*(continued)*

NOTE – Information provided in this book has been condensed. The publisher recommends that the product manufacturer be contacted, at the address or phone number provided, for more complete information on product safety, processing parameters, and applications not provided in this information sheet.

## XENOY 1101 *(continued)*

| Property | ASTM Method | English Units | SI Units |
|---|---|---|---|
| **Mechanical** | | | |
| Compressive Strength | D695 | | |
| Shear Strength | D732 | | |
| Izod Impact Strength | D256 | | |
| Notched, 1/8″ thick (3.2 mm) | | 13.3 ft-lb/in | 710 J/m |
| Notched @ −40°C | | 5.6 ft-lb/in | 300 J/m |
| Unnotched, 1/8″ thick (3.2 mm) | | | |
| Gardner Impact, 1/8″ thick (3.2 mm) | | >40 ft-lb | |
| Rockwell Hardness, R-scale | D785 | | |
| **Thermal** | | | |
| Heat Deflection Temp. | D648 | | |
| 66 psi (0.46 MPa) | | 223°F | 106°C |
| 264 psi (1.82 MPa) | | 210°F | 99°C |
| Coeff. of Therm. Expansion (linear) | E831 | | |
| Mold Direction × $10^{-5}$ | | | |
| Range: −40 100°F (−40 40°C) | | | |
| Range: 140 280°F (60 140°C) | | | |
| **Electrical** | | | |
| Dielectric Strength | D149 | | |
| 1/16″ (1.6 mm) | | 738 V/mil | 29.1 kV/mm |
| 1/8″ (1.82 MPa) | | 528 V/mil | 20.7 kV/mm |
| Dielectric Constant | D150 | | |
| 100 Hz | | 2.99 | 2.99 |
| $10^6$ Hz | | 3.04 | 3.04 |
| Dissipation Factor | D150 | | |
| 100 Hz | | .002 | .002 |
| $10^6$ Hz | | .019 | .019 |
| Volume Resistivity | D257 | 2.4 × $10^{16}$ohm-cm | 2.4 × $10^{16}$ohm-cm |
| **UL Properties** | | | |
| Flammability | UL94 | | |
| Arc Resistance | D495 | | |
| High Voltage Arc Tracking Rate | UL746A | | |
| High Ampere Arc Ignition | UL746A | | |
| Hot Wire Ignition | UL746A | | |
| Comparative Track Index (CTI) | UL746A | | |
| UL Temp. Index | UL746B | | |
| Elec. Properties | | | |
| Mech. Properties with Impact | | | |
| Mech. Properties without Impact | | | |

# General Electric Company

**Contact Address:**

General Electric Company
Plastics Group
One Plastics Avenue
Pittsfield, MA 01201

Phone: 1 (800) 845-0600

## XENOY 1102

**Product Information:**

XENOY is a polycarbonate based composite polymer alloy, also available unreinforced or with glass reinforcement, formulated to improve impact resistance. XENOY resins can be injection molded, extruded, blow molded, or thermoformed. XENOY has excellent physical strength, and is resistant to impact. Its excellent melt flow properties allows it to be processed over a wide range of temperature and pressure. XENOY has excellent chemical resistance, UV-stability, dimensional stability, low mold shrinkage, and low water absorbtion.

**Product Data:**

An unreinforced polycarbonate-polyester blend. It offers excellent low temperature impact, plus chemical resistance.

**XENOY 1102**

| Property | ASTM Method | English Units | International |
|---|---|---|---|
| **Physical** | | | |
| Specific Gravity | D792 | 1.20 | 1.20 |
| Specific Volume | D892 | 23.1 $in^3/lb$ | 833 $cm^3/kg$ |
| Water Absorption, 24 hours | D570 | 0.12% | 0.12% |
| Mold Shrinkage Flow Direction and Cross Flow Direction (125 mil) (3.175 mm) | | 8–10 in/in × $10^{-3}$ | 8–10 mm/mm × $10^{-3}$ |
| **Mechanical** | | | |
| Tensile Strength | D638 | 7900 psi | 54 MPa |
| Elongation at Break | D638 | 150% | 150% |
| Flexural Strength | D790 | 11,900 psi | 82 MPa |
| Flexural Modulus | D790 | 285,000 psi | 1964 MPa |

*(continued)*

NOTE – Information provided in this book has been condensed. The publisher recommends that the product manufacturer be contacted, at the address or phone number provided, for more complete information on product safety, processing parameters, and applications not provided in this information sheet.

## XENOY 1102 *(continued)*

| Property | ASTM Method | English Units | SI Units |
|---|---|---|---|
| **Mechanical** | | | |
| Compressive Strength | D695 | | |
| Shear Strength | D732 | | |
| Izod Impact Strength<br>Notched, 1/8″ thick (3.2 mm)<br>Notched @ RT<br>Notched @ 0°C<br>Notched @ −30°C | D256 | <br><br>15.0 ft-lb/in<br>13.0 ft-lb/in<br>12.0 ft-lb/in | <br><br>800 J/m<br>800 J/mN<br>640 J/m |
| Gardner Impact, 1/8″ thick (3.2 mm) | | | |
| Rockwell Hardness, R-scale | D785 | | |
| **Thermal** | | | |
| Heat Deflection Temp.<br>66 psi (0.46 MPa)<br>264 psi (1.82 MPa) | D648 | <br>230°F<br>195°F | <br>110°C<br>91°C |
| Coeff. of Therm. Expansion (linear) | E831 | | |
| Mold Direction × $10^{-5}$<br>Range: −40 100°F (−40 40°C)<br>Range: 140 280°F (60 140°C) | | | |
| **Electrical** | | | |
| Dielectric Strength<br>1/16″ (1.6 mm)<br>1/8″ (3.2 mm) | D149 | | |
| Dielectric Constant<br>100 Hz<br>$10^6$ Hz | D150 | | |
| Dissipation Factor<br>100 Hz<br>$10^6$ Hz | D150 | | |
| Volume Resistivity | D257 | | |
| **UL Properties** | | | |
| Flammability | UL94 | | |
| Arc Resistance | D495 | | |

*(continued)*

## XENOY 1102 *(continued)*

| Property | ASTM Method | English Units | SI Units |
|---|---|---|---|
| **UL Properties** | | | |
| High Voltage Arc Tracking Rate | UL746A | | |
| High Ampere Arc Ignition | UL746A | | |
| Hot Wire Ignition | UL746A | | |
| Comparative Track Index (CTI) | UL746A | | |
| UL Temp. Index<br>Elec. Properties<br>Mech. Properties with Impact<br>Mech. Properties without Impact | UL746B | | |

The values shown above are typical values that have been obtained using test bars molded from typical lots and are not intended for specification purposes. These values are for natural colors only. Addition of pigments may alter some values. Each user should make its own tests of its molded parts to determine the suitability of the material for its own use.

# General Electric Company

Contact Address:

General Electric Company
Plastics Group
One Plastics Avenue
Pittsfield, MA 01201

Phone: 1 (800) 845-0600

## XENOY 1104

Product Information:

XENOY is a polycarbonate based composite polymer alloy, also available unreinforced or with glass reinforcement, formulated to improve impact resistance. XENOY resins can be injection molded, extruded, blow molded, or thermoformed. XENOY has excellent physical strength, and is resistant to impact. Its excellent melt flow properties allows it to be processed over a wide range of temperature and pressure. XENOY has excellent chemical resistance, UV-stability, dimensional stability, low mold shrinkage, and low water absorbtion.

Product Data:

An unreinforced polycarbonate-polyester blend. It offers excellent low temperature impact, plus chemical resistance.

**XENOY 1104**

| Property | ASTM Method | English Units | SI Units |
|---|---|---|---|
| **Physical** | | | |
| Specific Gravity | D792 | 1.17 | 1.17 |
| Specific Volume | D892 | 23.7 in³/lb | 857 cm³/kg |
| Water Absorption, 24 hours | D570 | 0.13% | 0.13% |
| Mold Shrinkage<br>Flow Direction and<br>Cross Flow Direction<br>(125 mil) (3.175 mm) | | 6–9 in/in × $10^{-3}$ | 6–9 mm/mm × $10^{-3}$ |
| **Mechanical** | | | |
| Tensile Strength | D638 | 7200 psi | 50 MPa |
| Elongation at Break | D638 | 165% | 165% |
| Flexural Strength | D790 | 10,000 psi | 69 MPa |
| Flexural Modulus | D790 | 250,000 psi | 1724 MPa |

*(continued)*

NOTE – Information provided in this book has been condensed. The publisher recommends that the product manufacturer be contacted, at the address or phone number provided, for more complete information on product safety, processing parameters, and applications not provided in this information sheet.

## XENOY 1104 *(continued)*

| Property | ASTM Method | English Units | SI Units |
|---|---|---|---|
| **Mechanical** | | | |
| Compressive Strength | D695 | | |
| Shear Strength | D732 | | |
| Izod Impact Strength | D256 | | |
| Notched, 1/8″ thick (3.2 mm) | | 15 ft-lb/in | 800 J/m |
| Notched @ −40°C | | 10 ft-lb/in | 533 J/m |
| Unnotched, 1/8″ thick (3.2 mm) | | NB | NB |
| Instrumented Impact Test | D3763-85 | | |
| @ 73°F | @ Max | 425 in-lb | 47 J |
| @ −40°F | load | 495 in-lb | 56 J |
| Rockwell Hardness, R-scale | D785 | | |
| **Thermal** | | | |
| Heat Deflection Temp. | D648 | | |
| 66 psi (0.46 MPa) | | 240°F | 115°C |
| 264 psi (1.82 MPa) | | 205°F | 96°C |
| Coeff. of Therm. Expansion (linear) | E831 | | |
| Mold Direction $\times 10^{-5}$ | | | |
| Range: −40 100°F (−40 40°C) | | | |
| Range: 140 280°F (60 140°C) | | | |
| **Electrical** | | | |
| Dielectric Strength | D149 | | |
| 1/16″ (1.6 mm) | | 729 V/mil | 28.6 kV/mm |
| 1/8″ (3.2 mm) | | 505 V/mil | 19.8 kV/mm |
| Dielectric Constant | D150 | | |
| 100 Hz | | 2.93 | 2.93 |
| $10^6$ Hz | | 2.95 | 2.95 |
| Dissipation Factor | D150 | | |
| 100 Hz | | .002 | .002 |
| $10^6$ Hz | | .03 | .03 |
| Volume Resistivity | D257 | $9.5 \times 10^{16}$ ohm-cm | $9.5 \times 10^{16}$ ohm-cm |
| **UL Properties** | | | |
| Flammability | UL94 | | |
| Arc Resistance | D495 | | |

*(continued)*

## XENOY 1104 *(continued)*

| Property | ASTM Method | English Units | SI Units |
|---|---|---|---|
| **UL Properties** | | | |
| High Voltage Arc Tracking Rate | UL746A | | |
| High Ampere Arc Ignition | UL746A | | |
| Hot Wire Ignition | UL746A | | |
| Comparative Track Index (CTI) | UL746A | | |
| UL Temp. Index<br>Elec. Properties<br>Mech. Properties with Impact<br>Mech. Properties without Impact | UL746B | | |

The values shown above are typical values that have been obtained using test bars molded from typical lots and are not intended for specification purposes. These values are for natural colors only. Addition of pigments may alter some values. Each user should make its own tests of its molded parts to determine the suitability of the material for its own use.

# General Electric Company

## Contact Address:

General Electric Company
Plastics Group
One Plastics Avenue
Pittsfield, MA 01201

Phone: 1 (800) 845-0600

## XENOY 1200

### Product Information:

XENOY is a polycarbonate based composite polymer alloy, also available unreinforced or with glass reinforcement, formulated to improve impact resistance. XENOY resins can be injection molded, extruded, blow molded, or thermoformed. XENOY has excellent physical strength, and is resistant to impact. Its excellent melt flow properties allows it to be processed over a wide range of temperature and pressure. XENOY has excellent chemical resistance, UV-stability, dimensional stability, low mold shrinkage, and low water absorbtion.

### Product Data:

An unreinforced polycarbonate-polyester blend. It offers superior chemical resistance and low temperature impact.

**XENOY 1200**

| Property | ASTM Method | English Units | SI Units |
|---|---|---|---|
| **Physical** | | | |
| Specific Gravity | D792 | 1.20 | 1.20 |
| Specific Volume | D892 | 23.2 in$^3$/lb | 835 cm$^3$/kg |
| Water Absorption, 24 hours | D570 | 0.08% | 0.08% |
| Mold Shrinkage<br>Flow Direction and<br>Cross Flow Direction<br>(125 mil) (3.175 mm) | | 16–20<br>in/in $\times$ 10$^{-3}$ | 16–20<br>mm/mm $\times$ 10$^{-3}$ |

*(continued)*

NOTE—Information provided in this book has been condensed. The publisher recommends that the product manufacturer be contacted, at the address or phone number provided, for more complete information on product safety, processing parameters, and applications not provided in this information sheet.

## XENOY 1200 *(continued)*

| Property | ASTM Method | English Units | SI Units |
|---|---|---|---|
| **Mechanical** | | | |
| Tensile Strength | D638 | 6300 psi | 43 MPa |
| Elongation at Break | D638 | 175% | 175% |
| Flexural Strength | D790 | 9300 psi | 64 MPa |
| Flexural Modulus | D790 | 250,000 psi | 1724 MPa |
| Compressive Strength | D695 | 6700 psi | 46 MPa |
| Shear Strength | D732 | 6200 psi | 43 MPa |
| Izod Impact Strength | D256 | | |
| Notched, 1/8″ thick (3.2 mm) | | 15.0 ft-lb/in | 800 J/m |
| Notched @ −40°C | | 15.0 ft-lb/in | 640 J/m |
| Unnotched, 1/8″ thick (3.2 mm) | | 30 ft-lb/in | 1600 J/m |
| Gardner Impact, 1/8″ thick (3.2 mm) | | >40 ft-lb | 54 J |
| Rockwell Hardness, R-scale | D785 | 108 | 108 |
| **Thermal** | | | |
| Heat Deflection Temp. | D648 | | |
| 66 psi (0.46 MPa) | | 210°F | 99°C |
| 264 psi (1.82 MPa) | | 140°F | 60°C |
| Coeff. of Therm. Expansion (linear) | E831 | | |
| Mold Direction × $10^{-5}$ | | | |
| Range: −40 100°F (−40 40°C) | | 2.8 in/in/°F | 5.0 m/m/°C |
| Range: 140 280°F (60 140°C) | | 5.7 in/in/°F | 10.3 m/m/°C |
| **Electrical** | | | |
| Dielectric Strength | D149 | | |
| 1/16″ (1.6 mm) | | 710 V/mil | 28 kV/mm |
| 1/8″ (1.82 MPa) | | 485 V/mil | 19 kV/mm |
| Dielectric Constant | D150 | | |
| 100 Hz | | 3.10 | 3.10 |
| $10^6$ Hz | | 3.00 | 3.00 |
| Dissipation Factor | D150 | | |
| 100 Hz | | 0.0020 | 0.0020 |
| $10^6$ Hz | | 0.0200 | 0.0200 |
| Volume Resistivity | D257 | 5.50 × $10^{16}$ Ohm-cm | 5.50 × $10^{16}$ Ohm-cm |

*(continued)*

XENOY 1200 *(continued)*

| Property | ASTM Method | English Units | SI Units |
|---|---|---|---|
| **UL Properties** | | | |
| Flammability | UL94 | | |
| Arc Resistance | D495 | | |
| High Voltage Arc Tracking Rate | UL746A | | |
| High Ampere Arc Ignition | UL746A | | |
| Hot Wire Ignition | UL746A | | |
| Comparative Track Index (CTI) | UL746A | | |
| UL Temp. Index<br>Elec. Properties<br>Mech. Properties with Impact<br>Mech. Properties without Impact | UL746B | | |

# General Electric Company

Contact Address:

General Electric Company
Plastics Group
One Plastics Avenue
Pittsfield, MA 01201

Phone: 1 (800) 845-0600

## XENOY 1500

Product Information:

XENOY is a polycarbonate based composite polymer alloy, also available unreinforced or with glass reinforcement, formulated to improve impact resistance. XENOY resins can be injection molded, extruded, blow molded, or thermoformed. XENOY has excellent physical strength, and is resistant to impact. Its excellent melt flow properties allows it to be processed over a wide range of temperature and pressure. XENOY has excellent chemical resistance, UV-stability, dimensional stability, low mold shrinkage, and low water absorbtion.

Product Data:

An unreinforced polycarbonate-polyester blend. It offers excellent low temperature impact, plus chemical resistance.

**XENOY 1500**

| Property | ASTM Method | English Units | SI Units |
|---|---|---|---|
| **Physical** | | | |
| Specific Gravity | D792 | 1.22 | 1.22 |
| Specific Volume | D892 | 22.8 $in^3/lb$ | 825 $cm^3/kg$ |
| Water Absorption, 24 hours | D570 | 0.12% | 0.12% |
| Mold Shrinkage Flow Direction and Cross Flow Direction (125 mil) (3.175 mm) | | 6–9 in/in $\times 10^{-3}$ | 6–9 mm/mm $\times 10^{-3}$ |
| **Mechanical** | | | |
| Tensile Strength | D638 | 7900 psi | 54 MPa |
| Elongation at Break | D638 | 110% | 110% |
| Flexural Strength | D790 | 12,400 psi | 85 MPa |
| Flexural Modulus | D790 | 310,000 psi | 2137 MPa |

*(continued)*

NOTE – Information provided in this book has been condensed. The publisher recommends that the product manufacturer be contacted, at the address or phone number provided, for more complete information on product safety, processing parameters, and applications not provided in this information sheet.

## XENOY 1500 *(continued)*

| Property | ASTM Method | English Units | SI Units |
|---|---|---|---|
| **Mechanical** | | | |
| Compressive Strength | D695 | | |
| Shear Strength | D732 | | |
| Izod Impact Strength | D256 | | |
| Notched, 1/8″ thick (3.2 mm) | | 12 ft-lb/in | 640 J/m |
| Notched @ −40°C | | 3.2 ft-lb/in | 170 J/m |
| Unnotched, 1/8″ thick (3.2 mm) | | | |
| Instrumented Impact Test | D3763-85 | | |
| @ 73°F | @ Max | 450 in-lb | 51 J |
| @ −40°F | load | 499 in-lb | 56 J |
| **Thermal** | | | |
| Heat Deflection Temp. | D648 | | |
| 66 psi (0.46 MPa) | | 240°F | 115°C |
| 264 psi (1.82 MPa) | | 205°F | 96°C |
| Coeff. of Therm. Expansion (linear) | E831 | | |
| Mold Direction × $10^{-5}$ | | | |
| Range: −40 100°F (−40 40°C) | | | |
| Range: 140 280°F (60 140°C) | | | |
| **Electrical** | | | |
| Dielectric Strength | D149 | | |
| 1/16″ (1.6 mm) | | 710 V/mil | 27.9 kV/mm |
| 1/8″ (3.2 mm) | | 467 V/mil | 18.3 kV/mm |
| Dielectric Constant | D150 | | |
| 100 Hz | | 3.04 | 3.04 |
| $10^6$ Hz | | 3.14 | 3.14 |
| Dissipation Factor | D150 | | |
| 100 Hz | | .002 | .002 |
| $10^6$ Hz | | .006 | .006 |
| Volume Resistivity | D257 | $5.3 \times 10^{16}$ ohm-cm | $5.3 \times 10^{16}$ ohm-cm |
| **UL Properties** | | | |
| Flammability | UL94 | | |
| Arc Resistance | D495 | | |

*(continued)*

XENOY 1500 *(continued)*

| Property | ASTM Method | English Units | SI Units |
|---|---|---|---|
| **UL Properties** | | | |
| High Voltage Arc Tracking Rate | UL746A | | |
| High Ampere Arc Ignition | UL746A | | |
| Hot Wire Ignition | UL746A | | |
| Comparative Track Index (CTI) | UL746A | | |
| UL Temp. Index<br>Elec. Properties<br>Mech. Properties with Impact<br>Mech. Properties without Impact | UL746B | | |

The values shown above are typical values that have been obtained using test bars molded from typical lots and are not intended for specification purposes. These values are for natural colors only. Addition of pigments may alter some values. Each user should make its own tests of its molded parts to determine the suitability of the material for its own use.

# General Electric Company

## Contact Address:

General Electric Company
Plastics Group
One Plastics Avenue
Pittsfield, MA 01201

Phone: 1 (800) 845-0600

## XENOY 2230

### Product Information:

XENOY is a polycarbonate based composite polymer alloy, also available unreinforced or with glass reinforcement, formulated to improve impact resistance. XENOY resins can be injection molded, extruded, blow molded, or thermoformed. XENOY has excellent physical strength, and is resistant to impact. Its excellent melt flow properties allows it to be processed over a wide range of temperature and pressure. XENOY has excellent chemical resistance, UV-stability, dimensional stability, low mold shrinkage, and low water absorbtion.

### Product Data:

**XENOY 2230 Impact Modified Resin**

| Property | ASTM Method | English Units | SI Units |
|---|---|---|---|
| **Physical** | | | |
| Specific Gravity | D792 | 1.22 | |
| Specific Volume | D892 | 22.7 $in^3/lb$ | 820 $cm^3/kg$ |
| Water Absorption, 24 hours | D570 | .14% | .14% |
| Mold Shrinkage<br>Flow Direction and<br>Cross Flow Direction<br>(125 mil) (3.175 mm) | | 6–9 in/in $\times 10^{-3}$ | 6–9 mm/mm $\times 10^{-3}$ |
| **Mechanical** | | | |
| Tensile Strength | D638 | 8000 psi | 55 MPa |
| Elongation at Break | D638 | 120% | 120% |
| Flexural Strength | D790 | 12,500 psi | 86 MPa |
| Flexural Modulus | D790 | 325,000 psi | 2240 MPa |

*(continued)*

NOTE – Information provided in this book has been condensed. The publisher recommends that the product manufacturer be contacted, at the address or phone number provided, for more complete information on product safety, processing parameters, and applications not provided in this information sheet.

## XENOY 2230 Impact Modified Resin *(continued)*

| Property | ASTM Method | English Units | SI Units |
|---|---|---|---|
| **Mechanical** | | | |
| Compressive Strength | D695 | | |
| Shear Strength | D732 | | |
| Izod Impact Strength | D256 | | |
| Notched, 1/8″ thick (3.2 mm) | | 15 ft-lb/in | 800 J/m |
| Notched @ −40°C | | | |
| Unnotched, 1/8″ thick (3.2 mm) | | | |
| Gardner Impact, 1/8″ thick (3.2 mm) | | 40 ft-lb | 54 J |
| Rockwell Hardness, R-scale | D785 | 115 | 115 |
| **Thermal** | | | |
| Heat Deflection Temp. | D648 | | |
| @ 66 psi (0.46 MPa) | | 265°F | 129°C |
| @ 264 psi (1.82 MPa) | | 250°F | 121°C |
| Coeff. of Therm. Expansion (linear) | E831 | | |
| Mold Direction × $10^{-5}$ | | | |
| Range: −40→100°F (−40→40°C) | | | |
| Range: 140→280°F (60→140°C) | | | |
| **Electrical** | | | |
| Dielectric Strength | D149 | | |
| 1/16″ (1.6 mm) | | 705 V/mil | 27.7 kV/mm |
| 1/8″ (3.2 mm) | | 580 V/mil | 22.8 kV/mm |
| Dielectric Constant | D150 | | |
| 100 Hz | | 3.77 | 3.77 |
| $10^6$ Hz | | 3.73 | 3.73 |
| Dissipation Factor | D150 | | |
| 100 Hz | | .005 | .005 |
| $10^6$ Hz | | .026 | .026 |
| Volume Resistivity | D257 | $2.4 \times 10^{16}$ Ohm-cm | $2.4 \times 10^{16}$ Ohm-cm |

*(continued)*

**XENOY 2230 Impact Modified Resin *(continued)***

| Property | ASTM Method | English Units | SI Units |
|---|---|---|---|
| **UL Properties** | | | |
| Flammability | UL94 | HB/.031″ | |
| Arc Resistance | D495 | 99/.125″ sec | |
| High Voltage Arc Tracking Rate | UL746A | 2.6/.125″ in/min | |
| High Ampere Arc Ignition | UL746A | 26/.031 arcs | |
| Hot Wire Ignition | UL746A | 9/.031″ sec | |
| Comparative Track Index (CTI) | UL746A | 260/.125″ volts | |
| UL Temp. Index | UL746B | | |
| Electrical Properties | | 105/.031″ °C | |
| Mechanical Properties with Impact | | 105/.031″ °C | |
| Mechanical Properties without Impact | | 105/.031″ °C | |

# General Electric Company

## Contact Address:

General Electric Company
Plastics Group
One Plastics Avenue
Pittsfield, MA 01201

Phone: 1 (800) 845-0600

## XENOY 5220

### Product Information:

XENOY is a polycarbonate based composite polymer alloy, also available unreinforced or with glass reinforcement, formulated to improve impact resistance. XENOY resins can be injection molded, extruded, blow molded, or thermoformed. XENOY has excellent physical strength, and is resistant to impact. Its excellent melt flow properties allows it to be processed over a wide range of temperature and pressure. XENOY has excellent chemical resistance, UV-stability, dimensional stability, low mold shrinkage, and low water absorbtion.

### Product Data:

**XENOY 5220 Impact Modified Resin**

| Property | ASTM Method | English Units | SI Units |
|---|---|---|---|
| **Physical** | | | |
| Specific Gravity | D792 | 1.21 | |
| Specific Volume | D892 | 22.9 in³/lb | 829 cm³/kg |
| Water Absorption, 24 hours | D570 | .12% | |
| Mold Shrinkage<br>Flow Direction and<br>Cross Flow Direction<br>(125 mil) (3.175 mm) | | 8–10 in/in $\times$ $10^{-3}$ | |
| **Mechanical** | | | |
| Tensile Strength | D638 | 7700 psi | 53 MPa |
| Elongation at Break | D638 | 120% | 120% |
| Flexural Strength | D790 | 12,300 psi | 85 MPa |
| Flexural Modulus | D790 | 296,000 psi | 2040 MPa |

*(continued)*

NOTE—Information provided in this book has been condensed. The publisher recommends that the product manufacturer be contacted, at the address or phone number provided, for more complete information on product safety, processing parameters, and applications not provided in this information sheet.

### XENOY 5220 Impact Modified Resin *(continued)*

| Property | ASTM Method | English Units | SI Units |
|---|---|---|---|
| **Mechanical** | | | |
| Compressive Strength | D695 | | |
| Shear Strength | D732 | | |
| Izod Impact Strength | D256 | | |
| Notched, 1/8″ thick (3.2 mm) | | 13.3 ft-lb/in | 710 J/m |
| Notched @ −40°C | | 5.6 ft-lb/in | 300 J/m |
| Unnotched, 1/8″ thick (3.2 mm) | | NB | |
| Gardner Impact, 1/8″ thick (3.2 mm) | | 40 ft-lb | 54 J |
| Instrumented Impact Test | | | |
| @ 73°F (23°C) | | | |
| @ −40°F (−40°C) | | | |
| Rockwell Hardness, R-scale | D785 | 112 | 112 |
| **Thermal** | | | |
| Heat Deflection Temp. | D648 | | |
| @ 66 psi (0.46 MPa) | | 223°F | 106°C |
| @ 264 psi (1.82 MPa) | | 210°F | 99°C |
| Coeff. of Therm. Expansion (linear) | E831 | | |
| Mold Direction × $10^{-5}$ | | | |
| Range: −40→100°F (−40→40°C) | | | |
| Range: 140→280°F (60→140°C) | | | |
| **Electrical** | | | |
| Dielectric Strength | D149 | | |
| 1/16″ (1.6 mm) | | 738 V/mil | 29.1 kV/mm |
| 1/8″ (3.2 mm) | | 528 V/mil | 20.7 kV/mm |
| Dielectric Constant | D150 | | |
| 100 Hz | | 2.99 | |
| $10^6$ Hz | | 3.04 | |
| Dissipation Factor | D150 | | |
| 100 Hz | | .002 | .002 |
| $10^6$ Hz | | .019 | .019 |
| Volume Resistivity | D257 | $2.4 \times 10^{16}$ Ohm-cm | |

*(continued)*

**XENOY 5220 Impact Modified Resin *(continued)***

| Property | ASTM Method | English Units | SI Units |
|---|---|---|---|
| **UL Properties** | | | |
| Flammability | UL94 | HB/.031″ | |
| Arc Resistance | D495 | 126/.123″ sec | |
| High Voltage Arc Tracking Rate | UL746A | .80/.062″ in/min | |
| High Ampere Arc Ignition | UL746A | 200/.062 arcs | |
| Hot Wire Ignition | UL746A | 29/.123″ sec | |
| Comparative Track Index (CTI) | UL746A | 450/.123″ volts | |
| UL Temp. Index | UL746B | | |
| Electrical Properties | | 75/.062″ °C | |
| Mechanical Properties with Impact | | 75/.062″ °C | |
| Mechanical Properties without Impact | | 75/.062″ °C | |

# General Electric Company

## Contact Address:

General Electric Company
Plastics Group
One Plastics Avenue
Pittsfield, MA 01201

Phone: 1 (800) 845-0600

## XENOY DX5230

### Product Information:

XENOY is a polycarbonate based composite polymer alloy, also available unreinforced or with glass reinforcement, formulated to improve impact resistance. XENOY resins can be injection molded, extruded, blow molded, or thermoformed. XENOY has excellent physical strength, and is resistant to impact. Its excellent melt flow properties allows it to be processed over a wide range of temperature and pressure. XENOY has excellent chemical resistance, UV-stability, dimensional stability, low mold shrinkage, and low water absorbtion.

### Product Data:

**XENOY DX5230 Impact Modified Resin**

| Property | ASTM Method | English Units | SI Units |
|---|---|---|---|
| **Physical** | | | |
| Specific Gravity | D792 | 1.22 | |
| Specific Volume | D892 | 22.8 $in^3/lb$ | 825 $cm^3/kg$ |
| Water Absorption, 24 hours | D570 | .12% | .12% |
| Mold Shrinkage<br>Flow Direction and<br>Cross Flow Direction<br>(125 mil) (3.175 mm) | | 6–9 in/in $\times 10^{-3}$ | |
| **Mechanical** | | | |
| Tensile Strength | D638 | 7900 psi | 54 MPa |
| Elongation at Break | D638 | 110% | 110% |
| Flexural Strength | D790 | 12,400 psi | 85 MPa |
| Flexural Modulus | D790 | 310,000 psi | 2137 MPa |

*(continued)*

NOTE—Information provided in this book has been condensed. The publisher recommends that the product manufacturer be contacted, at the address or phone number provided, for more complete information on product safety, processing parameters, and applications not provided in this information sheet.

**XENOY DX5230 Impact Modified Resin *(continued)***

| Property | ASTM Method | English Units | SI Units |
|---|---|---|---|
| **Mechanical** | | | |
| Compressive Strength | D695 | | |
| Shear Strength | D732 | | |
| Izod Impact Strength<br>Notched, 1/8″ thick (3.2 mm)<br>Notched @ −40°C<br>Unnotched, 1/8″ thick (3.2 mm) | D256 | <br>12 ft-lb/in<br>3.2 ft-lb/in | <br>640 J/m<br>170 J |
| Gardner Impact, 1/8″ thick (3.2 mm) | | | |
| Instrumented Impact Test<br>@ 73°F (23°C)<br>@ −40°F (−40°C) | D3763-85 | <br>450 @ max load<br>499 @ max load | <br>51 @ max load<br>56 @ max load |
| Rockwell Hardness, R-scale | D785 | | |
| **Thermal** | | | |
| Heat Deflection Temp.<br>@ 66 psi (0.46 MPa)<br>@ 264 psi (1.82 MPa) | D648 | <br>240°F<br>205°F | <br>115°C<br>96°C |
| Coeff. of Therm. Expansion (linear) | E831 | | |
| Mold Direction × $10^{-5}$<br>Range: −40→100°F (−40→40°C)<br>Range: 140→280°F (60→140°C) | | | |
| **Electrical** | | | |
| Dielectric Strength<br>1/16″ (1.6 mm)<br>1/8″ (3.2 mm) | D149 | <br>710 V/mil<br>467 V/mil | <br>27.9 kV/mm<br>18.3 kV/mm |
| Dielectric Constant<br>100 Hz<br>$10^6$ Hz | D150 | <br>3.04<br>3.14 | <br>3.04<br>3.14 |
| Dissipation Factor<br>100 Hz<br>$10^6$ Hz | D150 | <br>.002<br>.006 | <br>.002<br>.006 |
| Volume Resistivity | D257 | $5.3 \times 10^{16}$ Ohm-cm | |

*(continued)*

**XENOY DX5230 Impact Modified Resin *(continued)***

| Property | ASTM Method | English Units | SI Units |
|---|---|---|---|
| **UL Properties** | | | |
| Flammability | UL94 | | |
| Arc Resistance | D495 | | |
| High Voltage Arc Tracking Rate | UL746A | | |
| High Ampere Arc Ignition | UL746A | | |
| Hot Wire Ignition | UL746A | | |
| Comparative Track Index (CTI) | UL746A | | |
| UL Temp. Index<br>Electrical Properties<br>Mechanical Properties with Impact<br>Mechanical Properties without Impact | UL746B | | |

# General Electric Company

## Contact Address:

General Electric Company
Plastics Group
One Plastics Avenue
Pittsfield, MA 01201

Phone: 1 (800) 845-0600

## XENOY DX5720

### Product Information:

XENOY is a polycarbonate based composite polymer alloy, also available unreinforced or with glass reinforcement, formulated to improve impact resistance. XENOY resins can be injection molded, extruded, blow molded, or thermoformed. XENOY has excellent physical strength, and is resistant to impact. Its excellent melt flow properties allows it to be processed over a wide range of temperature and pressure. XENOY has excellent chemical resistance, UV-stability, dimensional stability, low mold shrinkage, and low water absorbtion.

### Product Data:

**XENOY DX5720 Impact Modified Resin**

| Property | ASTM Method | English Units | SI Units |
|---|---|---|---|
| **Physical** | | | |
| Specific Gravity | D792 | 1.17 | |
| Specific Volume | D892 | 23.7 $in^3/lb$ | 857 $cm^3/kg$ |
| Water Absorption, 24 hours | D570 | .13% | .13% |
| Mold Shrinkage<br>Flow Direction and<br>Cross Flow Direction<br>(125 mil) (3.175 mm) | | 6–9 in/in $\times 10^{-3}$ | |
| **Mechanical** | | | |
| Tensile Strength | D638 | 7200 psi | 50 MPa |
| Elongation at Break | D638 | 165% | 165% |
| Flexural Strength | D790 | 10,000 psi | 69 MPa |
| Flexural Modulus | D790 | 250,000 psi | 1724 MPa |

*(continued)*

NOTE—Information provided in this book has been condensed. The publisher recommends that the product manufacturer be contacted, at the address or phone number provided, for more complete information on product safety, processing parameters, and applications not provided in this information sheet.

XENOY DX5720 Impact Modified Resin *(continued)*

| Property | ASTM Method | English Units | SI Units |
|---|---|---|---|
| **Mechanical** | | | |
| Compressive Strength | D695 | | |
| Shear Strength | D732 | | |
| Izod Impact Strength | D256 | | |
| Notched, 1/8″ thick (3.2 mm) | | 15 ft-lb/in | 800 J/m |
| Notched @ −40°C | | 10 ft-lb/in | 533 J |
| Unnotched, 1/8″ thick (3.2 mm) | | | |
| Gardner Impact, 1/8″ thick (3.2 mm) | Falling Dart | | |
| Instrumented Impact Test | D3763-85 | | |
| @ 73°F (23°C) | | 425 @ max load | 47 @ max load |
| @ −40°F (−40°C) | | 495 @ max load | 56 @ max load |
| Rockwell Hardness, R-scale | D785 | | |
| **Thermal** | | | |
| Heat Deflection Temp. | D648 | | |
| @ 66 psi (0.46 MPa) | | 240°F | 115°C |
| @ 264 psi (1.82 MPa) | | 205°F | 96°C |
| Coeff. of Therm. Expansion (linear) | E831 | | |
| Mold Direction × $10^{-5}$ | | | |
| Range: −40→100°F (−40→40°C) | | | |
| Range: 140→280°F (60→140°C) | | | |
| **Electrical** | | | |
| Dielectric Strength | D149 | | |
| 1/16″ (1.6 mm) | | 729 V/mil | 28.6 kV/mm |
| 1/8″ (3.2 mm) | | 505 V/mil | 19.8 kV/mm |
| Dielectric Constant | D150 | | |
| 100 Hz | | 2.93 | |
| $10^6$ Hz | | 2.95 | |
| Dissipation Factor | D150 | | |
| 100 Hz | | .0022 | |
| $10^6$ Hz | | .03 | |
| Volume Resistivity | D257 | $9.5 \times 10^{16}$ Ohm-cm | |

*(continued)*

**XENOY DX5720 Impact Modified Resin *(continued)***

| Property | ASTM Method | English Units | SI Units |
|---|---|---|---|
| **UL Properties** | | | |
| Flammability | UL94 | | |
| Arc Resistance | D495 | | |
| High Voltage Arc Tracking Rate | UL746A | | |
| High Ampere Arc Ignition | UL746A | | |
| Hot Wire Ignition | UL746A | | |
| Comparative Track Index (CTI) | UL746A | | |
| UL Temp. Index<br>Electrical Properties<br>Mechanical Properties with Impact<br>Mechanical Properties without Impact | UL746B | | |

# General Electric Company

Contact Address:

General Electric Company
Plastics Group
One Plastics Avenue
Pittsfield, MA 01201

Phone: 1 (800) 845-0600

## XENOY 6120

Product Information:

XENOY is a polycarbonate based composite polymer alloy, also available unreinforced or with glass reinforcement, formulated to improve impact resistance. XENOY resins can be injection molded, extruded, blow molded, or thermoformed. XENOY has excellent physical strength, and is resistant to impact. Its excellent melt flow properties allows it to be processed over a wide range of temperature and pressure. XENOY has excellent chemical resistance, UV-stability, dimensional stability, low mold shrinkage, and low water absorbtion.

Product Data:

**XENOY 6120 Impact Modified Resin**

| Property | ASTM Method | English Units | SI Units |
|---|---|---|---|
| **Physical** | | | |
| Specific Gravity | D792 | 1.25 | |
| Specific Volume | D892 | 22.3 in³/lb | 801 cm³/kg |
| Water Absorption, 24 hours | D570 | .08% | |
| Mold Shrinkage<br>Flow Direction<br>(0.125)<br>Cross Flow Direction<br>(0.125 mil) | | <br><br>16–21 in/in × $10^{-3}$<br><br>16–21 in/in × $10^{-3}$ | |
| **Mechanical** | | | |
| Tensile Strength | D638 | 6500 psi | 45 MPa |
| Elongation at Break | D638 | 175% | 175% |
| Flexural Strength | D790 | 10,500 psi | 72 MPa |
| Flexural Modulus | D790 | 275,000 psi | 1900 MPa |

*(continued)*

NOTE – Information provided in this book has been condensed. The publisher recommends that the product manufacturer be contacted, at the address or phone number provided, for more complete information on product safety, processing parameters, and applications not provided in this information sheet.

**XENOY 6120 Impact Modified Resin *(continued)***

| Property | ASTM Method | English Units | SI Units |
|---|---|---|---|
| **Mechanical** | | | |
| Compressive Strength | D695 | 8600 psi | 59 MPa |
| Shear Strength | D732 | 5300 psi | 36 MPa |
| Izod Impact Strength<br>Notched, 1/8″ thick (3.2 mm)<br>Unnotched, 1/8″ thick (3.2 mm) | D256 | <br>16 ft-lb/in<br>NB | <br>850 J/m |
| Gardner Impact, 1/8″ thick (3.2 mm) | | 40 ft-lb | 54 J |
| Rockwell Hardness, R-scale | D785 | 113 | 113 |
| **Thermal** | | | |
| Heat Deflection Temp.<br>@ 66 psi (0.46 MPa)<br>@ 264 psi (1.82 MPa) | D648 | <br>250°F<br>135°F | <br>121°C<br>57°C |
| Coeff. of Therm. Expansion (linear) | E831 | | |
| Mold Direction × $10^{-5}$<br>Range: −40→100°F (−40→40°C)<br>Range: 140→280°F (60→140°C) | | <br>2.8 in/in/°F<br>5.7 in/in/°F | <br>5.0 m/m/°C<br>10.3 m/m/°C |
| **Electrical** | | | |
| Dielectric Strength<br>1/16″ (1.6 mm)<br>1/8″ (3.2 mm) | D149 | <br>570 V/mil<br>350 V/mil | <br>22 kV/mm<br>14 kV/mm |
| Dielectric Constant<br>100 Hz<br>$10^6$ Hz | D150 | <br>3.4<br>3.3 | <br>3.4<br>3.3 |
| Dissipation Factor<br>100 Hz<br>$10^6$ Hz | D150 | <br>.003<br>.03 | <br>.003<br>.03 |
| Volume Resistivity | D257 | .50 × $10^{16}$<br>Ohm-cm | |

*(continued)*

**XENOY 6120 Impact Modified Resin (*continued*)**

| Property | ASTM Method | English Units | SI Units |
|---|---|---|---|
| **UL Properties** | | | |
| Flammability | UL94 | HB/.031″ | |
| Arc Resistance | D495 | 146/.123″ sec | |
| High Voltage Arc Tracking Rate | UL746A | .3/.123″ in/min | |
| High Ampere Arc Ignition | UL746A | 200/.123 arcs | |
| Hot Wire Ignition | UL746A | 37/.123″ sec | |
| Comparative Track Index (CTI) | UL746A | 600/.123″ volts | |
| UL Temp. Index | UL746B | | |
| Electrical Properties | | 120/.062″ °C | |
| Mechanical Properties with Impact | | 120/.062″ °C | |
| Mechanical Properties without Impact | | 140/.062″ °C | |

# General Electric Company

## Contact Address:

General Electric Company
Plastics Group
One Plastics Avenue
Pittsfield, MA 01201

Phone: 1 (800) 845-0600

## XENOY 6123

### Product Information:

XENOY is a polycarbonate based composite polymer alloy, also available unreinforced or with glass reinforcement, formulated to improve impact resistance. XENOY resins can be injection molded, extruded, blow molded, or thermoformed. XENOY has excellent physical strength, and is resistant to impact. Its excellent melt flow properties allows it to be processed over a wide range of temperature and pressure. XENOY has excellent chemical resistance, UV-stability, dimensional stability, low mold shrinkage, and low water absorbtion.

### Product Data:

**XENOY 6123 Impact Modified Resin**

| Property | ASTM Method | English Units | SI Units |
|---|---|---|---|
| **Physical** | | | |
| Specific Gravity | D792 | 1.24 | |
| Specific Volume | D892 | 22.4 in³/lb | 806 cm³/kg |
| Water Absorption, 24 hours | D570 | .11% | .11% |
| Mold Shrinkage<br>Flow Direction<br>(0.125)<br>Cross Flow Direction<br>(0.125 mil) | | <br><br>12–15 in/in × $10^{-3}$<br><br>12–15 in/in × $10^{-3}$ | |
| **Mechanical** | | | |
| Tensile Strength | D638 | 6800 psi | 47 MPa |
| Elongation at Break | D638 | 130% | 130% |
| Flexural Strength | D790 | 11,000 psi | 76 MPa |
| Flexural Modulus | D790 | 290,000 psi | 2000 MPa |

*(continued)*

NOTE – Information provided in this book has been condensed. The publisher recommends that the product manufacturer be contacted, at the address or phone number provided, for more complete information on product safety, processing parameters, and applications not provided in this information sheet.

**XENOY 6123 Impact Modified Resin *(continued)***

| Property | ASTM Method | English Units | SI Units |
|---|---|---|---|
| **Mechanical** | | | |
| Compressive Strength | D695 | 8810 psi | 61 MPa |
| Shear Strength | D732 | 6600 psi | 45 MPa |
| Izod Impact Strength<br>Notched, 1/8″ thick (3.2 mm)<br>Unnotched, 1/8″ thick (3.2 mm) | D256 | <br>15 ft-lb/in<br>NB | <br>800 J/m |
| Gardner Impact, 1/8″ thick (3.2 mm) | | 40 ft-lb | 54 J |
| Rockwell Hardness, R-scale | D785 | 116 | 116 |
| **Thermal** | | | |
| Heat Deflection Temp.<br>@ 66 psi (0.46 MPa)<br>@ 264 psi (1.82 MPa) | D648 | <br>240°F<br>188°F | <br>115°C<br>87°C |
| Coeff. of Therm. Expansion (linear) | E831 | | |
| Mold Direction $\times 10^{-5}$<br>Range: −40→100°F (−40→40°C)<br>Range: 140→280°F (60→140°C) | | <br>5.0 in/in/°F<br>8.0 in/in/°F | <br>9.0 m/m/°C<br>14.4 m/m/°C |
| **Electrical** | | | |
| Dielectric Strength<br>1/16″ (1.6 mm)<br>1/8″ (3.2 mm) | D149 | <br>575 V/mil<br>445 V/mil | <br>23 kV/mm<br>17 kV/mm |
| Dielectric Constant<br>100 Hz<br>$10^6$ Hz | D150 | <br>3.1<br>3.1 | <br>3.1<br>3.1 |
| Dissipation Factor<br>100 Hz<br>$10^6$ Hz | D150 | <br>.002<br>.02 | <br>.002<br>.02 |
| Volume Resistivity | D257 | $1.0 \times 10^{16}$ Ohm-cm | |

*(continued)*

**XENOY 6123 Impact Modified Resin *(continued)***

| Property | ASTM Method | English Units | SI Units |
|---|---|---|---|
| **UL Properties** | | | |
| Flammability | UL94 | HB/.062″ | |
| Arc Resistance | D495 | 146/.123″ sec | |
| High Voltage Arc Tracking Rate | UL746A | 2.6/.062″ in/min | |
| High Ampere Arc Ignition | UL746A | 200/.062 arcs | |
| Hot Wire Ignition | UL746A | 37/.123″ sec | |
| Comparative Track Index (CTI) | UL746A | 330/.123″ volts | |
| UL Temp. Index<br>Electrical Properties<br>Mechanical Properties with Impact<br>Mechanical Properties without Impact | UL746B | <br>120/.062″ °C<br>115/.062″ °C<br>125/.062″ °C | |

# General Electric Company

## Contact Address:

General Electric Company
Plastics Group
One Plastics Avenue
Pittsfield, MA 01201

Phone: 1 (800) 845-0600

## XENOY 6240

### Product Information:

XENOY is a polycarbonate based composite polymer alloy, also available unreinforced or with glass reinforcement, formulated to improve impact resistance. XENOY resins can be injection molded, extruded, blow molded, or thermoformed. XENOY has excellent physical strength, and is resistant to impact. Its excellent melt flow properties allows it to be processed over a wide range of temperature and pressure. XENOY has excellent chemical resistance, UV-stability, dimensional stability, low mold shrinkage, and low water absorbtion.

### Product Data:

**XENOY 6240—10% Glass Reinforced Impact Modified Resin**

| Property | ASTM Method | English Units | SI Units |
|---|---|---|---|
| **Physical** | | | |
| Specific Gravity | D792 | 1.30 | 1.30 |
| Specific Volume | D892 | 21.3 $in^3/lb$ | 770 $cm^3/kg$ |
| Water Absorption, 24 hours | D570 | .11% | .11% |
| Mold Shrinkage<br>Flow Direction<br>(0.125)<br>Cross Flow Direction<br>(0.125 mil) | | <br><br>7-9 in/in $\times 10^{-3}$<br><br>11-13 in/in $\times 10^{-3}$ | |
| **Mechanical** | | | |
| Tensile Strength | D638 | 9000 psi | 62 MPa |
| Elongation at Break | D638 | 4% | 4% |
| Flexural Strength | D790 | 14,000 psi | 96 MPa |
| Flexural Modulus | D790 | 400,000 psi | 2700 MPa |

*(continued)*

NOTE–Information provided in this book has been condensed. The publisher recommends that the product manufacturer be contacted, at the address or phone number provided, for more complete information on product safety, processing parameters, and applications not provided in this information sheet.

**XENOY 6240—10% Glass Reinforced Impact Modified Resin *(continued)***

| Property | ASTM Method | English Units | SI Units |
|---|---|---|---|
| **Mechanical** | | | |
| Compressive Strength | D695 | 11,400 psi | 78 MPa |
| Shear Strength | D732 | 6900 psi | 47 MPa |
| Izod Impact Strength | D256 | | |
| Notched, 1/8″ thick (3.2 mm) | | 3.5 ft-lb/in | 190 J/m |
| Unnotched, 1/8″ thick (3.2 mm) | | 14 ft-lb | 750 J/m |
| Gardner Impact, 1/8″ thick (3.2 mm) | | | |
| Rockwell Hardness, R-scale | D785 | 113 | 113 |
| **Thermal** | | | |
| Heat Deflection Temp. | D648 | | |
| @ 66 psi (0.46 MPa) | | 350°F | 176°C |
| @ 264 psi (1.82 MPa) | | 250°F | 121°C |
| Coeff. of Therm. Expansion (linear) | E831 | | |
| Mold Direction × $10^{-5}$ | | | |
| Range: −40→100°F (−40→40°C) | | 2.9 in/in/°F | 5.3 m/m/°C |
| Range: 140→280°F (60→140°C) | | 2.9 in/in/°F | 5.3 m/m/°C |
| **Electrical** | | | |
| Dielectric Strength | D149 | | |
| 1/16″ (1.6 mm) | | 640 V/mil | 25 kV/mm |
| 1/8″ (3.2 mm) | | 480 V/mil | 19 kV/mm |
| Dielectric Constant | D150 | | |
| 100 Hz | | 3.4 | 3.4 |
| $10^6$ Hz | | 3.3 | 3.3 |
| Dissipation Factor | D150 | | |
| 100 Hz | | .002 | .002 |
| $10^6$ Hz | | .01 | .01 |
| Volume Resistivity | D257 | .7 × $10^{16}$ Ohm-cm | |

*(continued)*

XENOY 6240—10% Glass Reinforced Impact Modified Resin *(continued)*

| Property | ASTM Method | English Units | SI Units |
|---|---|---|---|
| **UL Properties** | | | |
| Flammability | UL94 | | |
| Arc Resistance | D495 | | |
| High Voltage Arc Tracking Rate | UL746A | | |
| High Ampere Arc Ignition | UL746A | | |
| Hot Wire Ignition | UL746A | | |
| Comparative Track Index (CTI) | UL746A | | |
| UL Temp. Index<br>Electrical Properties<br>Mechanical Properties with Impact<br>Mechanical Properties without Impact | UL746B | | |

# General Electric Company

## Contact Address:

General Electric Company
Plastics Group
One Plastics Avenue
Pittsfield, MA 01201

Phone: 1 (800) 845-0600

# XENOY 6370

## Product Information:

XENOY is a polycarbonate based composite polymer alloy, also available unreinforced or with glass reinforcement, formulated to improve impact resistance. XENOY resins can be injection molded, extruded, blow molded, or thermoformed. XENOY has excellent physical strength, and is resistant to impact. Its excellent melt flow properties allows it to be processed over a wide range of temperature and pressure. XENOY has excellent chemical resistance, UV-stability, dimensional stability, low mold shrinkage, and low water absorbtion.

## Product Data:

**XENOY 6370—30% Glass Reinforced Impact Modified Resin**

| Property | ASTM Method | English Units | SI Units |
|---|---|---|---|
| **Physical** | | | |
| Specific Gravity | D792 | 1.44 | 1.44 |
| Specific Volume | D892 | 19.3 $in^3/lb$ | 695 $cm^3/kg$ |
| Water Absorption, 24 hours | D570 | .09% | .09% |
| Mold Shrinkage<br>Flow Direction<br>(0.125)<br>Cross Flow Direction<br>(0.125 mil) | | <br><br>4–5 in/in × $10^{-3}$<br><br>7–9 in/in × $10^{-3}$ | |
| **Mechanical** | | | |
| Tensile Strength | D638 | 13,300 psi | 92 MPa |
| Elongation at Break | D638 | 4% | 4% |
| Flexural Strength | D790 | 20,000 psi | 138 MPa |
| Flexural Modulus | D790 | 780,000 psi | 5374 MPa |

*(continued)*

NOTE – Information provided in this book has been condensed. The publisher recommends that the product manufacturer be contacted, at the address or phone number provided, for more complete information on product safety, processing parameters, and applications not provided in this information sheet.

**XENOY 6370—30% Glass Reinforced Impact Modified Resin *(continued)***

| Property | ASTM Method | English Units | SI Units |
|---|---|---|---|
| **Mechanical** | | | |
| Compressive Strength | D695 | 10,800 psi | 74 MPa |
| Shear Strength | D732 | 6540 psi | 45 MPa |
| Izod Impact Strength | D256 | | |
| Notched, 1/8″ thick (3.2 mm) | | 3.2 ft-lb/in | 170 J/m |
| Unnotched, 1/8″ thick (3.2 mm) | | 12 ft-lb | 640 J/m |
| Gardner Impact, 1/8″ thick (3.2 mm) | | | |
| Rockwell Hardness, R-scale | D785 | 109 | 109 |
| **Thermal** | | | |
| Heat Deflection Temp. | D648 | | |
| @ 66 psi (0.46 MPa) | | 400°F | 204°C |
| @ 264 psi (1.82 MPa) | | 300°F | 149°C |
| Coeff. of Therm. Expansion (linear) | E831 | | |
| Mold Direction × $10^{-5}$ | | | |
| Range: −40→100°F (−40→40°C) | | 1.5 in/in/°F | 2.7 m/m/°C |
| Range: 140→280°F (60→140°C) | | 1.1 in/in/°F | 2.0 m/m/°C |
| **Electrical** | | | |
| Dielectric Strength | D149 | | |
| 1/16″ (1.6 mm) | | 700 V/mil | 28 kV/mm |
| 1/8″ (3.2 mm) | | 530 V/mil | 21 kV/mm |
| Dielectric Constant | D150 | | |
| 100 Hz | | 4.0 | 4.0 |
| $10^6$ Hz | | 3.9 | 3.9 |
| Dissipation Factor | D150 | | |
| 100 Hz | | .003 | .003 |
| $10^6$ Hz | | .02 | .02 |
| Volume Resistivity | D257 | .48 × $10^{16}$ Ohm-cm | |

*(continued)*

**XENOY 6370—30% Glass Reinforced Impact Modified Resin *(continued)***

| Property | ASTM Method | English Units | SI Units |
|---|---|---|---|
| **UL Properties** | | | |
| Flammability | UL94 | HB/.061″ | |
| Arc Resistance | D495 | 130/.122″ sec | |
| High Voltage Arc Tracking Rate | UL746A | .83/.061″ in/min | |
| High Ampere Arc Ignition | UL746A | 29/.061″ arcs | |
| Hot Wire Ignition | UL746A | 36/.061″ sec | |
| Comparative Track Index (CTI) | UL746A | 300/.122″ volts | |
| UL Temp. Index | UL746B | | |
| Electrical Properties | | 140/.061″ | |
| Mechanical Properties with Impact | | 130/.061″ | |
| Mechanical Properties without Impact | | 140/.061″ | |

# General Electric Company

## Contact Address:

General Electric Company
Plastics Group
One Plastics Avenue
Pittsfield, MA 01201

Phone: 1 (800) 845-0600

## XENOY 6380

### Product Information:

XENOY is a polycarbonate based composite polymer alloy, also available unreinforced or with glass reinforcement, formulated to improve impact resistance. XENOY resins can be injection molded, extruded, blow molded, or thermoformed. XENOY has excellent physical strength, and is resistant to impact. Its excellent melt flow properties allows it to be processed over a wide range of temperature and pressure. XENOY has excellent chemical resistance, UV-stability, dimensional stability, low mold shrinkage, and low water absorbtion.

### Product Data:

**XENOY 6380—30% Glass Reinforced Impact Modified Resin**

| Property | ASTM Method | English Units | SI Units |
|---|---|---|---|
| **Physical** | | | |
| Specific Gravity | D792 | 1.51 | 1.51 |
| Specific Volume | D892 | 18.3 $in^3/lb$ | 658 $cm^3/kg$ |
| Water Absorption, 24 hours | D570 | .10% | .10% |
| Mold Shrinkage<br>Flow Direction<br>(0.125)<br>Cross Flow Direction<br>(0.125 mil) | | <br><br>4–6 in/in $\times 10^{-3}$<br><br>7–9 in/in $\times 10^{-3}$ | |
| **Mechanical** | | | |
| Tensile Strength | D638 | 12,000 psi | 83 MPa |
| Elongation at Break | D638 | | |
| Flexural Strength | D790 | 20,000 psi | 138 MPa |
| Flexural Modulus | D790 | 850,000 psi | 5856 MPa |

*(continued)*

NOTE—Information provided in this book has been condensed. The publisher recommends that the product manufacturer be contacted, at the address or phone number provided, for more complete information on product safety, processing parameters, and applications not provided in this information sheet.

**XENOY 6380—30% Glass Reinforced Impact Modified Resin *(continued)***

| Property | ASTM Method | English Units | SI Units |
|---|---|---|---|
| **Mechanical** | | | |
| Compressive Strength | D695 | 11,600 psi | 80 MPa |
| Shear Strength | D732 | 6400 psi | 44 MPa |
| Izod Impact Strength | D256 | | |
| Notched, 1/8″ thick (3.2 mm) | | 3.1 ft-lb/in | 165 J/m |
| Unnotched, 1/8″ thick (3.2 mm) | | 13.5 ft-lb | 721 J/m |
| Gardner Impact, 1/8″ thick (3.2 mm) | | | |
| Rockwell Hardness, R-scale | D785 | 110 | 110 |
| **Thermal** | | | |
| Heat Deflection Temp. | D648 | | |
| @ 66 psi (0.46 MPa) | | 415°F | 212°C |
| @ 264 psi (1.82 MPa) | | 330°F | 165°C |
| Coeff. of Therm. Expansion (linear) | E831 | | |
| Mold Direction × $10^{-5}$ | | | |
| Range: −40→100°F (−40→40°C) | | 3.5 in/in/°F | 6.4 m/m/°C |
| Range: 140→280°F (60→140°C) | | 6.5 in/in/°F | 11.7 m/m/°C |
| **Electrical** | | | |
| Dielectric Strength | D149 | | |
| 1/16″ (1.6 mm) | | 610 V/mil | 24 kV/mm |
| 1/8″ (3.2 mm) | | 520 V/mil | 20 kV/mm |
| Dielectric Constant | D150 | | |
| 100 Hz | | 4.2 | 4.2 |
| $10^6$ Hz | | 4.0 | 4.0 |
| Dissipation Factor | D150 | | |
| 100 Hz | | .003 | .003 |
| $10^6$ Hz | | .002 | .002 |
| Volume Resistivity | D257 | $4.6 \times 10^{16}$ Ohm-cm | |

*(continued)*

XENOY 6380—30% Glass Reinforced Impact Modified Resin *(continued)*

| Property | ASTM Method | English Units | SI Units |
|---|---|---|---|
| **UL Properties** | | | |
| Flammability | UL94 | HB/.062″ | |
| Arc Resistance | D495 | 68/.125″ sec | |
| High Voltage Arc Tracking Rate | UL746A | .8/.062″ in/min | |
| High Ampere Arc Ignition | UL746A | 27/.062″ arcs | |
| Hot Wire Ignition | UL746A | 27/.062″ sec | |
| Comparative Track Index (CTI) | UL746A | 390/.125″ volts | |
| UL Temp. Index | UL746B | | |
| Electrical Properties | | 75/.062″ | |
| Mechanical Properties with Impact | | 75/.062″ | |
| Mechanical Properties without Impact | | 75/.062″ | |

# General Electric Company

## Contact Address:

General Electric Company
Plastics Group
One Plastics Avenue
Pittsfield, MA 01201

Phone: 1 (800) 845-0600

# XENOY 6620

## Product Information:

XENOY is a polycarbonate based composite polymer alloy, also available unreinforced or with glass reinforcement, formulated to improve impact resistance. XENOY resins can be injection molded, extruded, blow molded, or thermoformed. XENOY has excellent physical strength, and is resistant to impact. Its excellent melt flow properties allows it to be processed over a wide range of temperature and pressure. XENOY has excellent chemical resistance, UV-stability, dimensional stability, low mold shrinkage, and low water absorbtion.

## Product Data:

**XENOY 6620 Impact Modified Resin**

| Property | ASTM Method | English Units | SI Units |
|---|---|---|---|
| **Physical** | | | |
| Specific Gravity | D792 | 1.20 | 1.20 |
| Specific Volume | D892 | 23.2 in$^3$/lb | 835 cm$^3$/kg |
| Water Absorption, 24 hours | D570 | .08% | .08% |
| Mold Shrinkage<br>Flow Direction<br>(0.125)<br>Cross Flow Direction<br>(0.125 mil) | | <br><br>14–20 in/in × 10$^{-3}$<br><br>14–20 in/in × 10$^{-3}$ | |
| **Mechanical** | | | |
| Tensile Strength | D638 | 6300 psi | 43 MPa |
| Elongation at Break | D638 | 175% | 175% |
| Flexural Strength | D790 | 9300 psi | 64 MPa |
| Flexural Modulus | D790 | 250,000 psi | 1724 MPa |

*(continued)*

NOTE—Information provided in this book has been condensed. The publisher recommends that the product manufacturer be contacted, at the address or phone number provided, for more complete information on product safety, processing parameters, and applications not provided in this information sheet.

**XENOY 6620 Impact Modified Resin *(continued)***

| Property | ASTM Method | English Units | SI Units |
|---|---|---|---|
| **Physical** | | | |
| Compressive Strength | D695 | 6700 psi | 46 MPa |
| Shear Strength | D732 | 6200 psi | 43 MPa |
| Izod Impact Strength | D256 | | |
| Notched, 1/8″ thick (3.2 mm) | | 15@ −30°C | |
| Unnotched, 1/8″ thick (3.2 mm) | | NB | |
| Gardner Impact, 1/8″ thick (3.2 mm) | | 40 ft-lb | 54 J |
| Rockwell Hardness, R-scale | D785 | 108 | 108 |
| **Thermal** | | | |
| Heat Deflection Temp. | D648 | | |
| @ 66 psi (0.46 MPa) | | 210°F | 99°C |
| @ 264 psi (1.82 MPa) | | 140°F | 60°C |
| Coeff. of Therm. Expansion (linear) | E831 | | |
| Mold Direction $\times 10^{-5}$ | | | |
| Range: −40→100°F (−40→40°C) | | 2.8 in/in/°F | 5.0 m/m/°C |
| Range: 140→280°F (60→140°C) | | 5.7 in/in/°F | 10.3 m/m/°C |
| **Electrical** | | | |
| Dielectric Strength | D149 | | |
| 1/16″ (1.6 mm) | | 710 V/mil | 28 kV/mm |
| 1/8″ (3.2 mm) | | 485 V/mil | 19 kV/mm |
| Dielectric Constant | D150 | | |
| 100 Hz | | 3.1 | 3.1 |
| $10^6$ Hz | | 3.0 | 3.0 |
| Dissipation Factor | D150 | | |
| 100 Hz | | .002 | .002 |
| $10^6$ Hz | | .02 | .02 |
| Volume Resistivity | D257 | $5.5 \times 10^{16}$ Ohm-cm | |

*(continued)*

**XENOY 6620 Impact Modified Resin *(continued)***

| Property | ASTM Method | English Units | SI Units |
|---|---|---|---|
| **UL Properties** | | | |
| Flammability | UL94 | HB/.062″ | |
| Arc Resistance | D495 | 130/.123″ sec | |
| High Voltage Arc Tracking Rate | UL746A | .67/.062″ in/min | |
| High Ampere Arc Ignition | UL746A | 115/.123″ arcs | |
| Hot Wire Ignition | UL746A | 15/.123″ sec | |
| Comparative Track Index (CTI) | UL746A | 600/.123″ volts | |
| UL Temp. Index | UL746B | | |
| Electrical Properties | | 75/.062″ | |
| Mechanical Properties with Impact | | 75/.062″ | |
| Mechanical Properties without Impact | | 75/.062″ | |

# MACH-1 Compounding

## Contact Address:

MACH-1 Compounding
775 East Highland Road
Macedonia, OH 44056

Phone: 1 (216) 467-8108

## CPE-24-2

### Product Information:

MACH-1 Compounding is a custom compounder of rubber and plastic materials. MACH-1 has a proprietary line of nitrile rubber blends with polyvinyl chloride called Bur-A-Loy®. Bur-A-Loy® products are used in the rubber industry and are compounded to make thermoset rubber products for the rubber roll, wire and cable, hose and other industries.

MACH-1 Compounding manufactures compounded thermoplastic CPE products, such as CPE 24-2 and CPE 30-2, for the wire and cable industry. Other thermoplastic resins used include PVC, PE, PP, ABS, TPE's, TPO's, TPU's, SBS, SEBS, CPVC, EVA, EMA, etc. We incorporate fillers, colors, curatives, plasticizers, rubbers and other additives into the resins mentioned.

MACH-1 Compounding is able to tailor its custom compounding service to composition and color, as well as to achieve effective user performance. Handling costs are substantially reduced and significant savings in time and money are achieved by designing the product to suit customer's existing equipment and processes.

MACH-1's production facilities are totally dedicated to custom mixing free of any conflicting finished products, giving you added assurance that your valued formulations and products are protected.

### Product Data:

**Requested Properties and Specifications: CPE-24-2**

| Properties | Specifications and Conditions | Test Results and Conditions |
|---|---|---|
| Durometer | Shore D—typ. | 48 |
| Tensile Strength | psi—typ. | 1400 |
| Elongation | %—typ. | 425 |
| 100% Modulus | psi—typ. | 1100 |
| Specific Gravity | gm/cc | 1.290 |
| Oxygen Index | LOI % | 31 |
| Low Temperature | ASTM D-746, °C | −29 |
| Moisture, oil & electric properties | | good |

Based on Dow formulation W546-82038-24-2.

NOTE—Information provided in this book has been condensed. The publisher recommends that the product manufacturer be contacted, at the address or phone number provided, for more complete information on product safety, processing parameters, and applications not provided in this information sheet.

# MACH-1 Compounding

## Contact Address:

MACH-1 Compounding
775 East Highland Road
Macedonia, OH 44056

Phone: 1 (216) 467-8108

## CPE-30-2

### Product Information:

MACH-1 Compounding is a custom compounder of rubber and plastic materials. MACH-1 has a proprietary line of nitrile rubber blends with polyvinyl chloride called Bur-A-Loy®. Bur-A-Loy® products are used in the rubber industry and are compounded to make thermoset rubber products for the rubber roll, wire and cable, hose and other industries.

MACH-1 Compounding manufactures compounded thermoplastic CPE products, such as CPE 24-2 and CPE 30-2, for the wire and cable industry. Other thermoplastic resins used include PVC, PE, PP, ABS, TPE's, TPO's, TPU's, SBS, SEBS, CPVC, EVA, EMA, etc. We incorporate fillers, colors, curatives, plasticizers, rubbers and other additives into the resins mentioned.

MACH-1 Compounding is able to tailor its custom compounding service to composition and color, as well as to achieve effective user performance. Handling costs are substantially reduced and significant savings in time and money are achieved by designing the product to suit customer's existing equipment and processes.

MACH-1's production facilities are totally dedicated to custom mixing free of any conflicting finished products, giving you added assurance that your valued formulations and products are protected.

### Product Data:

**Requested Properties and Specifications: CPE-30-2**

| Properties | Specifications and Conditions | Test Results and Conditions |
|---|---|---|
| Durometer | Shore D—typ. | 46 |
| Tensile Strength | psi—typ. | 1700 |
| Elongation | %—typ. | 500 |
| 100% Modulus | psi—typ. | 900 |
| Specific Gravity | gm/cc | 1.37 |
| Oxygen Index | LOI % | +32.0% |
| Low Temperature | ASTM D-746, °C | −35 |
| Moisture, oil & electric properties | | good |

Based on Dow formulation W546-82038-24-2.

NOTE—Information provided in this book has been condensed. The publisher recommends that the product manufacturer be contacted, at the address or phone number provided, for more complete information on product safety, processing parameters, and applications not provided in this information sheet.

# Mobay Chemical Corporation

## Contact Address:

Mobay Chemical Corporation
Plastics Division
Mobay Road
Pittsburgh, PA 15205-9741

Phone: 1 (412) 777-2000

## BAYBLEND DP-2-1439

### Product Information:

BAYBLEND DP-2-1439, -1440, and -1441 are unreinforced, flame retarded thermoplastic blends of Merlon polycarbonate and ABS resin. They represent new flame retardant technology that combines easy processability and excellent property performance. These injection molding grades of Bayblend are naturally opaque and are supplied in pellet form. Trial size quantities are available for evaluation (colors upon request).

### Markets & Applications:

Bayblend DP-2-1439, -1440, and -1441 resins have UL 94 V-O at 0.062 inch thickness as tested in Mobay's laboratories. Their non-blooming flame retardancy, good processability, and excellent toughness enabled them to be used for demanding applications such as:

*Business Machine*—Personal & point-of-sale computer housings; typewriter housings; general office equipment.

*Electrical/Electronic*—General wiring devices; plugs & receptacles; control housings; telecom system connectors.

### Processing:

Processing parameters for Bayblend DP-2-1439, -1440, and -1441 are noted below. Predrying prior to processing is essential to insure desired appearance and property performance.

### Typical Injection Molding Conditions:

| | Temperatures | | | |
|---|---|---|---|---|
| | °F | °C | | Pressures |
| Barrel—Rear | 430–445 | (220–230) | Injection | 10–16,000 psi (69–110 MPa) |
| Middle | 435–455 | (225–235) | Hold | 50–75% injection pressure |
| Front | 445–465 | (230–240) | Back | 50–100 psi (345–690 kPa) |
| Nozzle | 485–505 | (253–263) | | |
| Melt | 430–500 | (220–260) | Fill Rate | Slow to Moderate |
| Mold—Stationary | 120–175 | (50–80) | Cushion | 0.13–0.25 in (3–6 mm) |
| Moving | 120–175 | (50–80) | Screw Speed | 40–70 rpm |
| | | | Clamp | 3–5 ton/in$^2$ |

*Note:* Melt temperatures of 520°F/270°C should not be exceeded.

*Drying:* 175–195°F/80–90°C for 3–4 hours in a dehumidifying hopper dryer with a maximum dew point of 0°F (−18°C) to achieve a moisture content of less than 0.1%.

NOTE—Information provided in this book has been condensed. The publisher recommends that the product manufacturer be contacted, at the address or phone number provided, for more complete information on product safety, processing parameters, and applications not provided in this information sheet.

## Typical Physical Properties:

### BAYBLEND DP-2-1439 *(continued)*

| Property | ASTM Method (Others) | English Units | SI Units |
|---|---|---|---|
| **Physical** | | | |
| Density | D792 | 0.0422 lb/in³ | 1.17 gm/cm³ |
| Mold Shrinkage | D955 | 0.005–0.007 in/in | 0.5–0.07% |
| Rockwell Hardness (R-Scale) | D785 | R115 | |
| **Mechanical** | | | |
| Tensile Strength at Yield | D638 | 7700 lb/in² | 53 MPa |
| Elongation at Yield | D638 | 3.5% | |
| Tensile Strength at Break | D638 | 6500 lb/in² | 45 MPa |
| Elongation at Break | D638 | 70% | |
| Tensile Modulus | D638 | 377,000 lb/in² | 2.60 GPa |
| Flexural Strength | D790 | 12,600 lb/in² | 87 MPa |
| Flexural Modulus | D790 | 377,000 lb/in² | 2.60 GPa |
| Izod Impact Strength (Notched): 3.2mm (0.125 in) Thickness @ 23°C (73°F) | D256 | 5.6 ft-lb/in | 300 J/m |
| Gardner Impact Strength (1.27 mm/0.5 in diameter Dart) @ 23°C (73°F) | D3029 | 45 ft-lb | 33 J |
| **Thermal** | | | |
| Deflection Temperature under Load | D648 | | |
| 1.82 MPa (264 psi) Load | | 180°F | 80°C |
| 0.46 MPa (66 psi) Load | | 194°F | 90°C |
| Vicat Softening Temp., B/120 | D1525 | 194°F | 90°C |
| **Electrical** | | | |
| Dielectric Strength | D149 | >2000 V/mil | >80 kV/mm |
| Comparative Track Index | (IEC) | 300 | |
| **Flammability** | | | |
| Vertical @ Minimum Thickness V-O* Rating | (UL 94) | 0.062 in | 1.57 mm |
| **Rheology** | | | |
| Spiral Flow @ 2mm/0.079 in Thickness & 260°/500°F Melt Temp. | (Mobay) | 23 in | 580 mm |

*The flammability results are based on small scale laboratory tests for purposes of relative comparison and do not necessarily represent the hazard presented by this material or any other material under actual fire conditions. The tests were conducted in Mobay's laboratories.

# Mobay Chemical Corporation

## Contact Address:

Mobay Chemical Corporation
Plastics Division
Mobay Road
Pittsburgh, PA 15205-9741

Phone: 1 (412) 777-2000

## BAYBLEND DP-2-1440

### Product Information:

BAYBLEND DP-2-1439, -1440, and -1441 are unreinforced, flame retarded thermoplastic blends of Merlon polycarbonate and ABS resin. They represent new flame retardant technology that combines easy processability and excellent property performance. These injection molding grades of Bayblend are naturally opaque and are supplied in pellet form. Trial size quantities are available for evaluation (colors upon request).

### Markets & Applications:

Bayblend DP-2-1439, -1440, and -1441 resins have UL 94 V-O at 0.062 inch thickness as tested in Mobay's laboratories. Their non-blooming flame retardancy, good processability, and excellent toughness enabled them to be used for demanding applications such as:

*Business Machine*—Personal & point-of-sale computer housings; typewriter housings; general office equipment.

*Electrical/Electronic*—General wiring devices; plugs & receptacles; control housings; telecom system connectors.

### Processing:

Processing parameters for Bayblend DP-2-1439, -1440, and -1441 are noted below. Predrying prior to processing is essential to insure desired appearance and property performance.

### Typical Injection Molding Conditions:

| | Temperatures | | | |
|---|---|---|---|---|
| | °F | °C | | Pressures |
| Barrel—Rear | 430–445 | (220–230) | Injection | 10–16,000 psi (69–110 MPa) |
| Middle | 435–455 | (225–235) | Hold | 50–75% injection pressure |
| Front | 445–465 | (230–240) | Back | 50–100 psi (345–690 kPa) |
| Nozzle | 485–505 | (253–263) | | |
| Melt | 430–500 | (220–260) | Fill Rate | Slow to Moderate |
| Mold—Stationary | 120–175 | (50–80) | Cushion | 0.13–0.25 in (3–6 mm) |
| Moving | 120–175 | (50–80) | Screw Speed | 40–70 rpm |
| | | | Clamp | 3–5 ton/in$^2$ |

*Note:* Melt temperatures of 520°F/270°C should not be exceeded.

*Drying:* 175–195°F/80–90°C for 3–4 hours in a dehumidifying hopper dryer wiht a maximum dew point of 0°F (−18°C) to achieve a moisture content of less than 0.1%.

NOTE—Information provided in this book has been condensed. The publisher recommends that the product manufacturer be contacted, at the address or phone number provided, for more complete information on product safety, processing parameters, and applications not provided in this information sheet.

## Typical Physical Properties:

**BAYBLEND DP-2-1440**

| Property | ASTM Method (Others) | English Units | SI Units |
|---|---|---|---|
| **Physical** | | | |
| Density | D792 | 0.0422 lb/in³ | 1.18 gm/cm³ |
| Mold Shrinkage | D955 | 0.005–0.007 in/in | 0.5–0.07% |
| Rockwell Hardness (R-Scale) | D785 | R119 | |
| **Mechanical** | | | |
| Tensile Strength at Yield | D638 | 8000 lb/in² | 55 MPa |
| Elongation at Yield | D638 | 4.0% | |
| Tensile Strength at Break | D638 | 6500 lb/in² | 45 MPa |
| Elongation at Break | D638 | 70% | |
| Tensile Modulus | D638 | 377,000 lb/in² | 2.60 GPa |
| Flexural Strength | D790 | 13,000 lb/in² | 90 MPa |
| Flexural Modulus | D790 | 377,000 lb/in² | 2.60 GPa |
| Izod Impact Strength (Notched): 3.2mm (0.125 in) Thickness @ 23°C (73°F) | D256 | 7.5 ft-lb/in | 400 J/m |
| Gardner Impact Strength (1.27 mm/0.5 in diameter Dart) @ 23°C (73°F) | D3029 | 46 ft-lb | 34 J |
| **Thermal** | | | |
| Deflection Temperature under Load | D648 | | |
| 1.82 MPa (264 psi) Load | | 194°F | 90°C |
| 0.46 MPa (66 psi) Load | | 212°F | 100°C |
| Vicat Softening Temp., B/120 | D1525 | 212°F | 100°C |
| **Electrical** | | | |
| Dielectric Strength | D149 | >2000 V/mil | >80 kV/mm |
| Comparative Track Index | (IEC) | 275 | |
| **Flammability** | | | |
| Vertical @ Minimum Thickness V-O* Rating (as tested in Mobay's lab) | (UL 94) | 0.062 in | 1.57 mm |
| **Rheology** | | | |
| Spiral Flow @ 2mm/0.079 in Thickness & 260°/500°F Melt Temp. | (Mobay) | 21 in | 540 mm |

*The flammability results are based on small scale laboratory tests for purposes of relative comparison and do not necessarily represent the hazard presented by this material or any other material under actual fire conditions. The tests were conducted in Mobay's laboratories.

# Mobay Chemical Corporation

## Contact Address:

Mobay Chemical Corporation
Plastics Division
Mobay Road
Pittsburgh, PA 15205-9741

Phone: 1 (412) 777-2000

## BAYBLEND DP-2-1441

### Product Information:

BAYBLEND DP-2-1439, -1440, and -1441 are unreinforced, flame retarded thermoplastic blends of Merlon polycarbonate and ABS resin. They represent new flame retardant technology that combines easy processability and excellent property performance. These injection molding grades of Bayblend are naturally opaque and are supplied in pellet form. Trial size quantities are available for evaluation (colors upon request).

### Markets & Applications:

Bayblend DP-2-1439, -1440, and -1441 resins have UL 94 V-O at 0.062 inch thickness as tested in Mobay's laboratories. Their non-blooming flame retardancy, good processability, and excellent toughness enabled them to be used for demanding applications such as:

*Business Machine*—Personal & point-of-sale computer housings; typewriter housings; general office equipment.

*Electrical/Electronic*—General wiring devices; plugs & receptacles; control housings; telecom system connectors.

### Processing:

Processing parameters for Bayblend DP-2-1439, -1440, and -1441 are noted below. Predrying prior to processing is essential to insure desired appearance and property performance.

### Typical Injection Molding Conditions:

| | Temperatures | | | |
|---|---|---|---|---|
| | °F | °C | | Pressures |
| Barrel—Rear | 430–445 | (220–230) | Injection | 10–16,000 psi (69–110 MPa) |
| Middle | 435–455 | (225–235) | Hold | 50–75% injection pressure |
| Front | 445–465 | (230–240) | Back | 50–100 psi (345–690 kPa) |
| Nozzle | 485–505 | (253–263) | | |
| Melt | 430–500 | (220–260) | Fill Rate | Slow to Moderate |
| Mold—Stationary | 120–175 | (50–80) | Cushion | 0.13–0.25 in (3–6 mm) |
| Moving | 120–175 | (50–80) | Screw Speed | 40–70 rpm |
| | | | Clamp | 3-5 ton/in² |

*Note:* Melt temperatures of 520°F/270°C should not be exceeded.

*Drying:* 175–195°F/80–90°C for 3–4 hours in a dehumidifying hopper dryer wiht a maximum dew point of 0°F (−18°C) to achieve a moisture content of less than 0.1%.

NOTE—Information provided in this book has been condensed. The publisher recommends that the product manufacturer be contacted, at the address or phone number provided, for more complete information on product safety, processing parameters, and applications not provided in this information sheet.

## Typical Physical Properties:

**BAYBLEND DP-2-1441**

| Property | ASTM Method (Others) | English Units | SI Units |
|---|---|---|---|
| **Physical** | | | |
| Density | D792 | 0.0426 lb/in³ | 1.18 gm/cm³ |
| Mold Shrinkage | D955 | 0.005–0.007 in/in | 0.5–0.07% |
| Rockwell Hardness (R-Scale) | D785 | R121 | |
| **Mechanical** | | | |
| Tensile Strength at Yield | D638 | 8600 lb/in² | 59 MPa |
| Elongation at Yield | D638 | 4.5% | |
| Tensile Strength at Break | D638 | 7300 lb/in² | 50 MPa |
| Elongation at Break | D638 | 70% | |
| Tensile Modulus | D638 | 377,000 lb/in² | 2.60 GPa |
| Flexural Strength | D790 | 13,900 lb/in² | 96 MPa |
| Flexural Modulus | D790 | 377,000 lb/in² | 2.60 GPa |
| Izod Impact Strength (Notched):<br>3.2mm (0.125 in) Thickness<br>@ 23°C (73°F) | D256 | 9.4 ft-lb/in | 500 J/m |
| Gardner Impact Strength<br>(1.27 mm/0.5 in diameter Dart)<br>@ 23°C (73°F) | D3029 | 48 ft-lb | 35 J |
| **Thermal** | | | |
| Deflection Temperature under Load<br>1.82 MPa (264 psi) Load<br>0.46 MPa (66 psi) Load | D648 | <br>208°F<br>230°F | <br>98°C<br>110°C |
| Vicat Softening Temp., B/120 | D1525 | 230°F | 110°C |
| **Electrical** | | | |
| Dielectric Strength | D149 | >2000 V/mil | >80 kV/mm |
| Comparative Track Index | (IEC) | 275 | |
| **Flammability** | | | |
| Vertical @ Minimum Thickness<br>V-O* Rating<br>(as tested in Mobay's lab) | (UL 94) | 0.062 in | 1.57 mm |
| **Rheology** | | | |
| Spiral Flow @ 2mm/0.079 in<br>Thickness & 260°/500°F Melt Temp. | (Mobay) | 20 in | 500 mm |

*The flammability results are based on small scale laboratory tests for purposes of relative comparison and do not necessarily represent the hazard presented by this material or any other material under actual fire conditions. The tests were conducted in Mobay's laboratories.

# Mobay Chemical Corporation

## Contact Address:

Mobay Chemical Corporation
Plastics Division
Mobay Road
Pittsburgh, PA 15205-9741

Phone: 1 (412) 777-2000

## BAYBLEND DP-2-1482

### Product Information:

Bayblend is a thermoplastic blend of Merlon polycarbonate and ABS resins, which combines the good properties of both polymers. Bayblend DP-2-1482, DP-2-1483, and DP-2-1484 are based on technology commercially utilized extensively in the worldwide automotive market. Depending on type, the heat deflection temperatures range from 95 to 105°C (between that of ABS and PC). However, due to its unique two-phase morphology, Bayblend exhibits excellent creep behavior and, in a no-load state, molded articles show a much lower tendency to warp than competitive materials of similar heat deflection temperature. Stiffness and hardness are comparable with polycarbonate up to the softening range. The Gardner impact strength also approaches that of impact modified polycarbonate (down to −29°C).

As amorphous thermoplastics, these Bayblend developmental grades exhibit the high dimensional accuracy in service which is typical of this group of materials. Total shrinkage and shrinkage differential are low, while post-molding shrinkage is negligible. Low moisture absorption is another key Bayblend attribute.

These Bayblend grades can meet narrow tolerances and yet offer wide latitude for the designer. Molded parts with high gloss or textured surfaces can be produced in virtually any opaque color shades. Foil stamping, electroplating, and coating are achievable with standard methods. Due to its good electrical insulating properties, Bayblend is also highly suitable for protecting electrical apparatus. This combination of properties makes these Bayblend resins ideally suited for applications in the automotive and general transportation markets. Typical properties are summarized in the table.

Typical existing interior applications for this family of Bayblend materials include air vents, central consoles, decorative trim, glove compartment doors, grilles, electrical housings, instrument frames, instrument panels, buffer strips, speaker fascias, steering column shrouds, wheel nut caps, window surrounds, and mirror housings.

Exterior applications include cowl vent grilles, body side moldings, wheel covers, tail-light housings, head-lamp surrounds, logo plates, spoilers, and body components.

NOTE—Information provided in this book has been condensed. The publisher recommends that the product manufacturer be contacted, at the address or phone number provided, for more complete information on product safety, processing parameters, and applications not provided in this information sheet.

## Typical Physical Properties:

**BAYBLEND DP-2-1482**

| Property | ASTM Method (Others) | English Units | SI Units |
|---|---|---|---|
| **Physical** | | | |
| Density | D792 | 0.0397 lb/in³ | 1.10 Mg/m³ |
| Mold Shrinkage | D955 | 0.004–0.007 in/in | 0.004–0.007 mm/mm |
| Water Absorption (immersion)<br>Content at 24 hrs, 23°C (73°F)<br>Equilibrium 240 hrs, 23°C/73°F | D570 | <br>0.2%<br>0.7% | |
| Rockwell Hardness (R Scale) | D785 | 114 | |
| **Mechanical** | | | |
| Tensile Strength at Yield | D638 | 6900 lb/in² | 48 MPa |
| Elongation at Yield | D638 | 4% | |
| Tensile Strength at Break | D638 | 5800 lb/in² | 40 MPa |
| Elongation at Break | D638 | 40% | |
| Flexural Strength | D790 | 10,100 lb/in² | 70 MPa |
| Flexural Modulus | D790 | 276,000 lb/in² | 1.90 GPa |
| Izod Impact Strength (Notched)<br>3.2mm (0.125 in) Thickness<br>@ 23°C (73°F)<br>6.4mm (0.250 in) Thickness<br>@ 23°C (73°F) | D256 | <br><br>9.6 ft-lb/in<br><br>6.3 ft-lb/in | <br><br>512 J/m<br><br>336 J/m |
| Izod Impact Strength (Unnotched) | D256 | no break | |
| Gardner Impact Strength<br>(12.7mm/0.5 in diameter Dart)<br>23°C/73°F<br>−29°C/−20°F | (Mobay) | <br><br>28 ft-lb<br>22 ft-lb | <br><br>38 J<br>30 J |
| **Thermal** | | | |
| Deflection Temperature under Load<br>1.82 MPa (264 psi) Load<br>0.46 MPa (66 psi) Load | D648 | <br>203°F<br>100°F | <br>95°C<br>100°C |
| Coefficient of Linear Thermal Expansion | D696 | 44–47 E-6 in/in/°F | 80–85 E-6 mm/mm/°C |

# Mobay Chemical Corporation

## Contact Address:

Mobay Chemical Corporation
Plastics Division
Mobay Road
Pittsburgh, PA 15205-9741

Phone: 1 (412) 777-2000

# BAYBLEND DP-2-1483

## Product Information:

Bayblend is a thermoplastic blend of Merlon polycarbonate and ABS resins, which combines the good properties of both polymers. Bayblend DP-2-1482, DP-2-1483, and DP-2-1484 are based on technology commercially utilized extensively in the worldwide automotive market. Depending on type, the heat deflection temperatures range from 95 to 105°C (between that of ABS and PC). However, due to its unique two-phase morphology, Bayblend exhibits excellent creep behavior and, in a no-load state, molded articles show a much lower tendency to warp than competitive materials of similar heat deflection temperature. Stiffness and hardness are comparable with polycarbonate up to the softening range. The Gardner impact strength also approaches that of impact modified polycarbonate (down to −29°C).

As amorphous thermoplastics, these Bayblend developmental grades exhibit the high dimensional accuracy in service which is typical of this group of materials. Total shrinkage and shrinkage differential are low, while post-molding shrinkage is negligible. Low moisture absorption is another key Bayblend attribute.

These Bayblend grades can meet narrow tolerances and yet offer wide latitude for the designer. Molded parts with high gloss or textured surfaces can be produced in virtually any opaque color shades. Foil stamping, electroplating, and coating are achievable with standard methods. Due to its good electrical insulating properties, Bayblend is also highly suitable for protecting electrical apparatus. This combination of properties makes these Bayblend resins ideally suited for applications in the automotive and general transportation markets. Typical properties are summarized in the table.

Typical existing interior applications for this family of Bayblend materials include air vents, central consoles, decorative trim, glove compartment doors, grilles, electrical housings, instrument frames, instrument panels, buffer strips, speaker fascias, steering column shrouds, wheel nut caps, window surrounds, and mirror housings.

Exterior applications include cowl vent grilles, body side moldings, wheel covers, tail-light housings, head-lamp surrounds, logo plates, spoilers, and body components.

NOTE – Information provided in this book has been condensed. The publisher recommends that the product manufacturer be contacted, at the address or phone number provided, for more complete information on product safety, processing parameters, and applications not provided in this information sheet.

## Typical Physical Properties:

**BAYBLEND DP-2-1483**

| Property | ASTM Method (Others) | English Units | SI Units |
|---|---|---|---|
| **Physical** | | | |
| Density | D792 | 0.0405 $lb/in^3$ | 1.12 $Mg/m^3$ |
| Mold Shrinkage | D955 | 0.004–0.007 in/in | 0.004–0.007 mm/mm |
| Water Absorption (immersion)<br>Content at 24 hrs, 23°C (73°F)<br>Equilibrium 240 hrs, 23°C/73°F | D570 | <br>0.2%<br>0.6% | |
| Rockwell Hardness (R Scale) | D785 | 118 | |
| **Mechanical** | | | |
| Tensile Strength at Yield | D638 | 7100 $lb/in^2$ | 49 MPa |
| Elongation at Yield | D638 | 5% | |
| Tensile Strength at Break | D638 | 6000 $lb/in^2$ | 41 MPa |
| Elongation at Break | D638 | 45% | |
| Flexural Strength | D790 | 11,100 $lb/in^2$ | 77 MPa |
| Flexural Modulus | D790 | 310,000 $lb/in^2$ | 2.14 GPa |
| Izod Impact Strength (Notched)<br>3.2mm (0.125 in) Thickness<br>@ 23°C (73°F)<br>6.4mm (0.250 in) Thickness<br>@ 23°C (73°F) | D256 | <br><br>10.1 ft-lb/in<br><br>8.0 ft-lb/in | <br><br>539 J/m<br><br>427 J/m |
| Izod Impact Strength (Unnotched) | D256 | no break | |
| Gardner Impact Strength<br>(12.7mm/0.5 in diameter Dart)<br>23°C/73°F<br>−29°C/−20°F | (Mobay) | <br><br>29 ft-lb<br>33 ft-lb | <br><br>39 J<br>45 J |
| **Thermal** | | | |
| Deflection Temperature under Load<br>1.82 MPa (264 psi) Load<br>0.46 MPa (66 psi) Load | D648 | <br>212°F<br>248°F | <br>100°C<br>120°C |
| Coefficient of Linear<br>Thermal Expansion | D696 | 42–44 E-6<br>in/in/°F | 75–80 E-6<br>mm/mm/°C |

# Mobay Chemical Corporation

## Contact Address:

Mobay Chemical Corporation
Plastics Division
Mobay Road
Pittsburgh, PA 15205-9741

Phone: 1 (412) 777-2000

## BAYBLEND DP-2-1484

### Product Information:

Bayblend is a thermoplastic blend of Merlon polycarbonate and ABS resins, which combines the good properties of both polymers. Bayblend DP-2-1482, DP-2-1483, and DP-2-1484 are based on technology commercially utilized extensively in the worldwide automotive market. Depending on type, the heat deflection temperatures range from 95 to 105°C (between that of ABS and PC). However, due to its unique two-phase morphology, Bayblend exhibits excellent creep behavior and, in a no-load state, molded articles show a much lower tendency to warp than competitive materials of similar heat deflection temperature. Stiffness and hardness are comparable with polycarbonate up to the softening range. The Gardner impact strength also approaches that of impact modified polycarbonate (down to −29°C).

As amorphous thermoplastics, these Bayblend developmental grades exhibit the high dimensional accuracy in service which is typical of this group of materials. Total shrinkage and shrinkage differential are low, while post-molding shrinkage is negligible. Low moisture absorption is another key Bayblend attribute.

These Bayblend grades can meet narrow tolerances and yet offer wide latitude for the designer. Molded parts with high gloss or textured surfaces can be produced in virtually any opaque color shades. Foil stamping, electroplating, and coating are achievable with standard methods. Due to its good electrical insulating properties, Bayblend is also highly suitable for protecting electrical apparatus. This combination of properties makes these Bayblend resins ideally suited for applications in the automotive and general transportation markets. Typical properties are summarized in the table.

Typical existing interior applications for this family of Bayblend materials include air vents, central consoles, decorative trim, glove compartment doors, grilles, electrical housings, instrument frames, instrument panels, buffer strips, speaker fascias, steering column shrouds, wheel nut caps, window surrounds, and mirror housings.

Exterior applications include cowl vent grilles, body side moldings, wheel covers, tail-light housings, head-lamp surrounds, logo plates, spoilers, and body components.

NOTE—Information provided in this book has been condensed. The publisher recommends that the product manufacturer be contacted, at the address or phone number provided, for more complete information on product safety, processing parameters, and applications not provided in this information sheet.

## Typical Physical Properties:

### BAYBLEND DP-2-1484

| Property | ASTM Method (Others) | English Units | SI Units |
|---|---|---|---|
| **Physical** | | | |
| Density | D792 | 0.0412 lb/in$^3$ | 1.14 Mg/m$^3$ |
| Mold Shrinkage | D955 | 0.004–0.007 in/in | 0.004–0.007 mm/mm |
| Water Absorption (immersion)<br>Content at 24 hrs, 23°C (73°F)<br>Equilibrium 240 hrs, 23°C/73°F | D570 | <br>0.2%<br>0.6% | |
| Rockwell Hardness (R Scale) | D785 | 119 | |
| **Mechanical** | | | |
| Tensile Strength at Yield | D638 | 7600 lb/in$^2$ | 52 MPa |
| Elongation at Yield | D638 | 5% | |
| Tensile Strength at Break | D638 | 6300 lb/in$^2$ | 43 MPa |
| Elongation at Break | D638 | 50% | |
| Flexural Strength | D790 | 11,600 lb/in$^2$ | 80 MPa |
| Flexural Modulus | D790 | 300,000 lb/in$^2$ | 2.07 GPa |
| Izod Impact Strength (Notched)<br>3.2mm (0.125 in) Thickness<br>@ 23°C (73°F)<br>6.4mm (0.250 in) Thickness<br>@ 23°C (73°F) | D256 | <br><br>10.2 ft-lb/in<br><br>8.4 ft-lb/in | <br><br>544 J/m<br><br>448 J/m |
| Izod Impact Strength (Unnotched) | D256 | no break | |
| Gardner Impact Strength<br>(12.7mm/0.5 in diameter Dart)<br>23°C/73°F<br>−29°C/−20°F | (Mobay) | <br><br>29 ft-lb<br>34 ft-lb | <br><br>39 J<br>46 J |
| **Thermal** | | | |
| Deflection Temperature under Load<br>1.82 MPa (264 psi) Load<br>0.46 MPa (66 psi) Load | D648 | <br>221°F<br>257°F | <br>105°C<br>125°C |
| Coefficient of Linear Thermal Expansion | D696 | 39–42 E-6 in/in/°F | 70–75 E-6 mm/mm/°C |

# Mobay Chemical Corporation

Contact Address:

Mobay Chemical Corporation
Plastics Division
Mobay Road
Pittsburgh, PA 15205-9741

Phone: 1 (412) 777-2000

## BAYBLEND MC-2500

Product Information:

BAYBLEND thermoplastic resins are Merlon polycarbonate/ABS alloys formulated to provide properties supeior to those of the individual polymers themselves. They find application as engineering polymers used to make high quality products.

BAYBLEND MC-2500 resin, a uniquely formulated polymeric alloy of Merlon brand polycarbonate with ABS resin, possesses an attractive combination of engineering characteristics.

- high impact strength for abuse resistance
- resistance to deflection under load at high temperatures
- rigidity for use in structural components
- dimensional stability for molding to close tolerances
- cosmetic surface appearance in converted parts
- easy to mold at moderate stock temperatures
- attractive cost/performance characteristics

The structural properties of Bayblend MC-2500, being well above those of ABS, enable this resin to compete successfully with high performance engineering thermoplastic resins, on both technical and economical bases, and also to qualify for economical replacement of conventional metal components in many applications.

Bayblend MC-2500 resin, available in the form of pellets, is designed for injection molding in equipment that is conventionally recommended for polycarbonate or ABS. The flow characteristics of the resin enable processing at moderate stock temperatures in the range of 245–275 °C.

A cream colored, opaque appearance characterizes natural Bayblend MC-2500 resin, and it can be color-matched accurately to fulfill requirements of stylized applications. Examples of uses already filled by this product include automotive lamp housings, tractor grills, radio speaker grills, electrical switches, mobile home accessories, components of a coffee maker, guard for a chain saw, food serving tray, side marker, floor scrubber handle, and hand-held shower heads.

*(continued)*

NOTE—Information provided in this book has been condensed. The publisher recommends that the product manufacturer be contacted, at the address or phone number provided, for more complete information on product safety, processing parameters, and applications not provided in this information sheet.

## Typical Physical Properties:

**BAYBLEND MC-2500**

| Property | ASTM Method (Others) | English Units | SI Units |
|---|---|---|---|
| **General** | | | |
| Density | D792 | 0.0404 lb/in³ | 1.12 Mg/m³ |
| Specific Volume | D792 | 24.8 in³/lb | 89 E-5 m³/kg |
| Mold Shrinkage | D955 | 0.005–0.007 in/in | 0.005–0.007 mm/mm |
| Water Absorption (immersion) 24 hr | D570 | 0.21% | |
| **Mechanical** | | | |
| Tensile Strength at Yield | D638 | 8500 lb/in² | 59 MPa |
| Tensile Strength at Break | D638 | 7300 lb/in² | 50 MPa |
| Tensile Strain at Yield | D638 | 5% | |
| Tensile Strain at Break | D638 | 15% | |
| Tensile Modulus | D638 | 380,000 lb/in² | 2.59 GPa |
| Flexural Strength at 5% Strain | D790 | 13,000 lb/in² | 90 MPa |
| Flexural Modulus | D790 | 350,000 lb/in² | 2.43 GPa |
| Izod Impact Strength (Notched) 23°C [3.2mm (0.125 in) Specimen] | D256 | 10.5 ft-lb/in | 560 J/m |
| Impact Strength, Dart Drop, 25mm Diameter Dart [3.2mm (0.125 in) Specimen] | | 112 ft-lb | 152 J |
| Rockwell Hardness (R Scale) | D785 | 117 | |
| **Thermal** | | | |
| Deflection Temperature<br>1.82 MPa (264 psi) Load<br>0.46 MPa (66 psi) Load | D648 | <br>240°F<br>250°F | <br>116°C<br>121°C |
| Coefficient of Thermal Expansion | D696 | 35 E-6 1/°F | 63 E-6 1/°C |
| Flammability Rating<br>1.6–6.4 mm (.062–.250 in) thickness | (UL 94) | | HB |
| **Electrical** [3.2mm (0.125 in) Specimen] | | | |
| Dielectric Strength | D149 | >434 V/mil | >17 kV/mm |
| Volume Resistivity | D257 | 1.0 E + 17 ohm-cm | |

*(continued)*

## BAYBLEND MC-2500 *(continued)*

| Property | ASTM Method (Others) | English Units | SI Units |
|---|---|---|---|
| **Electrical** [3.2mm (0.125 in) Specimen] | | | |
| Dielectric Constant | D150 | | |
| 60 Hz | | 2.90 | |
| 1 MHz | | 2.81 | |
| Dissipation Factor | D150 | | |
| 60 Hz | | 0.0038 | |
| 1 MHz | | 0.0082 | |
| Arc Resistance | D495 | 91 s | |

# Mobay Chemical Corporation

Contact Address:

Mobay Chemical Corporation
Plastics Division
Mobay Road
Pittsburgh, PA 15205-9741

Phone: 1 (412) 777-2000

## BAYBLEND MC-3200

Product Information:

BAYBLEND thermoplastic resins are Merlon polycarbonate/ABS alloys formulated to provide properties superior to those of the individual polymers themselves. They find application as engineering polymers used to make high quality products.

Bayblend MC-3200 is a general purpose blend of Merlon FCR polycarbonate and ABS resin. Through a unique combination of polymers, this new grade exhibits improved processability and higher melt flow rates when compared to Bayblend MC-2500, without sacrificing property performance. Bayblend MC-3200 is especially suited for demanding injection molded applications, where part geometry and size present a challenge.

- high impact strength for abuse resistance
- resistance to deflection under load at high temperatures
- rigidity for use in structural components
- dimensional stability for molding to close tolerances
- cosmetic surface appearance in converted parts
- high melt flow rates at moderate stock temperatures
- attractive cost/performance characteristics

The structural properties of Bayblend MC-3200, being well above those of ABS, enable this resin to compete successfully with high performance engineering thermoplastic resins, on both technical and economical bases, and also to qualify for economical replacement of conventional metal components in many applications.

Bayblend MC-3200 resin, available in the form of pellets, is designed for injection molding in equipment that is conventionally recommended for polycarbonate or ABS. The flow characteristics of the resin enable processing at moderate stock temperatures in the range of 245–275 °C.

A cream colored, opaque appearance characterizes natural Bayblend MC-3200 resin, and it can be color-matched accurately to fulfill requirements of stylized applications. Examples of potential uses include automotive lamp housings, tractor grills, radio speaker grills, electrical switches, mobile home accessories, components of a coffee maker, guard for a chain saw, food serving tray, side marker, floor scrubber handle, and hand-held shower heads.

*(continued)*

NOTE – Information provided in this book has been condensed. The publisher recommends that the product manufacturer be contacted, at the address or phone number provided, for more complete information on product safety, processing parameters, and applications not provided in this information sheet.

## Typical Physical Properties:

**BAYBLEND MC-3200**

| Property | ASTM Method (Others) | English Units | SI Units |
|---|---|---|---|
| **General** | | | |
| Density | D792 | 0.0404 lb/in³ | 1.12 Mg/m³ |
| Specific Volume | D792 | 24.8 in³/lb | 89 E-5 m³/kg |
| Mold Shrinkage | D955 | 0.005–0.008 in/in | 0.005–0.008 mm/mm |
| Water Absorption (immersion) 24 hr | D570 | 0.21% | |
| **Mechanical** | | | |
| Tensile Strength at Yield | D638 | 8500 lb/in² | 59 MPa |
| Tensile Strength at Break | D638 | 7300 lb/in² | 50 MPa |
| Tensile Elongation at Yield | D638 | 5% | |
| Tensile Elongation at Break | D638 | 15% | |
| Tensile Modulus | D638 | 380,000 lb/in² | 2.59 GPa |
| Flexural Strength at 5% Strain | D790 | 13,000 lb/in² | 90 MPa |
| Flexural Modulus | D790 | 350,000 lb/in² | 2.43 GPa |
| Izod Impact Strength (Notched) 23°C [3.2mm (0.125 in) Specimen] | D256 | 8.5 ft-lb/in | 454 J/m |
| Gardner Impact Strength (2.5mm/0.100 in) Dart: 12.7mm (0.50 in) diameter | | 38 ft-lb | 51.5 J |
| Rockwell Hardness (R Scale) | D785 | 117 | |
| **Thermal** | | | |
| Deflection Temperature<br>1.82 MPa (264 psi) Load<br>0.46 MPa (66 psi) Load | D648 | <br>240°F<br>250°F | <br>116°C<br>121°C |
| Coefficient of Thermal Expansion | D696 | 35 E-6 1/°F | 63 E-6 1/°C |
| Flammability Rating 1.6–6.4 mm (.062–.250 in) thickness | (UL 94) | HB* | |
| **Electrical** [3.2mm (0.125 in) Specimen] | | | |
| Dielectric Strength | D149 | >434 V/mil | >17 kV/mm |
| Volume Resistivity | D257 | 1.0 E + 17 ohm-cm | |

*The flammability results are based on small scale laboratory tests and do not necessarily represent the hazard presented by this material or any other material under actual fire conditions. The tests were conducted in Mobay's laboratories.

*(continued)*

**BAYBLEND MC-3200 *(continued)***

| Property | ASTM Method (Others) | English Units | SI Units |
|---|---|---|---|
| **Electrical** [3.2mm (0.125 in) Specimen] | | | |
| Dielectric Constant | D150 | | |
| 60 Hz | | 2.90 | |
| 1 MHz | | 2.81 | |
| Dissipation Factor | D150 | | |
| 60 Hz | | 0.0038 | |
| 1 MHz | | 0.0082 | |
| Arc Resistance | D495 | 91 s | |
| **Rheological** | | | |
| Spiral Flow | | | |
| Length @ 2.54 mm (0.100 in) thickness | | | |
| for 249°C (480°F) Melt Temperature | | | |
| MC-3200 | | 21 in | 533 mm |
| MC-2500 | | 18 in | 457 mm |
| for 274°C (525°F) Melt Temperature | | | |
| MC-3200 | | 30 in | 762 mm |
| MC-2500 | | 26 in | 660 mm |

# Mobay Chemical Corporation

## Contact Address:

Mobay Chemical Corporation
Plastics Division
Mobay Road
Pittsburgh, PA 15205-9741

Phone: 1 (412) 777-2000

## BAYBLEND MH-6570

### Product Information:

BAYBLEND MH series resins are blends of Merlon polycarbonate and ABS resins developed to be color stable (minimal color change during normal part life) to fluorescent lighting and filtered sunlight in an office environment. In addition, Bayblend MH-6570, MH-6580, and MH-6595 resins offer high impact strength, rigidity, good dimensional stability, high heat resistance plus cosmetic surface appearance to facilitate decorating.

### Typical Physical Properties:

**BAYBLEND MH-6570**

| Property | ASTM Method (Others) | English Units | SI Units |
|---|---|---|---|
| **Physical** | | | |
| Density | D792 | 0.0444 lb/in$^3$ | 1.23 Mg/m$^3$ |
| Specific Volume | D792 | 22.52 in$^3$/lb | 81 E-5 m$^3$/kg |
| Mold Shrinkage | D955 | 0.005–0.007 in/in | 0.005–0.007 mm/mm |
| Water Absorption (immersion) 24 hr 23°C (73°F) | D570 | 0.14% | |
| Rockwell Hardness (R Scale) | D785 | R117 | |
| **Mechanical** [@ 23°C (73°F) and 50% rh] | | | |
| Tensile Strength at Yield | D638 | 8900 lb/in$^2$ | 61.4 MPa |
| Tensile Strength at Break | D638 | 7200 lb/in$^2$ | 49.6 MPa |
| Tensile Elongation at Yield | D638 | 5% | |
| Tensile Elongation at Break | D638 | 45% | |
| Flexural Strength 3.2mm (0.125 in) Thickness | D790 | 12,000 lb/in$^2$ | 82.7 MPa |

*(continued)*

NOTE – Information provided in this book has been condensed. The publisher recommends that the product manufacturer be contacted, at the address or phone number provided, for more complete information on product safety, processing parameters, and applications not provided in this information sheet.

**BAYBLEND MH-6570 *(continued)***

| Property | ASTM Method (Others) | English Units | SI Units |
|---|---|---|---|
| **Mechanical** [@ 23°C (73°F) and 50% rh] | | | |
| Flexural Modulus<br>3.2mm (0.125 in) Thickness | D790 | 354,000 lb/in² | 2,440 MPa |
| Izod Impact Strength (Notched):<br>3.2mm (0.125 in) Thickness<br>@ 23°C (73°F) | | 8 ft-lb/in | 427 J/m |
| Gardner Impact Strength<br>(3.2mm/0.125 in)<br>Dart: 15.88mm (0.625 in) diameter | (Mobay) | 34.9 ft-lb | 47.3 J |
| **Thermal** | | | |
| Deflection Temperature under Load<br>1.82 MPa (264 psi) Load | D648 | 216°F | 102°C |
| Vicat Softening Temperature | D1525 | 208°F | 98°C |
| **Flammability Rating** | | | |
| Vertical @ Minimum Thickness | (UL 94) | 0.062 in | 1.6 mm |
| 5V @ Minimum Thickness | | 0.125 in | 3.2 mm |

# Mobay Chemical Corporation

## Contact Address:

Mobay Chemical Corporation
Plastics Division
Mobay Road
Pittsburgh, PA 15205-9741

Phone: 1 (412) 777-2000

## BAYBLEND MH-6580

### Product Information:

BAYBLEND MH series resins are blends of Merlon polycarbonate and ABS resins developed to be color stable (minimal color change during normal part life) to fluorescent lighting and filtered sunlight in an office environment. In addition, Bayblend MH-6570, MH-6580, and MH-6595 resins offer high impact strength, rigidity, good dimensional stability, high heat resistance plus cosmetic surface appearance to facilitate decorating.

### Typical Physical Properties:

**BAYBLEND MH-6580**

| Property | ASTM Method (Others) | English Units | SI Units |
|---|---|---|---|
| **Physical** | | | |
| Density | D792 | 0.0444 lb/in³ | 1.23 Mg/m³ |
| Specific Volume | D792 | 22.52 in³/lb | 81 E-5 m³/kg |
| Mold Shrinkage | D955 | 0.005–0.007 in/in | 0.005–0.007 mm/mm |
| Water Absorption (immersion) 24 hr 23°C (73°F) | D570 | 0.14% | |
| Rockwell Hardness (R Scale) | D785 | R117 | |
| **Mechanical** [@ 23°C (73°F) and 50% rh] | | | |
| Tensile Strength at Yield | D638 | 8700 lb/in² | 60.0 MPa |
| Tensile Strength at Break | D638 | 7400 lb/in² | 51.0 MPa |
| Tensile Elongation at Yield | D638 | 5% | |
| Tensile Elongation at Break | D638 | 20% | |
| Flexural Strength 3.2mm (0.125 in) Thickness | D790 | 13,000 lb/in² | 89.6 MPa |

*(continued)*

NOTE – Information provided in this book has been condensed. The publisher recommends that the product manufacturer be contacted, at the address or phone number provided, for more complete information on product safety, processing parameters, and applications not provided in this information sheet.

**BAYBLEND MH-6580 *(continued)***

| Property | ASTM Method (Others) | English Units | SI Units |
|---|---|---|---|
| **Mechanical** [@ 23°C (73°F) and 50% rh] | | | |
| Flexural Modulus<br>3.2mm (0.125 in) Thickness | D790 | 360,000 lb/in$^2$ | 2,480 MPa |
| Izod Impact Strength (Notched):<br>3.2mm (0.125 in) Thickness<br>@ 23°C (73°F) | D256 | 8 ft-lb/in | 427 J/m |
| Gardner Impact Strength<br>(3.2mm/0.125 in)<br>Dart: 15.88mm (0.625 in) diameter | (Mobay) | 37.8 ft-lb | 51.3 J |
| **Thermal** | | | |
| Deflection Temperature under Load<br>1.82 MPa (264 psi) Load | D648 | 216°F | 102°C |
| Vicat Softening Temperature | D1525 | 220°F | 104°C |
| **Flammability Rating** | | | |
| Vertical @ Minimum Thickness | (UL 94) | 0.062 in | 1.6 mm |
| 5V @ Minimum Thickness | | 0.125 in | 3.2 mm |

# Mobay Chemical Corporation

## Contact Address:

Mobay Chemical Corporation
Plastics Division
Mobay Road
Pittsburgh, PA 15205-9741

Phone: 1 (412) 777-2000

## BAYBLEND MH-6595

### Product Information:

BAYBLEND MH series resins are blends of Merlon polycarbonate and ABS resins developed to be color stable (minimal color change during normal part life) to fluorescent lighting and filtered sunlight in an office environment. In addition, Bayblend MH-6570, MH-6580, and MH-6595 resins offer high impact strength, rigidity, good dimensional stability, high heat resistance plus cosmetic surface appearance to facilitate decorating.

### Typical Physical Properties:

**BAYBLEND MH-6595**

| Property | ASTM Method (Others) | English Units | SI Units |
|---|---|---|---|
| **Physical** | | | |
| Density | D792 | 0.0423 lb/in³ | 1.17 Mg/m³ |
| Specific Volume | D792 | 22.50 in³/lb | 79 E-5 m³/kg |
| Mold Shrinkage | D955 | 0.005–0.007 in/in | 0.005–0.007 mm/mm |
| Water Absorption (immersion) 24 hr 23°C (73°F) | D570 | 0.010% | |
| Rockwell Hardness (R Scale) | D785 | R119 | |
| **Mechanical** [@ 23°C (73°F) and 50% rh] | | | |
| Tensile Strength at Yield | D638 | 8100 lb/in² | 55.0 MPa |
| Tensile Strength at Break | D638 | 7500 lb/in² | 52.0 MPa |
| Tensile Elongation at Yield | D638 | 5% | |
| Tensile Elongation at Break | D638 | 40% | |
| Flexural Strength 3.2mm (0.125 in) Thickness | D790 | 13,200 lb/in² | 91.0 MPa |

*(continued)*

NOTE—Information provided in this book has been condensed. The publisher recommends that the product manufacturer be contacted, at the address or phone number provided, for more complete information on product safety, processing parameters, and applications not provided in this information sheet.

BAYBLEND MH-6595 *(continued)*

| Property | ASTM Method (Others) | English Units | SI Units |
|---|---|---|---|
| **Mechanical** [@ 23°C (73°F) and 50% rh] | | | |
| Flexural Modulus<br>3.2mm (0.125 in) Thickness | D790 | 360,000 lb/in$^2$ | 2,480 MPa |
| Izod Impact Strength (Notched):<br>3.2mm (0.125 in) Thickness<br>@ 23°C (73°F) | D256 | 9.1 ft-lb/in | 486 J/m |
| Gardner Impact Strength<br>(3.2mm/0.125 in)<br>Dart: 15.88mm (0.625 in) diameter | (Mobay) | 33.9 ft-lb | 46.0 J |
| **Thermal** | | | |
| Deflection Temperature under Load<br>1.82 MPa (264 psi) Load | D648 | 230°F | 110°C |
| Vicat Softening Temperature | D1525 | 264°F | 129°C |
| **Flammability Rating** | | | |
| Vertical @ Minimum Thickness | (UL 94) | 0.125 in | 3.2 mm |
| 5V @ Minimum Thickness | | | |

# Mobay Chemical Corporation

## Contact Address:

Mobay Chemical Corporation
Plastics Division
Mobay Road
Pittsburgh, PA 15205-9741

Phone: 1 (412) 777-2000

## BAYBLEND MJ-2500

### Product Information:

BAYBLEND thermoplastic resins are Merlon polycarbonate/ABS alloys formulated to provide properties superior to those of the individual polymers themselves. They find application as engineering polymers used to make high quality products.

BAYBLEND MJ-2500 resin, a unique polymeric alloy of Merlon brand polycarbonate with ABS Resin, possesses a fascinating combination of engineering characteristics.

- high impact strength for abuse resistance
- compliance with FDA Regulations for certain food contact applications
- resistance to caustic chemicals
- resistance to deflection under load at high temperature
- rigidity for shape retention
- dimensional stability for maintaining close tolerances
- cosmetic surface appearance in converted parts
- processing at moderate stock temperatures
- attractive cost/performance characteristics

The structural properties of Bayblend MJ-2500, being well above those of ABS, enable this resin to compete successfully with high performance engineering thermoplastic resins, on both technical and economical bases. In certain food and medical applications, Bayblend MJ-2500 qualifies as an economical replacement material for conventional metal components, glass or other thermoplastics.

Bayblend MJ-2500 resin, available in the form of pellets, is designed for injection molding, extrusion and thermoforming using equipment that is conventionally recommended for polycarbonate or ABS. The flow characteristics of the resin enable processing at moderate stock temperatures in the range of 245–265°C.

A cream colored, opaque appearance characterizes natural Bayblend MC-2500 resin, and it can be color-matched accurately to fulfill requirements of stylized applications.

*(continued)*

NOTE—Information provided in this book has been condensed. The publisher recommends that the product manufacturer be contacted, at the address or phone number provided, for more complete information on product safety, processing parameters, and applications not provided in this information sheet.

## Typical Physical Properties:

**BAYBLEND MJ-2500**

| Property | ASTM Method (Others) | English Units | SI Units |
|---|---|---|---|
| **General** | | | |
| Density | D792 | 0.0404 lb/in$^3$ | 1.12 Mg/m$^3$ |
| Specific Volume | D792 | 24.7 in$^3$/lb | 89 E-5 m$^3$/kg |
| Mold Shrinkage | D955 | 0.005–0.007 in/in | 0.005–0.007 mm/mm |
| Water Absorption (immersion) 24 hr | D570 | 0.24% | |
| **Mechanical** | | | |
| Tensile Strength at Yield | D638 | 8400 lb/in$^2$ | 58.0 MPa |
| Tensile Strength at Break | D638 | 7250 lb/in$^2$ | 50 MPa |
| Tensile Strain at Yield | D638 | 5% | |
| Tensile Strain at Break | D638 | 65% | |
| Flexural Strength at 5% Strain | D790 | 1.2 E + 4 lb/in$^2$ | 84 MPa |
| Flexural Modulus | D790 | 3.2 E + 5 lb/in$^2$ | 2.2 GPa |
| Izod Impact Strength (Notched):<br>23°C [3.2mm (0.125 in) Specimen] | D256 | 10.0 ft-lb/in | 538 J/m |
| Impact Strength, Dart Drop, 12.7mm Diameter Dart<br>[2.0mm (0.08 in) Specimen] | | 38 ft-lb | 51 J |
| Rockwell Hardness (R Scale) | D785 | 120 | |
| **Thermal** | | | |
| Deflection Temperature<br>1.82 MPa (264 psi) Load<br>0.46 MPa (66 psi) Load | D648 | 220°F<br>225°F | 104°C<br>107°C |
| Coefficient of Thermal Expansion | D696 | 37 E-6 1/°F | 67 E-6 1/°C |
| Flammability Rating<br>1.6–6.4 mm (.062–.250 in) thickness | (UL 94) | HB | |
| **Electrical** | | | |
| Dielectric Strength | D149 | >430 V/mil | >17 kV/mm |
| Volume Resistivity | D257 | 1.0 E + 17 ohm-cm | |

*(continued)*

BAYBLEND MJ-2500 *(continued)*

| Property | ASTM Method (Others) | English Units | SI Units |
|---|---|---|---|
| **Electrical** | | | |
| Dielectric Constant | D150 | | |
| 60 Hz | | 2.9 | |
| 1 MHz | | 2.8 | |
| Dissipation Factor | D150 | | |
| 60 Hz | | 0.004 | |
| 1 MHz | | 0.008 | |
| Arc Resistance | D495 | >50 | |

# Mobay Chemical Corporation

## Contact Address:

Mobay Chemical Corporation
Plastics Division
Mobay Road
Pittsburgh, PA 15205-9741

Phone: 1 (412) 777-2000

---

# MAKROBLEND DP-4-1350

## Product Information:

MAKROBLEND thermoplastic resins are Merlon Polycarbonate/PET blends formulated to be chemically resistant to a wide range of hydrocarbon solvents and to water or water vapor. The products molded from MAKROBLEND resins have excellent physical properties and maintain them over a wide range of environmental conditions.

## Description:

MAKROBLEND DP-4-1350 is a 10% glass fiber reinforced polycarbonate/PET polyester blend. This injection molding resin is naturally opaque in apearance and supplied in pellet form. Trial size quantities are available in natural and black resin (colors upon special request).

## Markets and Applications:

MAKROBLEND DP-4-1350 has nearly twice the stiffness of the unreinforced MAKROBLEND UT-1018 resin. It is also chemically resistant to a broad range of hydrocarbons, acids, bases, and solvents. This unique combination of properties makes MAKROBLEND DP-4-1350 ideal for use in products found in demanding environments, for example:

- industrial/mechanical – power tools; pump components; instrument enclosures
- automotive – passenger compartment components
- military – ammunition; battery cases; instrument housings
- lawn and garden – mower shrouds; trim equipment parts

## Processing:

Outstanding flow characteristics and processability make it possible to mold large parts from MAKROBLEND DP-4-1350 resin.

Drying the resin before processing is essential to optimize property performance and appearance in molded parts. Suggested processing conditions are listed below.

NOTE – Information provided in this book has been condensed. The publisher recommends that the product manufacturer be contacted, at the address or phone number provided, for more complete information on product safety, processing parameters, and applications not provided in this information sheet.

## Typical Injection Molding Conditions:

| | Temperatures °F | Temperatures °C | | Pressures |
|---|---|---|---|---|
| Barrel—Rear | 470–490 | (245–255) | Injection | 10–20,000 psi (69–138 MPa) |
| Middle | 480–500 | (250–260) | Hold | 50–75% injection pressure |
| Front | 490–510 | (255–265) | Back | 50–100 psi (344–688 kPa) |
| Nozzle | 495–515 | (258–268) | | |
| | | | Screw Speed | 50 to 70 rpm |
| Melt | 500–530 | (260–275) | Fill Rate | Moderate |
| Mold—Stationary | 165–200 | (75–95) | Cushion | 0.13–0.25 in (3.2–6.4 mm) |
| Moving | 165–200 | (75–95) | Clamp | 3–5 ton/in² |

*Drying:* 212–230°F (100–110°C) for 4–6 hours in a dehumidifying hopper dryer with a maximum dew point of 0°F (−18°C) to achieve a moisture content of less than 0.02%.

## Typical Physical Properties:

### MAKROBLEND DP-4-1350

| Property* | ASTM Method (Others) | English Units | SI Units |
|---|---|---|---|
| **Physical** | | | |
| Specific Gravity | D792 | 1.31 | |
| Mold Shrinkage | D955 | 0.1–0.6% | |
| **Mechanical** | | | |
| Tensile Strength at Yield | D638 | 11,000 lb/in² | 76 MPa |
| Elongation at Yield | D638 | 5% | |
| Tensile Strength at Break | D638 | 9000 lb/in² | 62 MPa |
| Elongation at Break | D638 | 5% | |
| Flexural Strength, Ultimate | D790 | 19,400 lb/in² | 134 MPa |
| Flexural Modulus | D790 | 560,000 lb/in² | 3862 MPa |
| Izod Impact Strength (Notched) @ 23°C (73°F) | D256 | 2.6 ft-lb/in | 139 J/m |
| **Flammability**** | | | |
| Vertical Rating @ Minimum Thickness | (UL 94) | 94 HB** 0.125 in | 3.2 mm |
| **Thermal** | | | |
| Deflection Temperature under Load 6.4 mm (0.250 in) thickness | D648 | | |
| @ 1.82 MPa (264 psi) load | | 237°F | 114°C |
| @ 0.46 MPa (66 psi) load | | 293°F | 145°C |

*These items are provided as general information only. They are approximate values and are not part of the product specification.

**The flammability results are based on small scale laboratory tests for purposes of relative comparison and do not necessarily represent the hazard presented by this material or any other material under actual fire conditions. The tests were conducted in Mobay's laboratories.

# Mobay Chemical Corporation

## Contact Address:

Mobay Chemical Corporation
Plastics Division
Mobay Road
Pittsburgh, PA 15205-9741

Phone: 1 (412) 777-2000

## MAKROBLEND DP-4-1351

### Product Information:

MAKROBLEND thermoplastic resins are Merlon Polycarbonate/PET blends formulated to be chemically resistant to a wide range of hydrocarbon solvents and to water or water vapor. The products molded from MAKROBLEND resins have excellent physical properties and maintain them over a wide range of environmental conditions.

### Description:

MAKROBLEND DP-4-1351 is a 20% glass fiber reinforced polycarbonate/PET polyester blend. Opaque in appearance, this injection molding resin is supplied in pellet form. Trial size quantities are available in natural and black resin (colors upon special request).

### Markets and Applications:

MAKROBLEND DP-4-1351 has nearly three times the stiffness of unreinforced MAKROBLEND UT-1018 resin, while retaining good impact resistance. It is also chemically resistant to a broad range of hydrocarbons, acids, bases, and solvents. These characteristics, in combination with good strength and heat resistance properties make MAKROBLEND DP-4-1351 ideal for use in industrial products requiring excellent dimensional stability and low warpage, for example:

- industrial/mechanical—pump components; power tools; military equipment; and other products formerly manufactured in metal
- automotive—covers and grilles; passenger compartment devices
- lawn and garden—tractor body shrouds; lawn mower components; trim equipment parts

### Processing:

Outstanding flow characteristics and processability make it possible to mold large and/or complicated parts with MAKROBLEND DP-4-1351.

Drying the resin before processing is essential to optimize property performance and appearance in molded parts. Suggested processing conditions are listed below.

NOTE—Information provided in this book has been condensed. The publisher recommends that the product manufacturer be contacted, at the address or phone number provided, for more complete information on product safety, processing parameters, and applications not provided in this information sheet.

## Typical Injection Molding Conditions:

| | Temperatures °F | °C | | Pressures |
|---|---|---|---|---|
| Barrel—Rear | 470–490 | (245–255) | Injection | 10–20,000 psi (69–138 MPa) |
| Middle | 480–500 | (250–260) | Hold | 50–75% injection pressure |
| Front | 490–510 | (255–265) | Back | 50–100 psi (344–688 kPa) |
| Nozzle | 495–515 | (258–268) | | |
| | | | Screw Speed | 50 to 75 rpm |
| Melt | 500–530 | (260–275) | Fill Rate | Moderate |
| Mold—Stationary | 165–200 | (75–95) | Cushion | 0.13–0.25 in (3.2–6.4 mm) |
| Moving | 165–200 | (75–95) | Clamp | 3–5 ton/in² |

*Drying:* 212–230°F (100–110°C) for 4–6 hours in a dehumidifying hopper dryer with a maximum dew point of 0°F (–18°C) to achieve a moisture content of less than 0.02%.

## Typical Physical Properties:

**MAKROBLEND DP-4-1351**

| Property* | ASTM Method (Others) | English Units | SI Units |
|---|---|---|---|
| **Physical** | | | |
| Specific Gravity | D792 | 1.37 | |
| Mold Shrinkage | D955 | 0.1–0.6% | |
| **Mechanical** | | | |
| Tensile Strength at Yield | D638 | 13,000 lb/in² | 90 MPa |
| Elongation at Yield | D638 | 5% | |
| Tensile Strength at Break | D638 | 12,000 lb/in² | 83 MPa |
| Elongation at Break | D638 | 5% | |
| Flexural Strength, Ultimate | D790 | 24,000 lb/in² | 166 MPa |
| Flexural Modulus | D790 | 850,000 lb/in² | 5862 MPa |
| Izod Impact Strength (Notched) @ 23°C (73°F) | D256 | 2.4 ft-lb/in | 128 J/m |
| **Flammability**** | | | |
| Vertical Rating @ Minimum Thickness | (UL 94) | 94 HB** 0.125 in | 3.2 mm |
| **Thermal** | | | |
| Deflection Temperature under Load 6.4 mm (0.250 in) thickness | D648 | | |
| @ 1.82 MPa (264 psi) load | | 252°F | 122°C |
| @ 0.46 MPa (66 psi) load | | 354°F | 179°C |

*These items are provided as general information only. They are approximate values and are not part of the product specification.
**The flammability results are based on small scale laboratory tests for purposes of relative comparison and do not necessarily represent the hazard presented by this material or any other material under actual fire conditions. The tests were conducted in Mobay's laboratories.

# Mobay Chemical Corporation

## Contact Address:

Mobay Chemical Corporation
Plastics Division
Mobay Road
Pittsburgh, PA 15205-9741

Phone: 1 (412) 777-2000

## MAKROBLEND DP-4-1354

### Product Information:

MAKROBLEND thermoplastic resins are Merlon Polycarbonate/PET blends formulated to be chemically resistant to a wide range of hydrocarbon solvents and to water or water vapor. The products molded from MAKROBLEND resins have excellent physical properties and maintain them over a wide range of environmental conditions.

### Description:

MAKROBLEND DP-4-1354 is a 9% mineral reinforced general purpose polycarbonate/PET polyester blend. This injection molding resin is opaque in appearance and is supplied in pellet form. Trial size quantities are available in natural and black resin (colors upon special request).

### Markets and Applications:

MAKROBLEND DP-4-1354 has improved stiffness when compared with MAKROBLEND UT-1018 resin while retaining good impact resistance. It also exhibits chemical resistant to a variety of reagents such as hydrocarbons, acids, bases and solvents. These unique characteristics, in combination with weatherability, low water absorption, and warpage resistance, make MAKROBLEND DP-4-1354 suited for industrial applications requiring dimensional stability and a good surface appearance, for example:

- industrial/mechanical—pump components; power tools
- consumer—furniture parts; kitchenware
- automotive—passenger compartment parts

### Processing:

Outstanding flow characteristics and processability make it possible to mold very large parts from MAKROBLEND DP-4-1354 resin.

Drying the resin before processing is essential to optimize property performance and appearance in molded parts. Suggested processing conditions are listed below.

NOTE—Information provided in this book has been condensed. The publisher recommends that the product manufacturer be contacted, at the address or phone number provided, for more complete information on product safety, processing parameters, and applications not provided in this information sheet.

## Typical Injection Molding Conditions:

| | Temperatures °F | °C | | Pressures |
|---|---|---|---|---|
| Barrel – Rear | 470–490 | (245–255) | Injection | 10–20,000 psi (69–138 MPa) |
| Middle | 480–500 | (250–260) | Hold | 50–75% injection pressure |
| Front | 490–510 | (255–265) | Back | 50–100 psi (344–688 kPa) |
| Nozzle | 495–515 | (258–268) | | |
| | | | Screw Speed | 50 to 75 rpm |
| Melt | 500–530 | (260–275) | Fill Rate | Moderate |
| Mold – Stationary | 165–200 | (75–95) | Cushion | 0.13–0.25 in (3.2–6.4 mm) |
| Moving | 165–200 | (75–95) | Clamp | 3-5 ton/in² |

*Drying:* 200–220°F (93–104°C) for 4–6 hours in a dehumidifying hopper dryer with a maximum dew point of 0°F (−18°C) to achieve a moisture content of less than 0.02%.

## Typical Physical Properties:

**MAKROBLEND DP-4-1354**

| Property* | ASTM Method (Others) | English Units | SI Units |
|---|---|---|---|
| **Physical** | | | |
| Specific Gravity | D792 | 1.27 | |
| Mold Shrinkage | D955 | 0.6–0.8% | |
| **Mechanical** | | | |
| Tensile Strength at Yield | D638 | 7800 lb/in² | 54 MPa |
| Elongation at Yield | D638 | 5% | |
| Tensile Strength at Break | D638 | 6600 lb/in² | 46 MPa |
| Elongation at Break | D638 | 25% | |
| Flexural Strength, Ultimate | D790 | 12,500 lb/in² | 86 MPa |
| Flexural Modulus | D790 | 400,000 lb/in² | 2759 MPa |
| Izod Impact Strength (Notched) @ 23°C (73°F) | D256 | no break | |
| Izod Impact Strength (Notched) @ 23°C (73°F) @ −29°C (−20°F) | D256 | 4.0 ft-lb/in 2.1 ft-lb/in | 214 J/m 114 J/m |
| Gardner Impact (Dart: 15.88mm/ (0.625 in) dia.) @ 23°C/73°F | (Mobay) | 58 ft-lb | 79 J |

*These items are provided as general information only. They are approximate values and are not part of the product specification.

*(continued)*

## MAKROBLEND DP-4-1354 *(continued)*

| Property* | ASTM Method (Others) | English Units | SI Units |
|---|---|---|---|
| **Flammability**** | | | |
| Vertical Rating | (UL 94) | 94 HB** | |
| @ Minimum Thickness | | 0.125 in | 3.2 mm |
| **Thermal** | | | |
| Deflection Temperature under Load | D648 | | |
| 6.4 mm (0.250 in) thickness) | | | |
| @ 1.82 MPa (264 psi) load | | 230°F | 110°C |
| @ 0.46 MPa (66 psi) load | | 262°F | 128°C |

**The flammability results are based on small scale laboratory tests for purposes of relative comparison and do not necessarily represent the hazard presented by this material or any other material under actual fire conditions. The tests were conducted in Mobay's laboratories.

# Mobay Chemical Corporation

## Contact Address:

Mobay Chemical Corporation
Plastics Division
Mobay Road
Pittsburgh, PA 15205-9741

Phone: 1 (412) 777-2000

## MAKROBLEND DP-4-1356

### Product Information:

MAKROBLEND thermoplastic resins are Merlon Polycarbonate/PET blends formulated to be chemically resistant to a wide range of hydrocarbon solvents and to water or water vapor. The products molded from MAKROBLEND resins have excellent physical properties and maintain them over a wide range of environmental conditions.

### Description:

MAKROBLEND DP-4-1356 is an opaque, impact-modified polycarbonate/PET polyester blend, which is supplied in pellet form. Trial size quantities are available in natural and black resin (colors upon special request).

### Markets and Applications:

MAKROBLEND DP-4-1356 was specifically developed for applications requiring impact strength, chemical resistance and good practical heat resistance. This combination of properties, in conjunction with low water absorption and good dimensional stability, make MAKROBLEND DP-4-1356 ideally suited for equipment found in industrial environments, for example:

- industrial/mechanical—hand-held radios; portable instruments; enclosures requiring good heat sag properties
- military—instrument housings; ammunition; weapon components
- automotive—passenger compartment components

### Processing:

Outstanding flow characteristics and processability make possible extrusion as well as injection molding of large parts.

Drying the resin before processing is essential to optimize property performance and appearance in molded parts. Suggested processing conditions are listed below.

*(continued)*

NOTE—Information provided in this book has been condensed. The publisher recommends that the product manufacturer be contacted, at the address or phone number provided, for more complete information on product safety, processing parameters, and applications not provided in this information sheet.

## Typical Injection Molding Conditions:

| | Temperatures °F | °C | | Pressures |
|---|---|---|---|---|
| Barrel–Rear | 470–490 | (245–255) | Injection | 10–20,000 psi (69–138 MPa) |
| Middle | 480–500 | (250–260) | Hold | 50–75% injection pressure |
| Front | 490–510 | (255–265) | Back | 50–100 psi (344–688 kPa) |
| Nozzle | 495–515 | (258–268) | | |
| | | | Screw Speed | 50 to 75 rpm |
| Melt | 500–550 | (260–290) | Fill Rate | Moderate |
| Mold–Stationary | 65–180 | (20–80) | Cushion | 0.13–0.25 in (3.2–6.4 mm) |
| Moving | 65–180 | (20–80) | Clamp | 3–5 ton/in² |

*Drying:* 212–230°F (100–110°C) for 4–6 hours in a dehumidifying hopper dryer with a maximum dew point of 0°F (–18°C) to achieve a moisture content of less than 0.02%.

## Typical Physical Properties:

**MAKROBLEND DP-4-1356**

| Property* | ASTM Method (Others) | English Units | SI Units |
|---|---|---|---|
| **Physical** | | | |
| Specific Gravity | D792 | 1.23 | |
| Mold Shrinkage | D955 | 0.6–0.8% | |
| **Mechanical** | | | |
| Tensile Strength at Yield | D638 | 8500 lb/in² | 59 MPa |
| Tensile Strength at Break | D638 | 8400 lb/in² | 58 MPa |
| Elongation at Break | D638 | 148% | |
| Flexural Strength, Ultimate | D790 | 13,000 lb/in² | 90 MPa |
| Flexural Modulus | D790 | 348,000 lb/in² | 2400 MPa |
| Izod Impact Strength (Notched) @ 23°C (73°F) | D256 | 20 ft-lb/in | 1068 J/m |
| Gardner Impact (Dart: 15.88mm/ (0.625 in) dia.) | (Mobay) | | |
| @ 23°C (73°F) | | 51 ft-lb | 69 J |
| @ –29°C (–20°F) | | 53 ft-lb | 72 J |

*These items are provided as general information only. They are approximate values and are not part of the product specification.

*(continued)*

**MAKROBLEND DP-4-1356 *(continued)***

| Property* | ASTM Method (Others) | English Units | SI Units |
|---|---|---|---|
| **Flammability**** | | | |
| Vertical Rating | (UL 94) | 94 HB** | |
| @ Minimum Thickness | | 0.125 in | 3.2 mm |
| **Thermal** | | | |
| Deflection Temperature under Load | D648 | | |
| 6.4 mm (0.250 in) thickness) | | | |
| @ 1.82 MPa (264 psi) load | | 261 °F | 127°C |
| @ 0.46 MPa (66 psi) load | | 279°F | 137°C |

**The flammability results are based on small scale laboratory tests for purposes of relative comparison and do not necessarily represent the hazard presented by this material or any other material under actual fire conditions. The tests were conducted in Mobay's laboratories.

# Mobay Chemical Corporation

## Contact Address:

Mobay Chemical Corporation
Plastics Division
Mobay Road
Pittsburgh, PA 15205-9741

Phone: 1 (412) 777-2000

## MAKROBLEND DP-4-1357

### Product Information:

MAKROBLEND thermoplastic resins are Merlon Polycarbonate/PET blends formulated to be chemically resistant to a wide range of hydrocarbon solvents and to water or water vapor. The products molded from MAKROBLEND resins have excellent physical properties and maintain them over a wide range of environmental conditions.

### Description:

MAKROBLEND DP-4-1357 is an unfilled, general purpose blend of MERLON polycarbonate and PET polyester resins. This injection molding resin is naturally opaque in appearance and supplied in pellet form. Trial size quantities are available in natural and black resin (colors upon special request).

### Markets and Applications:

MAKROBLEND DP-4-1357 has a high melt flow rate to facilitate molding small intricate parts or large housings. It also has excellent toughness, chemical resistance, and heat resistance. Specifically, DP-4-1357 is ideal for use in applications such as:

- telecommunications – radio housings; speakler grills; instrument housings
- lawn and garden – tractor shrouds and consoles; fertilizer housings

### Processing:

Processing parameters for MAKROBLEND DP-4-1357 are noted below. Predrying prior to processing is essential to ensure desired appearance and property performance.

### Typical Injection Molding Conditions:

| | Temperatures | | | |
|---|---|---|---|---|
| | °F | °C | | Pressures |
| Barrel – Rear | 470–490 | (245–255) | Injection | 10–20,000 psi (69–138 MPa) |
| Middle | 480–500 | (250–260) | Hold | 50–75% injection pressure |
| Front | 490–510 | (255–265) | Back | 50–100 psi (344–688 kPa) |
| Nozzle | 495–515 | (258–266) | | |
| Melt | 500–550 | (260–290) | Screw Speed | 50 to 75 rpm |
| | | | Fill Rate | Moderate |
| Mold – Stationary | 65–180 | (20–80) | Cushion | 0.13–0.25 in (3.2–6.4 mm) |
| Moving | 65–180 | (20–80) | Clamp | 3-5 ton/in² |

*Drying:* 212–230°F (100–110°C) for 4–6 hours in a dehumidifying hopper dryer with a maximum dew point of 0°F (–18°C) to achieve a moisture content of less than 0.02%.

## Typical Physical Properties:

### MAKROBLEND DP-4-1357

| Property* | ASTM Method (Others) | English Units | SI Units |
|---|---|---|---|
| **Physical** | | | |
| Specific Gravity | D792 | 1.22 | |
| Mold Shrinkage | D955 | 0.6–0.8% | |
| **Mechanical*** | | | |
| Tensile Strength at Yield | D638 | 7900 $lb/in^2$ | 54 MPa |
| Tensile Strength at Break | D638 | 8000 $lb/in^2$ | 55 MPa |
| Elongation at Break | D638 | 151% | |
| Flexural Strength, Ultimate | D790 | 12,700 $lb/in^2$ | 88 MPa |
| Flexural Modulus | D790 | 338,000 $lb/in^2$ | 2330 MPa |
| Izod Impact Strength, Notched @ 23°C (73°F) | D256 | 15 ft-lb/in | 800 J/m |
| **Flammability**** | | | |
| Vertical Rating @ Minimum Thickness | (UL 94) | 94 HB** 0.062 in | 1.6 mm |
| **Thermal** | | | |
| Deflection Temperature under Load 6.4 mm (0.250 in) thickness | D648 | | |
| @ 1.82 MPa (264 psi) load | | 248°F | 120°C |
| @ 0.46 MPa (66 psi) load | | 273°F | 134°C |

*These items are provided as general information only. They are approximate values and are not part of the product specification.

**The flammability results are based on small scale laboratory tests for purposes of relative comparison and do not necessarily represent the hazard presented by this material or any other material under actual fire conditions. The tests were conducted in Mobay's laboratories.

# Mobay Chemical Corporation

## Contact Address:

Mobay Chemical Corporation
Plastics Division
Mobay Road
Pittsburgh, PA 15205-9741

Phone: 1 (412) 777-2000

## MAKROBLEND UT-1018

### Product Information:

MAKROBLEND thermoplastic resins are Merlon Polycarbonate/PET blends formulated to be chemically resistant to a wide range of hydrocarbon solvents and to water or water vapor. The products molded from MAKROBLEND resins have excellent physical properties and maintain them over a wide range of environmental conditions.

MAKROBLEND UT-1018 is an opaque, impact-modified polycarbonate/PET injection molding resin. This product has been formulated for applications where resistance to gasoline and other hydrocarbons is of prime importance and where toughness, especially at very low temperatures, is required. Makroblend UT-1018 resin has excellent dimensional stability even in high moisture environments, plus excellent abrasion resistance.

Superior flowability and processability make it possible to mold very large parts. Potential applications for Makroblend UT-1018 include car bumpers, mirror housings and brackets, protective helmets, ski boot and binding components, snow blower components, lawn mower decks, geophones, furniture castors and conveyor components.

Suggested processing conditions are listed below for Makroblend UT-1018 resin. Drying the resin before processing is essential to optimize property performance and appearance in molded parts.

### Typical Injection Molding Conditions:

| | Temperatures | | | |
|---|---|---|---|---|
| | °F | °C | | Pressures |
| Barrel – Rear | 470–490 | (245–255) | Injection | 10–20,000 psi (69–138 MPa) |
| Middle | 480–500 | (250–260) | Hold | 50–75% injection pressure |
| Front | 490–510 | (255–265) | Back | 50–100 psi (344–688 kPa) |
| Nozzle | 495–515 | (258–268) | | |
| | | | Screw Speed | 50 to 75 rpm |
| Melt | 500–530 | (260–277) | Fill Rate | Slow to Moderate |
| Mold – Stationary | 65–165 | (18–74) | Cushion | 0.13–0.25 in (3.2–6.4 mm) |
| Moving | 65–165 | (18–74) | Clamp | 3-5 ton/in$^2$ |

*Drying:* 210–215°F (99–102°C) for 4–6 hours in a dehumidifying hopper dryer with a maximum dew point of 0°F (–18°C) to achieve a moisture content of less than 0.02%.

NOTE – Information provided in this book has been condensed. The publisher recommends that the product manufacturer be contacted, at the address or phone number provided, for more complete information on product safety, processing parameters, and applications not provided in this information sheet.

## Typical Physical Properties:

**MAKROBLEND UT-1018**

| Property | ASTM Method (Others) | English Units | SI Units |
|---|---|---|---|
| **Physical** | | | |
| Specific Gravity | D792 | 1.22 | |
| Hardness, Rockwell (R Scale) | D785 | 114 | |
| Mold Shrinkage<br>Flow<br>Cross Flow | D955 | <br>0.006 in/in<br>0.009 in/in | <br>0.006 mm/mm<br>0.009 mm/mm |
| Moisture Absorption, 24 hr Immersion | D510 | 0.16% | |
| Taber Abrasion | D1044 | 200 mg loss/<br>1000 cycles | |
| **Mechanical** | | | |
| Tensile Strength at Yield | D638 | 7000 lb/in² | 48 MPa |
| Elongation at Yield | D638 | 5% | |
| Tensile Strength at Break | D638 | 7600 lb/in² | 52 MPa |
| Elongation at Break | D638 | 165% | |
| Flexural Strength, Ultimate | D790 | 10,900 lb/in² | 75 MPa |
| Flexural Modulus | D790 | 300,000 lb/in² | 2070 MPa |
| Izod Impact Strength (Notched)<br>@ 23°C (73°F)<br>@ −29°C (−20°F) | D256 | <br>18 ft-lb/in<br>13 ft-lb/in | <br>970 J/m<br>700 J/m |
| Gardner Impact<br>(Dart: 15.88mm/ (0.625 in) dia.)<br>@ 23°C/73°F<br>−29°C/−20°F | (Mobay) | <br><br>42 ft-lb<br>37 ft-lb | <br><br>57 J<br>50 J |
| **Flammability**** | | | |
| Vertical—UL Rating @<br>Minimum Thickness Rating | (UL 94) | HB<br>0.063 in | HB<br>1.6 mm |
| **Thermal** | | | |
| Deflection Temperature under Load<br>(6.4 mm/0.250″ thickness)<br>1.82 MPa (264 psi) Load<br>0.46 MPa (66 psi) Load | D648 | <br><br>190°F<br>239°F | <br><br>88°C<br>115°C |
| **Rheological** | | | |
| Spiral Flow<br>Temp. 270°C (520°F)<br>Thickness 2.54 mm (0.100 in) | (Mobay) | <br><br>16 in | <br><br>40.6 cm |

*At 23°C/73°F and 50% rh for 3.2 mm (0.125 in) thickness.

**References to flammability characteristics and/or numerical flame spread rating are not intended to reflect hazards presented by this or any other material under actual fire conditions.

# Monsanto Chemical Company

## Contact Address:

Monsanto Chemical Company
800 North Lindbergh Blvd.
St. Louis, MO 63167

Phone: (314) 694-1000

# SANTOPRENE

## Product Data

SANTOPRENE® thermoplastic rubber, crosslinked EPDM in polypropylene, is a family of advanced elastomers which successfully combine the performance characteristics of vulcanized rubber with the processing ease of thermoplastics. It consists of fully cured, very small rubber particles, uniformly dispersed in a continuous matrix of thermoplastic polymer. Because it is thermoplastic, and has very good melt flow characteristics, SANTOPRENE® can be injection molded, extruded, blow molded, and calendered with the efficiency and economy associated with thermoplastics. Clean scrap can be recycled.

SANTOPRENE® rubber is available in a range of hardness from 55 Shore A to 50 Shore D, each grade having its own physical property specifications. The combination of excellent environmental resistance, and fluid resistance equivalent to neoprene, provides a series of thermoplastic rubbers suitable for a broad range of product applications.

## Typical Physical Properties:

**Typical mechanical properties of SANTOPRENE® thermoplastic rubber**

| Properties | ASTM Test Method | Test Units U.S. (SI) | Test Units °F (°C) | SANTOPRENE Rubber Grades | | | | | | |
|---|---|---|---|---|---|---|---|---|---|---|
| | | | | 201-55 101-55 | 201-64 101-64 | 201-73 101-73 | 201-80 101-80 | 201-87 101-87 | 203-40 103-40 | 203-50 103-50 |
| Hardness | D-2240 | 5 sec. Shore | 77 (25) | 55A | 64A | 73A | 80A | 87A | 40D | 50D |
| Specific Gravity | D-297 | — | 77 (25) | 0.97 | 0.97 | 0.98 | 0.97 | 0.96 | 0.95 | 0.94 |
| Tensile Strength | D-412 | psi (MPa) | 77 (25) | 640 (4.4) | 1000 (6.9) | 1200 (8.3) | 1600 (11.0) | 2300 (15.9) | 2750 (19.0) | 4000[2] (27.6) |
| Ultimate Elongation | D-412 | % | 77 (25) | 300 | 400 | 375 | 450 | 530 | 600 | 600[2] |
| 100% Modulus | D-412 | psi (MPa) | 77 (25) | 290 (2.0) | 340 (2.3) | 470 (3.2) | 700 (4.8) | 1000 (6.9) | 1250 (8.6) | 1450[2] (10.0) |

*(continued)*

NOTE – Information provided in this book has been condensed. The publisher recommends that the product manufacturer be contacted, at the address or phone number provided, for more complete information on product safety, processing parameters, and applications not provided in this information sheet.

### Typical mechanical properties of SANTOPRENE® thermoplastic rubber[1] *(continued)*

| Properties | ASTM Test Method | Test Units U.S. (SI) | Test Units °F (°C) | SANTOPRENE Rubber Grades | | | | | | |
|---|---|---|---|---|---|---|---|---|---|---|
| | | | | 201-55 101-55 | 201-64 101-64 | 201-73 101-73 | 201-80 101-80 | 201-87 101-87 | 203-40 103-40 | 203-50 103-50 |
| Tear Strength | D-624 | pli (kN/m) | 77 (25) | 108 (18.9) | 140 (24.5) | 159 (27.8) | 194 (34.0) | 278 (48.7) | 369 (64.6) | 514 (90.0) |
| | | pli (kN/m) | 212 (100) | 42 (7.3) | 58 (10.2) | 76 (13.3) | 75 (13.1) | 133 (23.3) | 203 (35.5) | 364 (63.7) |
| Tension Set | D-412 | % | 77 (25) | 6 | 10 | 14 | 20 | 33 | 48 | 61 |
| Compression Set, 168 Hrs | D-395 | % | 77 (25) | 20 | 20 | 24 | 29 | 36 | 44 | 47 |
| | | % | 212 (100) | 29 | 36 | 40 | 45 | 58 | 67 | 70 |
| Flex Fatigue[3] | | Cycles to fail | 77 (25) | >7.0 | >3.4 | >3.4 | >3.4 | — | — | — |
| | | | | million cycles | | | | | | |
| Brittle Point | D-746 | °F (°C) | — — | <−76 (<−60) | <−76 (−60) | −81 (−63) | −81 (−63) | −78 (−61) | −71 (−57) | −29 (−34) |

[1]See note disclaimer, page 392.
[2]Grades 103-50 and 203-50 have a yield point at 1650 psi (11.4 MPa) and 50% elongation. Stress-strain measurements for these grades are according to ASTM D-638.
[3]Monsanto Fatigue-To-Failure Test.

### Typical temperature conditions for injection molding of SANTOPRENE® rubber[1]

| Grade of SANTOPRENE Rubber | 201-55 101-55 | 201-64 101-64 | 201-73 101-73 | 201-80 101-80 | 201-87 101-87 | 203-40 103-40 | 203-50 103-50 |
|---|---|---|---|---|---|---|---|
| Rear Barrel °F (°C) | 360 ± 20 (182 ± 11) | 360 ± 20 (182 ± 11) | 340 ± 10 (171 ± 5) | 345 ± 10 (174 ± 5) | 350 ± 10 (177 ± 5) | 360 ± 20 (182 ± 11) | 365 ± 20 (185 ± 11) |
| Front Barrel °F (°C) | 360 ± 20 (182 ± 11) | (182 ± 11) | 340 ± 10 (171 ± 5) | 345 ± 10 (174 ± 5) | 350 ± 10 (177 ± 5) | 360 ± 20 (182 ± 11) | 365 ± 20 (185 ± 11) |
| Nozzle °F (°C) | 360 ± 20 (182 ± 11) | 360 ± 20 (182 ± 11) | 340 ± 10 (171 ± 5) | 345 ± 10 (174 ± 5) | 350 ± 10 (177 ± 5) | 360 ± 20 (182 ± 11) | 365 ± 20 (185 ± 11) |
| Melt °F (°C) | 380 ± 20 (193 ± 11) | 385 ± 20 (196 ± 11) | 355 ± 20 (179 ± 11) | 375 ± 20 (190 ± 11) | 390 ± 20 (199 ± 11) | 390 ± 20 (199 ± 11) | 390 ± 20 (199 ± 11) |
| Mold °F (°C) | 50-175 (10-80) | 50-175 (10-80) | 50-175 (10-80) | 50-175 (10-80) | 50-175 (10-80) | 50-175 (10-80) | 50-175 (10-80) |

[1]See note disclaimer, page 392.

*(continued)*

**Typical temperature conditions for extrusion of SANTOPRENE® rubber[1]**

| Grade of SANTOPRENE Rubber | 201-55<br>101-55 | 201-64<br>101-64 | 201-73<br>101-73 | 201-80<br>101-80 | 201-87<br>101-87 | 203-40<br>103-40 | 203-50<br>103-50 |
|---|---|---|---|---|---|---|---|
| **Feed Throat** | ---NO HEATING--- | | | | | | |
| Feed Zone<br>°F<br>(°C) | 370 ± 10<br>(188 ± 5) | 380 ± 10<br>(193 ± 5) | 380 ± 10<br>(193 ± 5) | 385 ± 10<br>(196 ± 5) | 395 ± 10<br>(202 ± 5) | 400 ± 10<br>(204 ± 5) | 400 ± 10<br>(204 ± 5) |
| Transition Zone<br>°F<br>(°C) | 370 ± 10<br>(188 ± 5) | 385 ± 10<br>(196 ± 5) | 385 ± 10<br>(196 ± 5) | 385 ± 10<br>(196 ± 5) | 395 ± 10<br>(202 ± 5) | 395 ± 10<br>(202 ± 5) | 400 ± 10<br>(204 ± 5) |
| Metering Zone<br>°F<br>(°C) | 380 ± 10<br>(193 ± 5) | 385 ± 10<br>(196 ± 5) | 385 ± 10<br>(196 ± 5) | 395 ± 10<br>(202 ± 5) | 395 ± 10<br>(202 ± 5) | 400 ± 10<br>(204 ± 5) | 400 ± 10<br>(210 ± 5) |
| Front Zone<br>°F<br>(°C) | 380 ± 10<br>(193 ± 5) | 390 ± 10<br>(199 ± 5) | 395 ± 10<br>(202 ± 5) | 395 ± 10<br>(202 ± 5) | 395 ± 10<br>(202 ± 5) | 400 ± 10<br>(204 ± 5) | 410 ± 10<br>(210 ± 5) |
| Gate/Die<br>°F<br>(°C) | 390 ± 10<br>(199 ± 5) | 400 ± 10<br>(204 ± 5) | 395 ± 10<br>(202 ± 5) | 395 ± 10<br>(202 ± 5) | 400 ± 10<br>(204 ± 5) | 410 ± 10<br>(210 ± 5) | 410 ± 10<br>(210 ± 5) |
| Melt<br>°F<br>(°C) | 400 ± 10<br>(204 ± 5) | 410 ± 10<br>(210 ± 5) | 410 ± 10<br>(210 ± 5) | 410 ± 10<br>(210 ± 5) | 420 ± 10<br>(216 ± 5) | 430 ± 10<br>(221 ± 5) | 430 ± 10<br>(221 ± 5) |

[1]See note disclaimer, page 392.

# Monsanto Chemical Company

## Contact Address:

Monsanto Chemical Company
800 North Lindbergh Blvd.
St. Louis, MO 63167

Phone: (314) 694-1000

## TRIAX 1120

### Product Information

TRIAX-1000 series resins are alloys of nylon and ABS. The 1120 and 1125 grades are formulated for injection molding, but either may also be blow-molded, extruded and thermoformed. These semi-crystalline thermoplastics exhibit good chemical resistance, high heat performance and good abrasion resistance. Both are economical candidates for applications requiring high toughness.

TRIAX 1120 is formulated to combine outstanding appearance with excellent impact resistance over a wide range of temperatures. The processing conditions, which include low stock temperatures, fast injection rates and a wide processing range, provide high economical productivity. Various finishing techniques including painting, flocking, hot stamping, and sonic welding are possible.

### Typical Physical Properties:

**TRIAX 1120**

| Property | ASTM Method | Test Conditions | English Units | Dry As Molded | 50% R.H. |
|---|---|---|---|---|---|
| **Mechanical** | | | | | |
| Tensile Strength | D-638 | 73°F<br>−40°F | Kpsi | 6.6<br>10.6 | 5.8<br>10.1 |
| Elongation at Fail | D-638 | 73°F<br>−40°F | % | 270<br>85 | 290<br>125 |
| Tensile Modulus | D-638 | 73°F | Kpsi | 275 | 170 |
| Flexural Modulus | D-790 | −40°F<br>73°F<br>158°F<br>170°F | Kpsi | 360<br>300<br>120<br>115 | 390<br>165<br>135<br>125 |
| Poisson Ratio | — | — | — | 0.38 | — |

*(continued)*

NOTE – Information provided in this book has been condensed. The publisher recommends that the product manufacturer be contacted, at the address or phone number provided, for more complete information on product safety, processing parameters, and applications not provided in this information sheet.

TRIAX 1120 *(continued)*

| Property | ASTM Method | Test Conditions | English Units | Dry As Molded | 50% R.H. |
|---|---|---|---|---|---|
| **Toughness** | | | | | |
| Gardner Impact<br>2″ Ring, 1/2″ Dart<br>100 mils | D-3029G | 73°F<br>32°F<br>−4°F<br>−40°F | in-lbs | >400<br>>400<br>>400<br>385 | >400<br>>400<br>>400<br>>400 |
| Driven Dart Impact<br>1-1/4″ Ring, 1/2″<br>Dart<br>100 mils | Monsanto | 73°F<br><br>−4°F<br>−40°F | in-lbs | 450<br><br>430<br>125 | 460<br><br>440<br>405 |
| Notched Izod<br>1/8″ Bar<br>1/4″ Bar | D-256 | 73°F | ft-lbs/in | 16<br>16 | 19<br>18 |
| Charpy Impact<br>Notched, 1/4″ Bar<br>Unnotched, 1/4″ Bar | D-256 | 73°F | ft-lbs/in$^2$ | 22<br>No break | 25<br>No break |
| Tensile Impact | D-1822<br>Type L | 73°F | ft-lbs/in$^2$ | 260 | No break |
| **Thermal** | | | | | |
| Minimum Molding Temperature | — | — | °F | 445 | — |
| Apparent Viscosity | Dayton Rheometer | 475°F<br>200 sec$^{-1}$<br>500 sec$^{-1}$<br>1000 sec$^{-1}$ | cp × 10$^3$ | <br>1600<br>890<br>560 | <br>—<br>—<br>— |
| Melt Flow Index, "I" | D-1238 | 445°F/3.8kg | g/10 min | 1.5 | — |
| Vicat Softening Temperature | D-1525 | 1kg<br>5kg | °F<br>°F | 387<br>230 | 338<br>— |
| Heat Deflection Temperature | D-648 | 66 psi | °F | 198 | 196 |
| Specific Heat | C-351 | 495°F | BTU/lb-°F | 0.60 | — |
| Coefficient of Linear Expansion | D-696 | −20° to 70°F<br>70°F to 180°F<br>180°F to 280°F | in/in/°F | $5.7 \times 10^{-5}$<br>$9.5 \times 10^{-5}$<br>$21 \times 10^{-5}$ | —<br>—<br>— |
| Thermal Conductivity | C-177 | — | BTU-in/<br>hr-ft$^2$-°F | 2.0 | — |

*(continued)*

TRIAX 1120 *(continued)*

| Property | ASTM Method | Test Conditions | English Units | Dry As Molded | 50% R.H. |
|---|---|---|---|---|---|
| **Miscellaneous** | | | | | |
| Specific Gravity | D-792 | 73°F | — | 1.06 | 1.06 |
| Water Absorption 50% R.H. | D-570 | 73°F, 24 hrs<br>73°F, 72 hrs<br>73°F, 90 days<br>73°F, Equilibrium | %<br>%<br>%<br>% | <0.1<br>0.6<br>1.1<br>1.6 | —<br>—<br>—<br>— |
| Water Immersion | D-570 | 73°F, 24 hrs<br>73°F, Saturation | %<br>% | 1.1<br>4.3 | —<br>— |
| Hardness<br>Rockwell<br>Shore | <br>D-785<br>D-676 | <br>73°F<br>73°F | <br>R Scale<br>D Scale | <br>95<br>75 | <br>80<br>68 |
| Mold Shrinkage | D-955 | Thickness,mils<br>50<br>125<br>200 | <br>mils/in | <br>8–10<br>10–12<br>13–15 | <br>—<br>—<br>— |
| Limiting Oxygen Index | D-2863 | — | % $O_2$ | 20 | 21 |

TRIAX 1120

| Property | ASTM Method | Test Conditions | SI Units | Dry As Molded | 50% R.H. |
|---|---|---|---|---|---|
| **Mechanical** | | | | | |
| Tensile Strength | D-638 | 73°F<br>−40°F | MPa | 46<br>73 | 40<br>70 |
| Elongation at Fail | D-638 | 73°F<br>−40°F | % | 270<br>85 | 290<br>125 |
| Tensile Modulus | D-638 | 73°F | MPa | 1900 | 1175 |
| Flexural Modulus | D-790 | −40°F<br>73°F<br>158°F<br>170°F | MPa | 2485<br>2070<br>830<br>795 | 2690<br>1140<br>920<br>860 |
| Poisson Ratio | — | — | — | 0.38 | — |

*(continued)*

TRIAX 1120 *(continued)*

| Property | ASTM Method | Test Conditions | SI Units | Dry As Molded | 50% R.H. |
|---|---|---|---|---|---|
| **Toughness** | | | | | |
| Gardner Impact<br>2″ Ring, 1/2″ Dart<br>100 mils | D-3029G | 73°F<br>32°F<br>−4°F<br>−40°F | J | >45<br>>45<br>>45<br>44 | >45<br>>45<br>>45<br>>45 |
| Driven Dart Impact<br>1-1/4″ Ring, 1/2″<br>Dart<br>100 mils | Monsanto | 73°F<br>−4°F<br>−40°F | J | 51<br>49<br>14 | 52<br>50<br>46 |
| Notched Izod<br>1/8″ Bar<br>1/4″ Bar | D-256 | 73°F | J/m | 850<br>850 | 1000<br>950 |
| Charpy Impact<br>Notched, 1/4″ Bar<br>Unnotched, 1/4″ Bar | D-256 | 73°F | $KJ/m^2$ | 48<br>No break | 54<br>No break |
| Tensile Impact | D-1822<br>Type L | 73°F | $KJ/m^2$ | 545 | No break |
| **Thermal** | | | | | |
| Minimum Molding Temperature | — | — | °C | 230 | — |
| Apparent Viscosity | Dayton Rheometer | 475°F<br>200 $sec^{-1}$<br>500 $sec^{-1}$<br>1000 $sec^{-1}$ | $cp \times 10^3$ | 1600<br>890<br>560 | —<br>—<br>— |
| Melt Flow Index, "I" | D-1238 | 445°F/3.8kg | g/10 min | 1.5 | — |
| Vicat Softening Temperature | D-1525 | 1kg<br>5kg | °C<br>°C | 197<br>110 | 170<br>— |
| Heat Deflection Temperature | D-648 | 66 psi | °C | 92 | 91 |
| Specific Heat | C-351 | 495°F | KJ/kg-°K | 2.5 | — |
| Coefficient of Linear Expansion | D-696 | −20° to 70°F<br>70°F to 180°F<br>180°F to 280°F | mm/m/°K | $10 \times 10^{-5}$<br>$17 \times 10^{-5}$<br>$38 \times 10^{-5}$ | —<br>—<br>— |
| Thermal Conductivity | C-177 | — | W/m-°K | 0.29 | — |

*(continued)*

TRIAX 1120 *(continued)*

| Property | ASTM Method | Test Conditions | SI Units | Dry As Molded | 50% R.H. |
|---|---|---|---|---|---|
| **Miscellaneous** | | | | | |
| Specific Gravity | D-792 | 73°F | — | 1.06 | 1.06 |
| Water Absorption<br>50% R.H. | D-570 | 73°F, 24 hrs<br>73°F, 72 hrs<br>73°F, 90 days<br>73°F, Equilibrium | %<br>%<br>%<br>% | <0.1<br>0.6<br>1.1<br>1.6 | —<br>—<br>—<br>— |
| Water Immersion | D-570 | 73°F, 24 hrs<br>73°F, Saturation | %<br>% | 1.1<br>4.3 | —<br>— |
| Hardness<br>Rockwell<br>Shore | <br>D-785<br>D-676 | <br>73°F<br>73°F | <br>R Scale<br>D Scale | <br>95<br>75 | <br>80<br>68 |
| Mold Shrinkage | D-955 | Thickness,mils<br>50<br>125<br>200 | mils/in | 8–10<br>10–12<br>13–15 | —<br>—<br>— |
| Limiting Oxygen Index | D-2863 | — | % $O_2$ | 20 | 21 |

TRIAX 1120 *(continued)*

| Property | ASTM Method | Test Conditions | Dry As Molded | 50% R.H. |
|---|---|---|---|---|
| **Miscellaneous** | | | | |
| Volume Resistivity | D-257 | $10^{14}$ohm-cm | 97 | 0.09 |
| Surface Resistivity | D-257 | $10^{14}$ohm | 83 | 0.32 |
| Insulation Resistance | D-257 | $10^{14}$ohm | 6.2 | 0.0028 |
| High Voltage Arc Resistance 1/8″ (3.2mm) | D-495 | seconds | 108 | 97 |
| Hot Wire Ignition 1/16″ (1.6mm) | UL-746A | seconds | 18 | — |
| High Voltage Arc Resistance to Ignition, 1/8″ (3.2mm) | UL-746A | seconds | 12 | — |
| Comparative Tracking Index | D-3638 | volts | 600+ | — |
| High Voltage Arc Tracking Rate | UL-746A | in/min | 0.4 | — |

*(continued)*

**TRIAX 1120 *(continued)***

| Property | ASTM Method | Test Conditions | Dry As Molded | 50% R.H. |
|---|---|---|---|---|
| **Miscellaneous** | | | | |
| Temperature Index | UL-746B | | | |
| Electrical | | °C | 60 | — |
| Mechanical with Impact | | °C | 60 | — |
| Mechanical without Impact | | °C | 60 | — |
| Dielectric Constant | D-150 | | | |
| 100Hz | | — | 3.9 | 5.7 |
| $10^3$Hz | | — | 3.8 | 4.8 |
| $10^5$Hz | | — | 3.5 | 3.5 |
| Dissipation Factor | D-150 | | | |
| 100Hz | | — | 0.015 | 0.073 |
| $10^3$Hz | | — | 0.014 | 0.065 |
| $10^5$Hz | | — | 0.024 | 0.039 |
| Dielectric Strength | D-149 | | | |
| Short Time | | volts/mil | 425 | 380 |
| Step-by-Step | | volts/mil | 350 | 340 |
| High Current Arc Resistance | UL-746A | Arc, seconds | 200+ | — |
| UL Flammability 1/16″ (1.6mm) | UL-94 | — | HB | HB |

# Monsanto Chemical Company

## Contact Address:

Monsanto Chemical Company
800 North Lindbergh Blvd.
St. Louis, MO 63167

Phone: (314) 694-1000

## TRIAX 1125

### Product Information

TRIAX-1000 series resins are alloys of nylon and ABS. The 1120 and 1125 grades are formulated for injection molding, but either may also be blow-molded, extruded and thermoformed. These semi-crystalline thermoplastics exhibit good chemical resistance, high heat performance and good abrasion resistance. Both are economical candidates for applications requiring high toughness.

TRIAX 1125 offers high impact strength, and economical production rates for abuse resistant parts. It offers better abrasion resistance than many other engineering thermoplastics, and is resistant to many aggressive chemicals. It is an excellent candidate for textured surfaces where it accurately reproduces the mold finish.

### Typical Physical Properties:

**TRIAX 1125**

| Property | ASTM Method | Test Conditions | English Units | Dry As Molded | 50% R.H. |
|---|---|---|---|---|---|
| **Mechanical** | | | | | |
| Tensile Strength | D-638 | 73°F<br>−40°F | Kpsi | 6.8<br>10.4 | 5.9<br>10.4 |
| Elongation at Fail | D-638 | 73°F<br>−40°F | % | 270<br>120 | 270<br>120 |
| Tensile Modulus | D-638 | 73°F | Kpsi | 274 | 185 |
| Flexural Modulus | D-790 | −40°F<br>73°F<br>158°F<br>170°F | Kpsi | 350<br>310<br>130<br>120 | 340<br>150<br>120<br>110 |
| Poisson Ratio | — | — | — | 0.38 | — |

*(continued)*

NOTE – Information provided in this book has been condensed. The publisher recommends that the product manufacturer be contacted, at the address or phone number provided, for more complete information on product safety, processing parameters, and applications not provided in this information sheet.

## TRIAX 1125 *(continued)*

| Property | ASTM Method | Test Conditions | English Units | Dry As Molded | 50% R.H. |
|---|---|---|---|---|---|
| **Toughness** | | | | | |
| Gardner Impact<br>2″ Ring, 1/2″ Dart<br>100 mils | D-3029G | 73°F<br>32°F<br>−4°F<br>−40°F | in-lbs | >400<br>260<br>210<br>50 | >400<br>>400<br>250<br>50 |
| Driven Dart Impact<br>1-1/4″ Ring, 1/2″<br>Dart<br>100 mils | Monsanto | 73°F<br>−4°F<br>−40°F | in-lbs | 450<br>250<br>5 | 450<br>300<br>5 |
| Notched Izod<br>1/8″ Bar<br>1/4″ Bar | D-256 | 73°F | ft-lbs/in | 16<br>16 | 18<br>18 |
| Charpy Impact<br>Notched, 1/4″ Bar<br>Unnotched, 1/4″ Bar | D-256 | 73°F | ft-lbs/in² | 20<br>No break | 22<br>No break |
| Tensile Impact | D-1822<br>Type L | 73°F | ft-lbs/in² | 260 | No break |
| **Thermal** | | | | | |
| Minimum Molding Temperature | — | — | °C | 243 | — |
| Apparent Viscosity | Dayton Rheometer | 475°F<br>200 sec⁻¹<br>500 sec⁻¹<br>1000 sec⁻¹ | $cp \times 10^3$ | 2450<br>1060<br>690 | —<br>—<br>— |
| Melt Flow Index,<br>"I"<br>"T" | D-1238<br>D-1238 | 445°F/3.8kg<br>482°F/2.16kg | g/10 min<br>g/10 min | 0.1<br>1.3 | —<br>— |
| Vicat Softening Temperature | D-1525 | 1kg<br>5kg | °C<br>°C | 180<br>100 | 155<br>— |
| Heat Deflection Temperature | D-648 | 66 psi | °C | 94 | 92 |
| Specific Heat | C-351 | 495°F | KJ/Kg-°K | 2.4 | — |
| Coefficient of Linear Expansion | D-696 | −20° to 70°F<br>70°F to 180°F<br>180°F to 280°F | mm/m/°K | $10 \times 10^{-5}$<br>$17 \times 10^{-5}$<br>$38 \times 10^{-5}$ | —<br>—<br>— |
| Thermal Conductivity | C-177 | — | W/m-°K | 0.31 | — |

*(continued)*

TRIAX 1125 *(continued)*

| Property | ASTM Method | Test Conditions | English Units | Dry As Molded | 50% R.H. |
|---|---|---|---|---|---|
| **Miscellaneous** | | | | | |
| Specific Gravity | D-792 | 73°F | — | 1.06 | 1.06 |
| Water Absorption 50% R.H. | D-570 | 73°F, 24 hrs<br>73°F, 72 hrs<br>73°F, 90 days<br>73°F, Equilibrium | %<br>%<br>%<br>% | <0.1<br>0.6<br>1.1<br>1.6 | —<br>—<br>—<br>— |
| Water Immersion | D-570 | 73°F, 24 hrs<br>73°F, Saturation | %<br>% | 1.1<br>4.2 | —<br>— |
| Hardness<br>Rockwell<br>Shore | <br>D-785<br>D-676 | <br>73°F<br>73°F | <br>R Scale<br>D Scale | <br>99<br>77 | <br>75<br>68 |
| Mold Shrinkage | D-955 | Thickness,mils<br>50<br>125<br>200 | <br>mils/in | <br>8–10<br>12–14<br>16–20 | <br>—<br>—<br>— |
| Limiting Oxygen Index | D-2863 | — | % $O_2$ | 20 | 21 |

TRIAX 1125

| Property | ASTM Method | Test Conditions | SI Units | Dry As Molded | 50% R.H. |
|---|---|---|---|---|---|
| **Mechanical** | | | | | |
| Tensile Strength | D-638 | 73°F<br>−40°F | MPa | 47<br>72 | 41<br>72 |
| Elongation at Fail | D-638 | 73°F<br>−40°F | % | 270<br>120 | 270<br>120 |
| Tensile Modulus | D-638 | 73°F | MPa | 1890 | 1280 |
| Flexural Modulus | D-790 | −40°F<br>73°F<br>158°F<br>170°F | MPa | 2415<br>2140<br>900<br>830 | 2345<br>1035<br>830<br>760 |
| Poisson Ratio | — | — | — | 0.38 | — |

*(continued)*

**TRIAX 1125 *(continued)***

| Property | ASTM Method | Test Conditions | SI Units | Dry As Molded | 50% R.H. |
|---|---|---|---|---|---|
| **Toughness** | | | | | |
| Gardner Impact<br>2″ Ring, 1/2″ Dart<br>100 mils | D-3029G | 73°F<br>32°F<br>−4°F<br>−40°F | J | >45<br>29<br>24<br>6 | >45<br>>45<br>29<br>6 |
| Driven Dart Impact<br>1-1/4″ Ring, 1/2″<br>Dart<br>100 mils | Monsanto | 73°F<br><br>−4°F<br>−40°F | J | 51<br><br>29<br>0.6 | 51<br><br>34<br>0.6 |
| Notched Izod<br>1/8″ Bar<br>1/4″ Bar | D-256 | 73°F | J/m | 850<br>850 | 950<br>950 |
| Charpy Impact<br>Notched, 1/4″ Bar<br>Unnotched, 1/4″ Bar | D-256 | 73°F | $KJ/m^2$ | 44<br>No break | 48<br>No break |
| Tensile Impact | D-1822<br>Type L | 73°F | $KJ/m^2$ | 545 | No break |
| **Thermal** | | | | | |
| Minimum Molding Temperature | — | — | °C | 243 | — |
| Apparent Viscosity | Dayton Rheometer | 475°F<br>200 $sec^{-1}$<br>500 $sec^{-1}$<br>1000 $sec^{-1}$ | $cp \times 10^3$ | <br>2450<br>1060<br>690 | <br>—<br>—<br>— |
| Melt Flow Index,<br>"I"<br>"T" | <br>D-1238<br>D-1238 | <br>445°F/3.8kg<br>482°F/2.16kg | <br>g/10 min<br>g/10 min | <br>0.1<br>1.3 | <br>—<br>— |
| Vicat Softening Temperature | D-1525 | 1kg<br>5kg | °C<br>°C | 180<br>100 | 155<br>— |
| Heat Deflection Temperature | D-648 | 66 psi | °C | 94 | 92 |
| Specific Heat | C-351 | 495°F | KJ/kg-°K | 2.4 | — |
| Coefficient of Linear Expansion | D-696 | −20° to 70°F<br>70°F to 180°F<br>180°F to 280°F | mm/m/°K | $10 \times 10^{-5}$<br>$17 \times 10^{-5}$<br>$38 \times 10^{-5}$ | —<br>—<br>— |
| Thermal Conductivity | C-177 | — | W/m-°K | 0.31 | — |

*(continued)*

TRIAX 1125 *(continued)*

| Property | ASTM Method | Test Conditions | SI Units | Dry As Molded | 50% R.H. |
|---|---|---|---|---|---|
| **Miscellaneous** | | | | | |
| Specific Gravity | D-792 | 73°F | — | 1.06 | 1.06 |
| Water Absorption 50% R.H. | D-570 | 73°F, 24 hrs<br>73°F, 72 hrs<br>73°F, 90 days<br>73°F, Equilibrium | %<br>%<br>%<br>% | <0.1<br>0.6<br>1.1<br>1.6 | —<br>—<br>—<br>— |
| Water Immersion | D-570 | 73°F, 24 hrs<br>73°F, Saturation | %<br>% | 1.1<br>4.2 | —<br>— |
| Hardness<br>Rockwell<br>Shore | <br>D-785<br>D-676 | <br>73°F<br>73°F | <br>R Scale<br>D Scale | <br>99<br>77 | <br>75<br>68 |
| Mold Shrinkage | D-955 | Thickness,mils<br>50<br>125<br>200 | mils/in | <br>8–10<br>12–14<br>16–20 | <br>—<br>—<br>— |
| Limiting Oxygen Index | D-2863 | — | % $O_2$ | 20 | 21 |

TRIAX 1125 *(continued)*

| Property | ASTM Conditions | Units | Dry As Molded | 50% R.H. |
|---|---|---|---|---|
| **Miscellaneous** | | | | |
| Volume Resistivity | D-257 | $10^{14}$ohm-cm | 6.4 | 0.09 |
| Surface Resistivity | D-257 | $10^{14}$ohm | 2.0 | 0.82 |
| Insulation Resistance | D-257 | $10^{14}$ohm | 1.2 | 0.036 |
| High Voltage Arc Resistance 1/8″ (3.2mm) | D-495 | seconds | 105 | 105 |
| Hot Wire Ignition 1/16″ (1.6mm) | UL-746A | seconds | 14 | — |
| High Voltage Arc Resistance to Ignition, 1/8″ (3.2mm) | UL-746A | seconds | 54 | — |
| Comparative Tracking Index | D-3638 | volts | 600+ | — |
| High Voltage Arc Tracking Rate | UL-746A | in/min | Zero | — |
| Temperature Index<br>Electrical<br>Mechanical with Impact<br>Mechanical without Impact | UL-746B | <br>°C<br>°C<br>°C | <br>60<br>60<br>60 | <br>—<br>—<br>— |

*(continued)*

TRIAX 1125 *(continued)*

| Property | ASTM Conditions | Units | Dry As Molded | 50% R.H. |
|---|---|---|---|---|
| **Miscellaneous** | | | | |
| Dielectric Constant | D-150 | | | |
| 100Hz | | — | 4.8 | 6.3 |
| $10^3$Hz | | — | 4.5 | 5.0 |
| $10^5$Hz | | — | 3.7 | 3.8 |
| Dissipation Factor | D-150 | | | |
| 100Hz | | — | 0.05 | 0.088 |
| $10^3$Hz | | — | 0.05 | 0.074 |
| $10^5$Hz | | — | 0.0003 | 0.044 |
| Dielectric Strength | D-149 | | | |
| Short Time | | volts/mil | 450 | 450 |
| Step-by-Step | | volts/mil | 350 | 380 |
| High Current Arc Resistance | UL-746A | Arc, seconds | 200+ | — |
| UL Flammability 1/16″ (1.6mm) | UL-94 | — | HB | HB |

# Republic Plastics Company

## Contact Address:

Republic Plastics Company
2663 Eaton Rapids Road
Lansing, MI 48911

Phone: 1 (517) 887-0087

---

## AB-6013

### Description:

Republic Plastics Company is a custom blender of polymer blends and alloys. Most have a polyolefin base. Their ETA alloys are thermoplastic elastomers. RTA alloys are essentially rigid. ETA and RTA polymers can be injection molded, extruded, blow molded, and thermoformed. They have excellent resistance to environmental forces, and are recommended for outdoor applications. They can be pigmented using olefinic color concentrates, and painted using flexible polymers and top coats.

ETA and RTA alloys have excellent mechanical properties; combining high strength, toughness, impact strength, tear strength, and stress crack resistance over a wide temperature range. The products find application in a wide range of automotive parts and components as well as products for building materials, recreational vehicles, lawn and garden equipment, outdoor furniture and appliances.

ETA alloys can be processed on plunged or reciprocating screw injection molding machines, as well as extruders and other equipment for manufacturing thermoplastic products.

Product data sheets are representative of a wider range of specialty polymer alloys available from the Republic Plastics Company. Polymer alloys for specific product applications are available in-house or from modified formulations.

### Product Data:

AB-6013 is an Elastomeric Thermoplastic Alloy designed specifically as an acoustical barrier material to abate low frequency diesel engine noise. AB-6013 can be used as an effective noise barrier in applications such as wheel wells, engine side dash mat, transmission tunnel, and any other application where a flexible, chemical resistant, molded noise barrier is required. Test samples were conditioned 96 hours at 77°F ± 5°F prior to testing.

*(continued)*

NOTE – Information provided in this book has been condensed. The publisher recommends that the product manufacturer be contacted, at the address or phone number provided, for more complete information on product safety, processing parameters, and applications not provided in this information sheet.

## Typical Physical Properties:

**AB 6013 Injection Molding/Acoustical Barrier Grade**

| Property | ASTM Test Methods | Units | Typical Value |
|---|---|---|---|
| Color | | | Black |
| Specific Gravity | D792 | | 1.80 |
| Flexural Modulus<br>73°F | D790<br>.05″/min. | psi | 12,000 |
| Flammability Rating | SAE J-369 | in/min | <1 |
| Shrinkage (Tool) | | mils/in | 8.0 ± 1.0 |
| Tensile, Ultimate Strength<br>(Die C) | D412<br>2″/min. | psi | 375 |
| Elongation<br>(Die C) | D412<br>2″/min. | % | 300 |
| Tear Strength<br>(Die C) | D624<br>2″/min. | p/in | 100 |
| Hardness | D2240 | Shore A | 95 ± 3 |
| Melt Index | D1238<br>Cond. L | g/10 min. | 4.5 ± 30% |

# Republic Plastics Company

## Contact Address:

Republic Plastics Company
2663 Eaton Rapids Road
Lansing, MI 48911

Phone: 1 (517) 887-0087

## AB-6023

### Product Data:

AB-6023 is an Elastomeric Thermoplastic Alloy designed specifically as an acoustical barrier material to abate low frequency diesel engine noise. AB-6023 can be used as an effective noise barrier in applications such as wheel wells, engine side dash mat, transmission tunnels, speaker housing and any other applications where a flexible, chemical resistant, molded noise barrier is required.

### Typical Physical Properties:

**AB-6023 Injection Molding/Acoustical Barrier Grade**

| Property | ASTM Method | Units | Original Properties | Heat Aged 72 hrs. @ 250°F |
|---|---|---|---|---|
| Color | | | Black | Black |
| Specific Gravity | D792 | | 1.95 | 1.95 |
| Flexibility<br>73°F<br>−10° | 1″ Mandrel Bend | | Pass<br>Pass | Pass<br>Pass |
| Flexural Modulus<br>73°F | D790<br>.05″/min. | psi | 17,000 | — |
| Tensile, Ultimate<br>(2″/min.) | D412<br>(Die C) | psi | 700 | 700 |
| Elongation, Ultimate<br>(2″/min.) | D412<br>(Die C) | % | 350 | 300 |
| Tear Strength<br>(2″/min.) | D624<br>(Die C) | p/in | 250 | 230 |
| Hardness (Inst.) | D2240 | Shore D | 45 | 46 |
| Flammability | SAE J-369 | in/min | <1 | <1 |

*(continued)*

NOTE – Information provided in this book has been condensed. The publisher recommends that the product manufacturer be contacted, at the address or phone number provided, for more complete information on product safety, processing parameters, and applications not provided in this information sheet.

**AB-6023 Injection Molding/Acoustical Barrier Grade *(continued)***

| **Chemical Resistance (4 hr. Immersion @ 77°F)** | **% Weight Change** | **% Tear Strength Loss** |
|---|---|---|
| Water | 0 | 0 |
| Anti-Freeze | 0 | 0 |
| #2 Diesel Fuel | +3.9 | −10.0 |
| Brake Fluid | 0 | 0 |
| Windshield Washer Fluid | 0 | 0 |
| Transmission Fluid | +0.5 | 0 |

# Republic Plastics Company

## Contact Address:

Republic Plastics Company
2663 Eaton Rapids Road
Lansing, MI 48911

Phone: 1 (517) 887-0087

## AB-6053

### Product Data:

AB-6053 is an Elastomeric Thermoplastic Alloy designed specifically as an acoustical barrier material to abate low frequency diesel engine noise. AB-6053 can be used as an effective noise barrier in applications such as wheel wells, engine side dash mat, transmission tunnel, and any other application where a flexible, chemical resistant, molded noise barrier is required. Test samples were conditioned 96 hours at 77°F ± 5°F prior to testing.

### Typical Physical Properties:

**AB-6053 Injection Molding/Acoustical Barrier Grade**

| Property | ASTM Test Methods | Units | AB-6053 |
|---|---|---|---|
| Color | | | Black |
| Specific Gravity | D792 | | 1.90 ± .05 |
| Flexibility<br>70°F<br>−10°F | 1″ Mandrel<br>Bend | | <br>Pass<br>Pass |
| Flexural Modulus<br>73°F | D790<br>.05″/min. | <br>MPa<br>psi | <br>310<br>45,000 |
| Flammability Rating | SAE J-369 | in/min | <1 |
| Shrinkage<br>(Direction of Flow) | | mls/in | 9.0 ± 1.0 |
| Tensile Strength<br>(Die C) | D638<br>2″/min. | <br>MPa<br>psi | <br>5.5<br>800 |
| Elongation Ultimate<br>(Die C) | D638<br>2″/min. | <br>% min. | <br>200 |

*(continued)*

NOTE – Information provided in this book has been condensed. The publisher recommends that the product manufacturer be contacted, at the address or phone number provided, for more complete information on product safety, processing parameters, and applications not provided in this information sheet.

**AB-6053 Injection Molding/Acoustical Barrier Grade *(continued)***

| Property | ASTM Test Methods | Units | AB-6053 |
|---|---|---|---|
| Tear Strength (Die C) | D624 2"/min. | KN/m<br>p/in | 35<br>200 |
| Hardness | D2240 | Shore A | 95 ± 5 |
| Melt Index | D1238 Cond. L Proc. A | g/10 min. | 1.0 |

# Republic Plastics Company

## Contact Address:

Republic Plastics Company
2663 Eaton Rapids Road
Lansing, MI 48911

Phone: 1 (517) 887-0087

## AB-6103

### Product Data:

AB-6103 is an Elastomeric Thermoplastic Alloy designed specifically as an acoustical barrier material to abate low frequency diesel engine noise. AB-6103 can be used as an effective noise barrier in applications such as wheel wells, engine side dash mat, transmission tunnels, speaker housing and any other application where a flexible, chemical resistant, molded noise barrier is required. Test samples were conditioned 96 hours at 77°F ± 5°F prior to testing.

### Typical Physical Properties:

**AB-6103 Injection Molding/Acoustical Barrier Grade**

| Property | ASTM Method | Units | Original Properties | Heat Aged 72 hrs. @ 250°F |
|---|---|---|---|---|
| Color | | | Black | Black |
| Specific Gravity | D792 | | 1.95 | 1.95 |
| Flexibility<br>73°F<br>−10°F | 1″ Mandrel<br>Bend | | <br>Pass<br>Pass | <br>Pass<br>Pass |
| Gardner Impact<br>73°F<br>−10°F | <br>1/2″ radius<br>dart | <br>in-lbs<br>in-lbs | <br>>24<br>>12 | <br>>24<br>>12 |
| Flexural Modulus<br>73°F | D790<br>.05″/min. | psi | 100,000 | 100,000 |
| Flammability Rating | SAE J-369 | in/min | <1 | |
| Shrinkage<br>(Direction of Flow) | | mils/in | 8.0 | 10.0 |
| Tensile, Ultimate Strength<br>(Die C) | D412<br>2″/min. | psi | 1000 | 1000 |

*(continued)*

NOTE – Information provided in this book has been condensed. The publisher recommends that the product manufacturer be contacted, at the address or phone number provided, for more complete information on product safety, processing parameters, and applications not provided in this information sheet.

**AB-6103 Injection Molding/Acoustical Barrier Grade** ***(continued)***

| Property | ASTM Method | Units | Original Properties | Heat Aged 72 hrs. @ 250°F |
|---|---|---|---|---|
| Elongation Ultimate (Die C) | D412 2″/min. | % | 200 | 180 |
| Tear Strength (Die C) | D624 2″/min. | p/in | 200 | 160 |
| Hardness | D2240 | Shore A | 90 | 93 |
| Melt Index | D1238 Cond L. | g/10 min | 2.0 | |

# Republic Plastics Company

## Contact Address:

Republic Plastics Company
2663 Eaton Rapids Road
Lansing, MI 48911

Phone: 1 (517) 887-0087

## AB-7023

### Product Data:

AB-7023 is an Elastomeric Thermoplastic Alloy designed specifically as an acoustical barrier material to abate low frequency diesel engine noise. AB-7023 can be used as an effective noise barrier in applications such as wheel wells, engine side dash mat, transmission tunnels, speaker housing and any other application where a flexible, chemical resistant, molded noise barrier is required. Test samples were conditioned 96 hours at 77°F ± 5°F prior to testing.

### Typical Physical Properties:

**AB-7023 Injection Molding/Acoustical Barrier Grade**

| Property | ASTM Test Methods | Units | Typical Value |
|---|---|---|---|
| Color | | | Black |
| Specific Gravity | D792 | | 1.20 |
| Flexural Modulus 73°F | D790 .05″/min. | psi | 20,000 |
| Flammability Rating | SAE J-369 | in/min | <1 |
| Shrinkage (Tool) | | mils/in | 11.5 ± 1.0 |
| Tensile, Ultimate Strength (Die C) | D412 2″/min. | psi | 600 |
| Elongation (Die C) | D412 2″/min. | % | 400 |
| Tear Strength (Die C) | D624 2″/min. | p/in | 200 |
| Hardness | D2240 | Shore D | 39 ± 3 |
| Melt Index | D1238 Cond. L | g/10 min. | 5.5 ± 30% |

NOTE – Information provided in this book has been condensed. The publisher recommends that the product manufacturer be contacted, at the address or phone number provided, for more complete information on product safety, processing parameters, and applications not provided in this information sheet.

# Republic Plastics Company

## Contact Address:

Republic Plastics Company
2663 Eaton Rapids Road
Lansing, MI 48911

Phone: 1 (517) 887-0087

## AB-7103

### Product Data:

AB-7103 is an Elastomeric Thermoplastic Alloy designed specifically as an acoustical barrier material to abate low frequency diesel engine noise. AB-7103 can be used as an effective noise barrier in applications such as wheel wells, engine side dash mats, transmission tunnels, speaker housings and any other application where a flexible, chemical resistant, molded noise barrier is required.

### Typical Physical Properties:

**AB-7103 Injection Molding/Acoustical Barrier Grade**

| Property | ASTM Method | Units | Data |
|---|---|---|---|
| Color | | | Black |
| Specific Gravity | D792 | | 1.2 |
| Melt Flow Index | D1238 (L) | g/10 min. | 9 ± 30% |
| Tear Strength | D624 | p/in | 400 |
| Tensile Strength, Ultimate | D412 | psi | 1400 |
| Elongation, Ultimate | D412 | % | 290 |
| Flexural Modulus | D790 | psi | 115,000 |
| Shrinkage | | mls/in | 12.0 |
| Hardness | Shore D | | 64 |

NOTE – Information provided in this book has been condensed. The publisher recommends that the product manufacturer be contacted, at the address or phone number provided, for more complete information on product safety, processing parameters, and applications not provided in this information sheet.

# Republic Plastics Company

## Contact Address:

Republic Plastics Company
2663 Eaton Rapids Road
Lansing, MI 48911

Phone: 1 (517) 887-0087

## ETA-3011/3010

### Product Data:

Elastomeric Thermoplastic Alloy, ETA-3011 and ETA-3010 are designed for ease of molding with optimum part appearance and performance. Test results are from ⅛″ injection molded plaques.

### Typical Physical Properties:

**ETA-3011/3010 Injection Molding Grade**

| Property | ASTM Test Methods | Units | ETA-3011/3010 |
|---|---|---|---|
| Color | | | Black/Natural |
| Specific Gravity | D792 | | 0.91 ± .01 |
| Gardner Impact −20°F | | in-lb | >320 |
| Tensile, Ultimate (Die C) | D638 2″/min. | psi | 1200 |
| Flexural Modulus 73°C (min.) | D790 | psi | 13,000 |
| Elongation, Ultimate (Die C) | D638 2″/min. | % | 500 |
| Tear Strength (Die C) | D624 2″/min. | p/in | 300 |
| Hardness | D2240 | Shore A<br>Shore D | 92 ± 3<br>34 ± 3 |
| Melt Flow | D1238 Cond. L | g/10 min. | 4.0 |
| Shrinkage (Tool) | | mils/in. | 9.0 ± 1.0 |

NOTE – Information provided in this book has been condensed. The publisher recommends that the product manufacturer be contacted, at the address or phone number provided, for more complete information on product safety, processing parameters, and applications not provided in this information sheet.

# Republic Plastics Company

## Contact Address:

Republic Plastics Company
2663 Eaton Rapids Road
Lansing, MI 48911

Phone: 1 (517) 887-0087

## ETA-3031

### Product Data:

Elastomeric Thermoplastic Alloy, ETA-3031 is designed for ease of molding with optimum part appearance and performance. Test results are from ⅛″ injection molded plaques.

### Typical Physical Properties:

**ETA-3031 Injection Molding Grade**

| Property | ASTM Test Methods | Units | ETA-3031 |
|---|---|---|---|
| Color | | | |
| Specific Gravity | D792 | | 0.910 |
| Gardner Impact −20°F<br>−29°C | | in-lb<br>J | >320<br>>36 |
| Tensile, Ultimate (Die C) | D638<br>2″/min. | psi<br>MPa | 2000<br>14 |
| Flexural Modulus 73°C (min.) | D790 | psi<br>MPa | 27,000<br>186 |
| Elongation, Ultimate (Die C) | D638<br>2″/min. | % | 600 |
| Tear Strength (Die C) | D624<br>2″/min. | p/in<br>KN/m | 450<br>79 |
| Hardness | D2240 | Shore D | 48 |
| Melt Flow | D1238<br>Cond. L | g/10 min. | 3.5 |
| Shrinkage (Tool) | | mils/in. | 8.0 ± 1.0 |
| Notched Izod<br>Brittle Point | D746 | °C | no break<br>< −50 |

NOTE—Information provided in this book has been condensed. The publisher recommends that the product manufacturer be contacted, at the address or phone number provided, for more complete information on product safety, processing parameters, and applications not provided in this information sheet.

# Republic Plastics Company

## Contact Address:

Republic Plastics Company
2663 Eaton Rapids Road
Lansing, MI 48911

Phone: 1 (517) 887-0087

---

## ETA-3041C

### Product Data:

Elastomeric Thermoplastic Alloy, ETA-3041C is designed for ease of molding with optimum part appearance and performance. Test results are from ⅛″ injection molded plaques.

### Typical Physical Properties:

**ETA-3041C Injection Molding Grade**

| Property | ASTM Test Methods | Units | ETA-3041C |
|---|---|---|---|
| Color | | | Black |
| Specific Gravity | D792 | | .91 |
| Brittle Point | D746 | °C | < −50 |
| Gardner Impact<br>73°F<br>−40°F | | <br>in-lb<br>in-lb | <br>> 320<br>> 320 |
| Flexural Modulus<br>73°F<br>160°F<br>−20°F | D790 | <br>psi<br>psi<br>psi | <br>37,000<br>12,000<br>130,000 |
| Tensile, Ultimate<br>(Die C) | D638<br>2″/min. | psi | 2000 |
| Elongation, Ultimate<br>(Die C) | D638<br>2″/min. | % | 500 |
| Tear Strength<br>(Die C) | D624<br>2″/min. | p/in | 500 |
| Hardness | D2240 | Shore D | 49 ± 2.0 |
| Melt Flow | D1238<br>Cond. L | g/10 min. | 5.0 |
| Shrinkage | | mils/in. | 9.0 ± 1.0 |

NOTE – Information provided in this book has been condensed. The publisher recommends that the product manufacturer be contacted, at the address or phone number provided, for more complete information on product safety, processing parameters, and applications not provided in this information sheet.

# Republic Plastics Company


Contact Address:

Republic Plastics Company
2663 Eaton Rapids Road
Lansing, MI 48911

Phone: 1 (517) 887-0087


## ETA-3041E

### Product Data:

Elastomeric Thermoplastic Alloys (ETA) are Thermoplastic Elastomers designed for optimum strength down to −40°F. ETA extrusion grade polymers are specifically formulated for processing by profile extrusion, sheet extrusion and thermoforming. Test results are from ⅛″ injection molded plaques.

### Typical Physical Properties:

**ETA-3041E Blow Molding/Extrusion Grade**

| Property | ASTM Test Methods | Units | ETA-3041E |
|---|---|---|---|
| Color | | | Black |
| Specific Gravity | D792 | | .915 |
| Brittle Point | D746 | °C | < −70 |
| Gardner Impact<br>73°F<br>−20°F | | <br>in-lb<br>in-lb | <br>>160<br>>160 |
| Izod Impact, Notched<br>73°F<br>−20°F | D256 | <br>ft lb/in<br>ft lb/in | <br>no break<br>no break |
| Flexural Modulus<br>73°F<br>160°F<br>−40°F | D790 | <br>psi<br>psi<br>psi | <br>34,000<br>16,000<br>142,000 |
| Tensile Yield Strength<br>(Die C) | D638<br>2″/min. | psi | 2600 |
| Elongation, Ultimate<br>(Die C) | D638<br>2″/min. | % | 550 |
| Tear Strength<br>(Die C) | D624<br>2″/min. | p/in | 650 |
| Hardness | D2240 | Shore D | 53 |
| Melt Index | D1238 | g/10 min. | 0.5 max. |

NOTE—Information provided in this book has been condensed. The publisher recommends that the product manufacturer be contacted, at the address or phone number provided, for more complete information on product safety, processing parameters, and applications not provided in this information sheet.

# Republic Plastics Company

## Contact Address:

Republic Plastics Company
2663 Eaton Rapids Road
Lansing, MI 48911

Phone: 1 (517) 887-0087

## ETA-3051

### Product Data:

Elastomeric Thermoplastic Alloy, ETA-3051 is designed for ease of molding with optimum part appearance and performance. Test results are from ⅛″ injection molded plaques.

### Typical Physical Properties:

**ETA-3051 Injection Molding Grade**

| Property | ASTM Test Methods | Units | ETA-3051 |
|---|---|---|---|
| Color | | | Black |
| Specific Gravity | D792 | | .90 ± .01 |
| Brittle Point | D746 | °C | < −50 |
| Gardner Impact<br>73°F<br>−40°F | | <br>in-lb<br>in-lb | <br>>320<br>>320 |
| Flexural Modulus<br>73°F<br>160°F<br>−20°F | D790 | <br>psi<br>psi<br>psi | <br>60,000<br>30,000<br>160,000 |
| Tensile, Ultimate<br>(Die C) | D638<br>2″/min. | psi | 1500 |
| Elongation, Ultimate<br>(Die C) | D638<br>2″/min. | % | 400 |
| Tear Strength<br>(Die C) | D624<br>2″/min. | p/in | 450 |
| Hardness | D2240 | Shore D | 58 ± 5 |
| Melt Flow Index | D1238<br>Cond. L | g/10 min. | 4.5 ± 30% |

*(continued)*

NOTE – Information provided in this book has been condensed. The publisher recommends that the product manufacturer be contacted, at the address or phone number provided, for more complete information on product safety, processing parameters, and applications not provided in this information sheet.

ETA-3051 Injection Molding Grade *(continued)*

| Property | ASTM Test Methods | Units | ETA-3051 |
|---|---|---|---|
| Shrinkage (Tool) | | mils/in. | 15.0 ± 1.0 |
| Shrinkage (After Bake) | | mils/in. | 17.5 ± 1.0 |
| Notched Izod | D256 | | |
| −20°F | | ft-lb | 17.0 |
| R.T. | | ft-lb | No Break |

# Republic Plastics Company

## Contact Address:

Republic Plastics Company
2663 Eaton Rapids Road
Lansing, MI 48911

Phone: 1 (517) 887-0087

---

## ETA-3055

### Product Data:

Elastomeric Thermoplastic Alloy, ETA-3055 is designed for ease of molding with optimum part appearance and performance. Test results are from ⅛″ injection molded plaques.

### Typical Physical Properties:

**ETA-3055 Injection Molding Grade**

| Property | ASTM Test Methods | Units | ETA-3055 |
|---|---|---|---|
| Color | | | Black |
| Specific Gravity | D792 | | .91 ±.01 |
| Brittle Point | D746 | °C | < −50 |
| Gardner Impact<br>73°F<br><br>−40°F | | <br>in-lb<br>J<br>in-lb<br>J | <br>>320<br>36<br>>320<br>36 |
| Flexural Modulus<br>73°F | D790 | <br>psi<br>MPa | <br>55,000<br>380 |
| Tensile, Ultimate<br>(Die C) | D638<br>2″/min. | <br>psi<br>MPa | <br>2000<br>13.5 |
| Elongation, Ultimate<br>(Die C) | D638<br>2″/min. | <br>% | <br>500 |
| Tear Strength<br>(Die C) | D624<br>2″/min. | <br>p/in<br>KN/M | <br>450<br>79 |

*(continued)*

NOTE—Information provided in this book has been condensed. The publisher recommends that the product manufacturer be contacted, at the address or phone number provided, for more complete information on product safety, processing parameters, and applications not provided in this information sheet.

**ETA-3055 Injection Molding Grade *(continued)***

| Property | ASTM Test Methods | Units | ETA-3055 |
|---|---|---|---|
| Hardness | D2240 | Shore D | 55 ± 3.0 |
| Melt Flow Index | D1238<br>Cond. L | g/10 min. | 8.0 ± 30% |
| Shrinkage (Tool)<br>(After Bake) | | mils/in.<br>mils/in. | 10.5 ± 1.0<br>14.0 ± 1.0 |
| Rheometric Impact, Ult. E<br>−29°C | 2.2 M/S | J | 50 |

# Republic Plastics Company

## Contact Address:

Republic Plastics Company
2663 Eaton Rapids Road
Lansing, MI 48911

Phone: 1 (517) 887-0087

## ETA-3081

### Product Data:

Elastomeric Thermoplastic Alloy, ETA-3081 is designed for ease of molding with optimum part appearance and performance. Test results are from ⅛″ injection molded plaques.

### Typical Physical Properties:

**ETA-3081 Injection Molding Grade**

| Property | ASTM Test Methods | Units | ETA-3081 |
|---|---|---|---|
| Color | | | |
| Specific Gravity | D792 | | 0.910 |
| Gardner Impact −20°F<br>−29°C | | in-lb<br>J | >320<br>>36 |
| Tensile, Ultimate<br>(Die C) | D638<br>2″/min. | psi<br>MPa | 2000<br>14 |
| Flexural Modulus<br>73°C (min.) | D790 | psi<br>MPa | 80,000<br>552 |
| Elongation, Ultimate<br>(Die C) | D638<br>2″/min. | % | 500 |
| Tear Strength<br>(Die C) | D624<br>2″/min. | p/in<br>KN/m | 500<br>86 |
| Hardness | D2240 | Shore D | 57 |
| Melt Flow | D1238<br>Cond. L | g/10 min. | 4.0 |
| Shrinkage (Tool) | | mils/in. | 11.0 ± 1.0 |
| Notched Izod<br>Brittle Point | <br>D746 | <br>°C | no break<br>< −50 |

NOTE—Information provided in this book has been condensed. The publisher recommends that the product manufacturer be contacted, at the address or phone number provided, for more complete information on product safety, processing parameters, and applications not provided in this information sheet.

# Republic Plastics Company

## Contact Address:

Republic Plastics Company
2663 Eaton Rapids Road
Lansing, MI 48911

Phone: 1 (517) 887-0087

# ETA-3081E

## Product Data:

Elastomeric Thermoplastic Alloys (ETA) are Thermoplastic Elastomers designed for optimum strength down to −40°F. ETA extrusion grade polymers are specifically formulated for processing by profile extrusion, sheet extrusion and thermoforming. Test results are from ⅛″ injection molded plaques.

## Typical Physical Properties:

**ETA-3081E Blow Molding/Extrusion Grade**

| Property | ASTM Test Methods | Units | ETA-3081E |
|---|---|---|---|
| Color | | | Black |
| Specific Gravity | D792 | | .915 |
| Brittle Point | D746 | °C | <−70 |
| Gardner Impact<br>73°F<br>−20°F | | <br>in-lb<br>in-lb | <br>>160<br>>160 |
| Izod Impact, Notched<br>73°F<br>−20°F | D256 | <br>ft lb/in<br>ft lb/in | <br>no break<br>no break |
| Flexural Modulus 73°F | D790 | psi | 65,000 |
| Tensile Yield Strength (Die C) | D638 2″/min. | psi | 2500 |
| Elongation, Ultimate (Die C) | D638 2″/min. | % | 600 |
| Tear Strength (Die C) | D624 2″/min. | p/in | 600 |
| Hardness | D2240 | Shore D | 57 |
| Melt Index | D1238 | g/10 min. | 0.85 |

NOTE—Information provided in this book has been condensed. The publisher recommends that the product manufacturer be contacted, at the address or phone number provided, for more complete information on product safety, processing parameters, and applications not provided in this information sheet.

# Republic Plastics Company

## Contact Address:

Republic Plastics Company
2663 Eaton Rapids Road
Lansing, MI 48911

Phone: 1 (517) 887-0087

## ETA-3104

### Product Data:

ETA-3104 is an Elastomeric Thermoplastic Alloy designed for automotive rubstrips and other applications requiring good heat resistance and dimensional stability. ETA-3104 is available in black or natural which can be pigmented by the use of color concentrates. Test results are from ⅛″ injection molded plaques.

### Typical Physical Properties:

**ETA-3104 Injection Molding Grade**

| Property | ASTM Test Methods | Units | ETA-3104 |
|---|---|---|---|
| Color | | | Black |
| Specific Gravity | D792 | | .92 ± .01 |
| Flexural Modulus<br>73°F<br>160°F<br>−20°F | D790<br>.05″/min.<br>.05″/min.<br>.05″/min. | <br>psi<br>psi<br>psi | <br>80,000<br>25,000<br>260,000 |
| Tensile Strength (Die C) | D412<br>2″/min. | psi | 2000 |
| Elongation, Ultimate (Die C) | D412<br>2″/min. | % | 450 |
| Tear Strength (Die C) | D624<br>2″/min. | p/in | 550 |
| Hardness | D2240 | Shore D | 58 ± 5.0 |
| Melt Index | D1238<br>Cond. L | g/10 min. | 3.5 |
| Shrinkage (Direction of Flow) | | mils/in. | 10.0 ± 1.0 |
| GFDI @ −20°F | | in-lb | >320 |

NOTE—Information provided in this book has been condensed. The publisher recommends that the product manufacturer be contacted, at the address or phone number provided, for more complete information on product safety, processing parameters, and applications not provided in this information sheet.

# Republic Plastics Company

## Contact Address:

Republic Plastics Company
2663 Eaton Rapids Road
Lansing, MI 48911

Phone: 1 (517) 887-0087

## ETA-3121

### Product Data:

Elastomeric Thermoplastic Alloy, ETA-3121 is designed for ease of molding with optimum part appearance and performance. Test results are from ⅛″ injection molded plaques.

### Typical Physical Properties:

**ETA-3121 Injection Molding Grade**

| Property | ASTM Test Methods | Units | ETA-3121 |
|---|---|---|---|
| Color | | | Black |
| Specific Gravity | D792 | | 0.95 |
| Gardner Impact<br>73°F<br>−20°F | | <br>in-lb<br>in-lb | <br>320<br>300 |
| Flexural Modulus<br>73°F | D790 | psi | 135,000 min. |
| Tensile, Ultimate<br>(Die C) | D638<br>2″/min. | psi | 2000 |
| Elongation, Ultimate<br>(Die C) | D638<br>2″/min. | % | 200 |
| Tear Strength<br>(Die C) | D624<br>2″/min. | p/in | 600 |
| Hardness | D2240 | Shore D | 66 ± 3.0 |
| Melt Flow | D1238<br>Cond. L | g/10 min. | 5.0 ± 30% |
| Shrinkage (Tool)<br>(After bake at 250°F) | | mils/in.<br>mils/in. | 10.5 ± 1.0<br>13.0 ± 1.0 |

NOTE – Information provided in this book has been condensed. The publisher recommends that the product manufacturer be contacted, at the address or phone number provided, for more complete information on product safety, processing parameters, and applications not provided in this information sheet.

# Republic Plastics Company

## Contact Address:

Republic Plastics Company
2663 Eaton Rapids Road
Lansing, MI 48911

Phone: 1 (517) 887-0087

## ETA-3125

### Product Data:

Elastomeric Thermoplastic Alloy, ETA-3125 is designed for ease of molding with optimum part appearance and performance. Test results are from ⅛″ injection molded specimens.

### Typical Physical Properties:

**ETA-3125 Injection Molding Grade**

| Property | ASTM Test Methods | Units | ETA-3125 |
|---|---|---|---|
| Color | | | Black |
| Specific Gravity | D792 | | .98 ± .02 |
| Gardner Impact<br>73°F<br><br>−20°F | | <br>in-lb<br>J<br>in-lb<br>J | <br>>160<br>>18.1<br>>120<br>>13.0 |
| Flexural Modulus<br>73°F | D790 | <br>psi<br>MPa | <br>175,000<br>1200 |
| Tensile, Ultimate<br>(Die C) (min.) | D638<br>2″/min. | <br>psi<br>MPa | <br>2000<br>14 |
| Elongation, Ultimate<br>(Die C) (min.) | D638<br>2″/min. | % | 40 |
| Tear Strength<br>(Die C) (min.) | D624<br>2″/min. | <br>p/in<br>KN/M | <br>500<br>87 |
| Hardness | D2240 | Shore D | 69 ± 5 |

*(continued)*

NOTE—Information provided in this book has been condensed. The publisher recommends that the product manufacturer be contacted, at the address or phone number provided, for more complete information on product safety, processing parameters, and applications not provided in this information sheet.

ETA-3125 Injection Molding Grade *(continued)*

| Property | ASTM Test Methods | Units | ETA-3125 |
|---|---|---|---|
| Melt Flow | D1238 Cond. L | g/10 min. | 5.0 ± 30% |
| Shrinkage (Tool)<br>(After bake at 250°F) | | mils/in.<br>mils/in. | 9.0 – 10.0<br>11.0 – 12.0 |
| HDT @ .0455 MPa<br>HDT @ 1.82 MPa | D648 | °C<br>°C | 100<br>60 |
| Coefficient of Linear Thermal Expansion | R.T. – 180°F | in/in/°F<br>in/in/°C | $3.0 \times 10^{-5}$<br>$5.2 \times 10^{-5}$ |

# Republic Plastics Company

## Contact Address:

Republic Plastics Company
2663 Eaton Rapids Road
Lansing, MI 48911

Phone: 1 (517) 887-0087

## ETA-3131

### Product Data:

Elastomeric Thermoplastic Alloy, ETA-3131 is designed for ease of molding with optimum part appearance and performance. Test results are from ⅛″ injection molded specimens.

### Typical Physical Properties:

**ETA-3131 Injection Molding Grade**

| Property | ASTM Test Methods | Units | ETA-3131 |
|---|---|---|---|
| Color | | | |
| Specific Gravity | D792 | | 0.900 |
| Gardner Impact −20°F | | in-lb<br>J | >320<br>>36 |
| Tensile, Ultimate<br>(Die C) | D638<br>2″/min. | psi<br>MPa | 2500<br>17 |
| Flexural Modulus<br>73°C (min.) | D790 | psi<br>MPa | 130,000<br>897 |
| Elongation, Ultimate<br>(Die C) | D638<br>2″/min. | % | 400 |
| Tear Strength<br>(Die C) | D624<br>2″/min. | p/in<br>KN/m | 650<br>114 |
| Hardness | D2240 | Shore D | 65 |
| Melt Flow | D1238<br>Cond. L | g/10 min. | 4.0 |
| Shrinkage (Tool) | | mils/in. | 15.0 ± 1.0 |
| Notched Izod<br>Brittle Point | D746 | °C | no break<br>< −40 |

NOTE – Information provided in this book has been condensed. The publisher recommends that the product manufacturer be contacted, at the address or phone number provided, for more complete information on product safety, processing parameters, and applications not provided in this information sheet.

# Republic Plastics Company

## Contact Address:

Republic Plastics Company
2663 Eaton Rapids Road
Lansing, MI 48911

Phone: 1 (517) 887-0087

## ETA-3131E

### Product Data:

Elastomeric Thermoplastic Alloys (ETA) are Thermoplastic Elastomers designed for optimum strength down to −20°F. ETA extrusion grade polymers are specifically formulated for processing by profile extrusion, sheet extrusion and thermoforming. Test results are from ⅛″ injection molded plaques.

### Typical Physical Properties:

**ETA-3131E Blow Molding/Extrusion Grade**

| Property | ASTM Test Methods | Units | ETA-3131E |
|---|---|---|---|
| Color | | | Black |
| Specific Gravity | D792 | | .91 |
| Brittle Point | D746 | °C | <−70 |
| Gardner Impact<br>73°F<br>−20°F | | <br>in-lb<br>in-lb | <br>>160<br>>160 |
| Flexural Modulus<br>73°F<br>160°F<br>−40°F | D790 | <br>psi<br>psi<br>psi | <br>130,000<br>40,000<br>390,000 |
| Tensile Yield Strength<br>(Die C) | D638<br>2″/min. | psi | 3000 |
| Elongation, Ultimate<br>(Die C) | D638<br>2″/min. | % | 400 |
| Tear Strength<br>(Die C) | D624<br>2″/min. | p/in | 700 |
| Hardness | D2240 | Shore D | 65 |
| Melt Flow Index | D1238<br>Cond. L | g/10 min. | 1.5 max. |

NOTE – Information provided in this book has been condensed. The publisher recommends that the product manufacturer be contacted, at the address or phone number provided, for more complete information on product safety, processing parameters, and applications not provided in this information sheet.

# Republic Plastics Company

## Contact Address:

Republic Plastics Company
2663 Eaton Rapids Road
Lansing, MI 48911

Phone: 1 (517) 887-0087

## ETA-3163

### Product Data:

Elastomeric Thermoplastic Alloy, ETA-3163 is designed for ease of molding with optimum part appearance and performance. Test results are from ⅛″ injection molded specimens.

### Typical Physical Properties:

**ETA-3163 Injection Molding Grade**

| Property | ASTM Test Methods | Units | ETA-3163 |
|---|---|---|---|
| Color | | | Black |
| Specific Gravity | D792 | | .95 |
| Gardner Impact<br>73°F<br>−40°F | | <br>in-lb<br>in-lb | <br>320<br>320 |
| Tensile, Ultimate<br>(Die C) | D638<br>2″/min. | psi | 2100 |
| Flexural Modulus<br>73°F | D790 | psi | 150,000 |
| Elongation, Ultimate<br>(Die C) | D638<br>2″/min. | % | 200 |
| Tear Strength<br>(Die C) | D624<br>2″/min. | p/in | 600 |
| Hardness | D2240 | Shore D | 66 ± 3.0 |
| Melt Flow | D1238<br>Cond. L | g/10 min. | 5.0 ± 30% |
| Shrinkage (Tool)<br>(After bake at 250°F) | | mils/in.<br>mils/in. | 10.5 ± 1.0<br>12.5 ± 1.0 |

*(continued)*

NOTE – Information provided in this book has been condensed. The publisher recommends that the product manufacturer be contacted, at the address or phone number provided, for more complete information on product safety, processing parameters, and applications not provided in this information sheet.

ETA-3163 Injection Molding Grade *(continued)*

| Property | ASTM Test Methods | Units | ETA-3163 |
|---|---|---|---|
| Notched Izod | D256 | ft-lb/in | 11.0 NB |
| Heat Deflection Temperature @ 66 psi | D648 | °C | 95 |
| Coefficient of Linear Thermal Expansion | R.T. to −20°F | in/in/°F | $4.5 \times 10^{-5}$ |
| | R.T. to 180°F | in/in/°F | $4.0 \times 10^{-5}$ |

# Republic Plastics Company

## Contact Address:

Republic Plastics Company
2663 Eaton Rapids Road
Lansing, MI 48911

Phone: 1 (517) 887-0087

# ETA-3183

## Product Data:

Elastomeric Thermoplastic Alloy, ETA-3183 is designed for ease of molding with optimum part appearance and performance. Test results are from ⅛″ injection molded specimens.

## Typical Physical Properties:

**ETA-3183 Injection Molding Grade**

| Property | ASTM Test Methods | Units | ETA-3183 |
|---|---|---|---|
| Color | | | Black |
| Specific Gravity | D792 | | .97 |
| Gardner Impact<br>73°F<br>−20°F | | <br>in-lb<br>in-lb | <br>240<br>>320 |
| Tensile, Ultimate<br>(Die C) | D638<br>2″/min. | <br>% | <br>2200 |
| Flexural Modulus<br>73°F | D790 | <br>psi | <br>180,000 |
| Elongation, Ultimate<br>(Die C) | D638<br>2″/min. | <br>% | <br>200 |
| Tear Strength<br>(Die C) | D624<br>2″/min. | <br>p/in | <br>500 |
| Hardness | D2240 | Shore D | 66 ± 3.0 |
| Melt Flow | D1238<br>Cond. L | g/10 min. | 4.0 |
| Shrinkage (Tool)<br>(After bake at 250°F) | | mils/in.<br>mils/in. | 9.5 – 10.5<br>12.0 – 13.0 |
| Heat Deflection Temperature @ 66 psi | D648 | °C | 100 |
| Coefficient of Linear Thermal Expansion | R. T. to 180°F | in/in/°C | $6.0 \times 10^{-5}$ |

NOTE—Information provided in this book has been condensed. The publisher recommends that the product manufacturer be contacted, at the address or phone number provided, for more complete information on product safety, processing parameters, and applications not provided in this information sheet.

# Republic Plastics Company

Contact Address:

Republic Plastics Company
2663 Eaton Rapids Road
Lansing, MI 48911

Phone: 1 (517) 887-0087

## RTA-3184

Product Data:

RTA-3184 is an injection molding grade rigid Thermoplastic Alloy, based upon a modified polyolefin polymer. RTA-3184 is designed for interior applications requiring a stiff, yet impact resistant material with good dimensional stability. It is supplied in black or natural, and can be pigmented with color concentrates or painted. Test results are from ⅛″ injection molded plaques.

Typical Physical Properties:

**RTA-3184 Injection Molding Grade**

| Property | ASTM Test Methods | Units | RTA-3184 |
|---|---|---|---|
| Color | | | Black/Natural |
| Specific Gravity | D792 | | 0.96/0.93 |
| Gardner Impact<br>73°F (1/8″)<br>−20°F (1/8″) | | <br>in-lb<br>in-lb | <br>100<br>15 |
| Flexural Modulus<br>73°F | D790<br>.05″/min. | psi | 180,000 |
| Tensile Strength<br>(Die C) 1/8″ | D412<br>2″/min. | psi | 3200 |
| Elongation Ultimate<br>(Die C) 1/8″ | D412<br>2″/min. | % | 300 |
| Tear Strength<br>(Die C) 1/8″ | D624<br>2″/min. | p/in | 700 |
| Hardness | D2240 | Shore D | 69 |
| Melt Index | D1238<br>Cond. L | g/10 min. | 6.5 ± 30% |
| Shrinkage<br>(Direction of Flow) | | mils/in. | 13.0/16.0 |

NOTE—Information provided in this book has been condensed. The publisher recommends that the product manufacturer be contacted, at the address or phone number provided, for more complete information on product safety, processing parameters, and applications not provided in this information sheet.

# Republic Plastics Company

## Contact Address:

Republic Plastics Company
2663 Eaton Rapids Road
Lansing, MI 48911

Phone: 1 (517) 887-0087

## RTA-3223

### Product Data:

Elastomeric Thermoplastic Alloy, RTA-3223 is designed for ease of molding with optimum part appearance and performance. Test results are from ⅛″ injection molded plaques.

### Typical Physical Properties:

**RTA-3223 Injection Molding Grade**

| Property | ASTM Test Methods | Units | RTA-3223 |
|---|---|---|---|
| Color | | | Black/Natural |
| Specific Gravity | D792 | | .99 ± .02 |
| Gardner Impact<br>73°F<br><br>−20°F | | <br>in-lb<br><br>in-lb | <br>280<br>(100% Pass)<br>240<br>(100% Pass) |
| Tensile, Ultimate<br>(Die C) | D638<br>2″/min. | <br>psi<br>MPa | <br>2300<br>16 |
| Flexural Modulus<br>73°F | D790 | <br>psi<br>MPa | <br>230,000<br>1587 |
| Elongation, Ultimate<br>(Die C) | D638<br>2″/min. | <br>% | <br>50 |
| Tear Strength<br>(Die C) | D624<br>2″/min. | <br>p/in<br>KN/m | <br>500<br>88 |
| Hardness | D2240 | Shore D | 69 ± 3.0 |

*(continued)*

NOTE—Information provided in this book has been condensed. The publisher recommends that the product manufacturer be contacted, at the address or phone number provided, for more complete information on product safety, processing parameters, and applications not provided in this information sheet.

**RTA-3223 Injection Molding Grade *(continued)***

| Property | ASTM Test Methods | Units | RTA-3223 |
|---|---|---|---|
| Melt Flow | D1238<br>Cond. L | g/10 min. | 5.0 ± 30% |
| Shrinkage (Tool)<br>(After bake at 250°F) | | mils/in.<br>mils/in. | 8.0 ± 1.0<br>11.0 ± 1.0 |
| Heat Deflection<br>Temperature @ 66 psi | D648 | °C | 110 |

# Republic Plastics Company

## Contact Address:

Republic Plastics Company
2663 Eaton Rapids Road
Lansing, MI 48911

Phone: 1 (517) 887-0087

## RTA-3263

### Product Data:

Elastomeric Thermoplastic Alloy, RTA-3263 is designed for ease of molding with optimum part appearance and performance. Test results are from ⅛″ injection molded plaques.

### Typical Physical Properties:

**RTA-3263 Injection Molding Grade**

| Property | ASTM Test Methods | Units | RTA-3263 |
|---|---|---|---|
| Color | | | Black/Natural |
| Specific Gravity | D792 | | 1.0 ± .02 |
| Gardner Impact<br>73°F<br><br>−20°F | | <br>in-lb<br><br>in-lb | <br>280<br>(100% Pass)<br>240<br>(100% Pass) |
| Tensile, Ultimate<br>(Die C) | D638<br>2″/min. | <br>psi<br>MPa | <br>2300<br>16 |
| Flexural Modulus<br>73°F | D790 | <br>psi<br>MPa | <br>250,000<br>1725 |
| Elongation, Ultimate<br>(Die C) | D638<br>2″/min. | <br>% | <br>50 |
| Tear Strength<br>(Die C) | D624<br>2″/min. | <br>p/in<br>KN/m | <br>550<br>96 |
| Hardness | D2240 | Shore D | 69 ± 3.0 |

*(continued)*

NOTE—Information provided in this book has been condensed. The publisher recommends that the product manufacturer be contacted, at the address or phone number provided, for more complete information on product safety, processing parameters, and applications not provided in this information sheet.

**RTA-3263 Injection Molding Grade *(continued)***

| Property | ASTM Test Methods | Units | RTA-3263 |
|---|---|---|---|
| Melt Flow | D1238 Cond. L | g/10 min. | 5.0 ± 30% |
| Shrinkage (Tool)<br>(After bake at 250°F) | | mils/in.<br>mils/in. | 8.0 ± 1.0<br>11.0 ± 1.0 |
| Heat Deflection Temperature @ 66 psi | D648 | °C | 110 |

# Republic Plastics Company

## Contact Address:

Republic Plastics Company
2663 Eaton Rapids Road
Lansing, MI 48911

Phone: 1 (517) 887-0087

## ETA-5012

### Product Data:

Elastomeric Thermoplastic Alloys (ETA) are Thermoplastic Elastomers designed for optimum strength down to −40°F. ETA extrusion grade polymers are specifically formulated for processing by profile extrusion, sheet extrusion and thermoforming. Test results are from ⅛″ injection molded plaques.

### Typical Physical Properties:

**ETA-5012 Blow Molding/Extrusion Grade**

| Property | ASTM Test Methods | Units | ETA-5012 |
|---|---|---|---|
| Color | | | Black |
| Specific Gravity (Max.) | D792 | | .92 |
| Brittle Point | D746 | °F | < −50 |
| Gardner Impact<br>73°F<br>−20°F | | <br>in-lb<br>in-lb | <br>>160<br>>160 |
| Flexural Modulus<br>73°F | D790 | <br>psi | <br>10,000 |
| Izod Impact Notched | D256 | ft lb/in | no break |
| Tensile Yield Strength<br>(Die C) | D638 | <br>psi | <br>1800 |
| Elongation Ultimate<br>(Die C) | D638 | <br>% | <br>650 |
| Tear Strength<br>(Die C) | D624 | <br>p/in | <br>400 |
| Hardness | D2240 | Shore D | 35 ± 3.0 |
| Melt Index (Max.) | D1238 | g/10 min. | 0.3 max. |

NOTE—Information provided in this book has been condensed. The publisher recommends that the product manufacturer be contacted, at the address or phone number provided, for more complete information on product safety, processing parameters, and applications not provided in this information sheet.

# Republic Plastics Company

## Contact Address:

Republic Plastics Company
2663 Eaton Rapids Road
Lansing, MI 48911

Phone: 1 (517) 887-0087

## ETA-5021

### Product Data:

Elastomeric Thermoplastic Alloys (ETA) are Thermoplastic Elastomers designed for optimum strength down to −40°F. ETA extrusion grade polymers are specifically formulated for processing by profile extrusion, sheet extrusion, thermoforming and blow molding. Test results are taken from ⅛″ injection molded plaques.

### Typical Physical Properties:

**ETA-5021 Blow Molding/Extrusion Grade**

| Property | ASTM Test Methods | Units | ETA-5021 |
|---|---|---|---|
| Color | | | Black |
| Specific Gravity | D792 | | .90 |
| Brittle Point | D746 | °F | < −40 |
| Gardner Impact<br>73°F<br>−20°F | | <br>in-lb<br>in-lb | <br>>160<br>>160 |
| Flexural Modulus<br>73°F<br>160°F<br>−40°F | D790<br>.05″/min. | <br>psi<br>psi<br>psi | <br>20,000<br>7,000<br>75,000 |
| Izod Impact Notched<br>73°F<br>−20°F | D256 | <br>ft lb/in<br>ft lb/in | <br>no break<br>no break |
| Tensile Yield Strength<br>(Die C) | D412<br>2″/min. | <br>psi | <br>2000 |
| Elongation Ultimate<br>(Die C) | D412<br>2″/min. | <br>% | <br>600 |

NOTE—Information provided in this book has been condensed. The publisher recommends that the product manufacturer be contacted, at the address or phone number provided, for more complete information on product safety, processing parameters, and applications not provided in this information sheet.

**ETA-5021 Blow Molding/Extrusion Grade *(continued)***

| Property | ASTM Test Methods | Units | ETA-5021 |
|---|---|---|---|
| Tear Strength (Die C) | D624 2″/min. | p/in | 450 |
| Hardness | D2240 | Shore D | 42 ± 3.0 |
| Melt Index | D1238 Cond. L | g/10 min. | .25 max. |

# Republic Plastics Company

## Contact Address:

Republic Plastics Company
2663 Eaton Rapids Road
Lansing, MI 48911

Phone: 1 (517) 887-0087

## ETA-5031

### Product Data:

Elastomeric Thermoplastic Alloys (ETA) are Thermoplastic Elastomers designed for optimum strength down to −40°F. ETA extrusion grade polymers are specifically formulated for processing by profile extrusion, sheet extrusion and thermoforming. Test results are from ⅛″ injection molded plaques.

### Typical Physical Properties:

**ETA-5031 Blow Molding/Extrusion Grade**

| Property | ASTM Test Methods | Units | ETA-5031 |
|---|---|---|---|
| Color | | | Black |
| Specific Gravity (Max.) | D792 | | .90 |
| Brittle Point | D746 | °F | −60 |
| Gardner Impact | | | |
| 73°F | | in-lb | 160 |
| −20°F | | in-lb | 160 |
| Flexural Modulus | D790 | | |
| 73°F | .05″/min. | psi | 25,000 |
| 160°F | | psi | 4,000 |
| −40°F | | psi | 40,000 |
| Izod Impact Notched | D256 | | |
| 73°F | | ft lb/in | no break |
| −20°F | | ft lb/in | no break |
| Tensile Yield Strength (Die C) | D412 2″/min. | psi | 2000 |
| Elongation Ultimate (Die C) | D412 2″/min. | % | 550 |
| Tear Strength (Die C) | D624 2″/min. | p/in | 500 |
| Hardness | D2240 | Shore D | 48 ± 3.0 |
| Melt Index | D1238 Cond. L | g/10 min. | .25 max. |

NOTE – Information provided in this book has been condensed. The publisher recommends that the product manufacturer be contacted, at the address or phone number provided, for more complete information on product safety, processing parameters, and applications not provided in this information sheet.

# Republic Plastics Company

## Contact Address:

Republic Plastics Company
2663 Eaton Rapids Road
Lansing, MI 48911

Phone: 1 (517) 887-0087

## ETA-5081

### Product Data:

Elastomeric Thermoplastic Alloys (ETA) are Thermoplastic Elastomers designed for optimum strength down to −40°F. ETA extrusion grade polymers are specifically formulated for processing by profile extrusion, sheet extrusion and thermoforming. Test results are from ⅛″ injection molded plaques.

### Typical Physical Properties:

**ETA-5081 Blow Molding/Extrusion Grade**

| Property | ASTM Test Methods | Units | ETA-5081 |
|---|---|---|---|
| Color | | | Black |
| Specific Gravity | D792 | | .90 |
| Brittle Point | D746 | °F | −40 |
| Gardner Impact<br>73°F<br>−20°F | | <br>in-lb<br>in-lb | <br>160<br>160 |
| Flexural Modulus<br>73°F<br>160°F<br>−40°F | D790<br>.05″/min. | <br>psi<br>psi<br>psi | <br>80,000<br>33,000<br>320,000 |
| Izod Impact Notched<br>73°F<br>−20°F | D256 | <br>ft lb/in<br>ft lb/in | <br>no break<br>no break |
| Tensile Yield Strength<br>(Die C) | D412<br>2″/min. | psi | 3300 |
| Elongation Ultimate<br>(Die C) | D412<br>2″/min. | % | 600 |
| Tear Strength<br>(Die C) | D624<br>2″/min. | p/in | 600 |

*(continued)*

NOTE – Information provided in this book has been condensed. The publisher recommends that the product manufacturer be contacted, at the address or phone number provided, for more complete information on product safety, processing parameters, and applications not provided in this information sheet.

ETA-5081 Blow Molding/Extrusion Grade *(continued)*

| Property | ASTM Test Methods | Units | ETA-5081 |
|---|---|---|---|
| Hardness | D2240 | Shore D | 60 |
| Melt Index | D1238 Cond. L | g/10 min. | .8 |
| Heat Sag | | inches | 0.25 |

# Republic Plastics Company

**Contact Address:**

Republic Plastics Company
2663 Eaton Rapids Road
Lansing, MI 48911

Phone: 1 (517) 887-0087

## T RUBBER 1011

**Product Data:**

T RUBBERS are Elastomeric Alloys designed to serve as an alternative to thermoset molded rubber. T RUBBERS offer rubber-like properties, yet are easily processed utilizing conventional thermoplastic equipment. T RUBBER 1011 is specifically formulated for injection molding.

**Typical Physical Properties:**

**T RUBBER 1011 Blow Molding/Extrusion Grade**

| Property | ASTM Test Methods | Units | T RUBBER 1011 |
|---|---|---|---|
| Color | | | Black |
| Specific Gravity | D792 | | .95 ± .01 |
| Brittle Point | D746 | °C | < −50 |
| Flexural Modulus | D790 | psi<br>MPa | 10,000<br>69 |
| Tensile, Ultimate (Die C) | D638<br>2″/min. | psi<br>MPa | 1000<br>6.9 |
| Elongation, Ultimate (Die C) | D638<br>2″/min. | % | 500 |
| Tear Strength (Die C) | D624<br>2″/min. | p/in<br>KN/m | 200<br>35 |
| Hardness | D2240 | Shore A<br>Shore D | 90 ± 3.0<br>35 ± 3.0 |
| Shrinkage (Tool) | | mils/in. | 12.5 ± 1.0 |

NOTE – Information provided in this book has been condensed. The publisher recommends that the product manufacturer be contacted, at the address or phone number provided, for more complete information on product safety, processing parameters, and applications not provided in this information sheet.

# Republic Plastics Company

Contact Address:

Republic Plastics Company
2663 Eaton Rapids Road
Lansing, MI 48911

Phone: 1 (517) 887-0087

## T RUBBER 1011-E

Product Data:

T RUBBERS are Elastomeric Alloys designed to serve as an alternative to thermoset molded rubber. T RUBBERS offer rubber-like properties, yet are easily processed utilizing conventional thermoplastic equipment. T RUBBER 1011E is specifically formulated for injection molding.

Typical Physical Properties:

**T RUBBER 1011-E Blow Molding/Extrusion Grade**

| Property | ASTM Test Methods | Units | T RUBBER 1011E |
|---|---|---|---|
| Color | | | Black |
| Specific Gravity | D792 | | .95 ± .01 |
| Brittle Point | D746 | °C | < −50 |
| Flexural Modulus | D790 | psi<br>MPa | 17,000<br>117.3 |
| Tensile, Ultimate (Die C) | D638<br>2″/min. | psi<br>MPa | 1500<br>10.4 |
| Elongation, Ultimate (Die C) | D638<br>2″/min. | % | 450 |
| Tear Strength (Die C) | D624<br>2″/min. | p/in<br>KN/m | 350<br>61.3 |
| Hardness | D2240 | Shore A<br>Shore D | 93 ± 3.0<br>41 ± 3.0 |
| Shrinkage (Tool) | | mils/in. | 13.0 ± 1.0 |

NOTE—Information provided in this book has been condensed. The publisher recommends that the product manufacturer be contacted, at the address or phone number provided, for more complete information on product safety, processing parameters, and applications not provided in this information sheet.

# Research Polymers International

## Contact Address:

Research Polymers International
1205 Avenue H East
Grand Prairie, TX 75050

Phone: 1 (214) 641-7152

## RPI 136EP

### Product Information:

RPI 136EP is a 30% glass fiber reinforced polyolefin copolymer alloy characterized by high heat distortion, good heat aging, low specific gravity, and uniform processability.

RPI 136EP applications include injection molded parts that require high heat distortion and dimensional stability coupled with cost performance value. Examples include small tool housings and automobile under the hood parts.

### Processing:

Target melt temperature range from 420–480°F (215–250°C), depending on part geometry, cycle times and machine size. Highly filled materials can attract moisture, especially in humid environments. For best results dry at 180°F (85°C) for two hours.

### Product Data:

**RPI 136EP**

| Property | Typical Value | Units | ASTM Method |
|---|---|---|---|
| **Physical** | | | |
| Specific Gravity | 1.14 | | D792 |
| Water Absorption (24 hr.) | 0.03 | % | D570 |
| Mold Shrinkage | 0.005 | in/in<br>mm/mm | D955 |
| Melt Flow Rate | — | g/10 min | D1238 |
| **Mechanical** | | | |
| Tensile Strength | 13,500<br>93 | psi<br>MPa | D638 |
| Elongation @ Break | 3 | % | D638 |
| Flexural Strength | 11,000<br>76 | psi<br>MPa | D790 |

*(continued)*

NOTE – Information provided in this book has been condensed. The publisher recommends that the product manufacturer be contacted, at the address or phone number provided, for more complete information on product safety, processing parameters, and applications not provided in this information sheet.

**RPI 136EP *(continued)***

| Property | Typical Value | Units | ASTM Method |
|---|---|---|---|
| **Mechanical** | | | |
| Flexural Modulus | 720,000<br>4,970 | psi<br>MPa | D790 |
| Izod Impact Strength<br>Notched 1/8″<br><br>Notched 1/4″ | <br>1.8<br>0.10<br>—<br>— | <br>ft lb/in<br>KJ/m<br>ft lb/in<br>KJ/m | D256 |
| Gardner Impact 1/8″<br>23°C<br><br>−30°C | <br>60<br>6.8<br>—<br>— | <br>in-lb<br>J<br>in-lb<br>J | |
| **Thermal** | | | |
| Heat Deflection Temp.<br>@ 66 psi<br><br>@ 264 psi | <br>340<br>171<br>330<br>166 | <br>°F<br>°C<br>°F<br>°C | D648 |
| CLTE ($10^{-5}$) | 2.0<br>3.6 | in/in/°F<br>m/m/°C | D696 |
| UL Flammability | — | | UL 94 |

# Research Polymers International

## Contact Address:

Research Polymers International
1205 Avenue H East
Grand Prairie, TX 75050

Phone: 1 (214) 641-7152

## RPI 503EP

### Product Information:

RPI 503EP is a polycarbonate alloy designed for high heat resistance, good impact strength, and chemical resistance. RPI 503EP is superior to polycarbonate type resins in resistance to cleaning compounds and solvents.

RPI 503EP is particularly suited for extrusion applications where autoclaving in the presence of cleaning compounds causes failure in polycarbonate. Extruded sheet can be used in vacuum forming operations.

### Processing:

Material should be dried prior to run for 5 to 6 hours at 220°F (100°C). Target melt temperature range from 460–510°F (240–265°C), depending on geometry and machine size. Material should be purged at end of run or for extended shutdown.

### Product Data:

**RPI 503EP**

| Property | Typical Value | Units | ASTM Method |
|---|---|---|---|
| **Physical** | | | |
| Specific Gravity | 1.18 | | D792 |
| Water Absorption (24 hr.) | 0.4 | % | D570 |
| Mold Shrinkage | 0.005–0.006 | in/in<br>mm/mm | D955 |
| Melt Flow Rate | — | g/10 min | D1238 |
| **Mechanical** | | | |
| Tensile Strength | 9,100<br>63 | psi<br>MPa | D638 |
| Elongation @ Break | 110 | % | D638 |
| Flexural Strength | 12,300<br>85 | psi<br>MPa | D790 |

*(continued)*

NOTE – Information provided in this book has been condensed. The publisher recommends that the product manufacturer be contacted, at the address or phone number provided, for more complete information on product safety, processing parameters, and applications not provided in this information sheet.

RPI 503EP *(continued)*

| Property | Typical Value | Units | ASTM Method |
|---|---|---|---|
| **Mechanical** | | | |
| Flexural Modulus | 360,000<br>2,480 | psi<br>MPa | D790 |
| Izod Impact Strength<br>Notched 1/8″<br><br>Notched 1/4″ | <br>19<br>1.0<br>15<br>0.8 | <br>ft lb/in<br>KJ/m<br>ft lb/in<br>KJ/m | D256 |
| Gardner Impact 1/8″<br>23°C<br><br>−30°C | <br>>320<br>36<br>>320<br>36 | <br>in-lb<br>J<br>in-lb<br>J | |
| **Thermal** | | | |
| Heat Deflection Temp.<br>@ 66 psi<br><br>@ 264 psi | <br>255<br>124<br>245<br>118 | <br>°F<br>°C<br>°F<br>°C | D648 |
| CLTE ($10^{-5}$) | 3.4<br>6.1 | in/in/°F<br>m/m/°C | D696 |
| UL Flammability | — | | UL 94 |

# Research Polymers International

## Contact Address:

Research Polymers International
1205 Avenue H East
Grand Prairie, TX 75050

Phone: 1 (214) 641-7152

## RPI 504EP

### Product Information:

RPI 504EP is a polycarbonate alloy designed for high heat resistance, good impact strength, and chemical resistance. RPI 504EP is superior to polycarbonate type resins in resistance to cleaning compounds and solvents.

RPI 504EP applications include injection molded parts, specifically where autoclaving in the presence of cleaning compounds causes failure in polycarbonate.

### Processing:

Material should be dried prior to molding for 5 to 6 hours at 220°F (100°C). Target melt temperature range from 460–510°F (240–265°C), depending on part geometry, cycle times and machine size. Material should be purged at end of run or for extended shutdown.

### Product Data:

**RPI 504EP**

| Property | Typical Value | Units | ASTM Method |
|---|---|---|---|
| **Physical** | | | |
| Specific Gravity | 1.18 | | D792 |
| Water Absorption (24 hr.) | 0.4 | % | D570 |
| Mold Shrinkage | 0.005–0.006 | in/in<br>mm/mm | D955 |
| Melt Flow Rate | — | g/10 min | D1238 |
| **Mechanical** | | | |
| Tensile Strength | 9,200<br>64 | psi<br>MPa | D638 |
| Elongation @ Break | 100 | % | D638 |
| Flexural Strength | 12,500<br>86 | psi<br>MPa | D790 |

*(continued)*

NOTE – Information provided in this book has been condensed. The publisher recommends that the product manufacturer be contacted, at the address or phone number provided, for more complete information on product safety, processing parameters, and applications not provided in this information sheet.

RPI 504EP *(continued)*

| Property | Typical Value | Units | ASTM Method |
|---|---|---|---|
| **Mechanical** | | | |
| Flexural Modulus | 360,000<br>2,480 | psi<br>MPa | D790 |
| Izod Impact Strength<br>Notched 1/8″<br><br>Notched 1/4″ | <br>18<br>1.0<br>15<br>0.8 | <br>ft lb/in<br>KJ/m<br>ft lb/in<br>KJ/m | D256 |
| Gardner Impact 1/8″<br>23°C<br><br>−30°C | <br>>320<br>36<br>>320<br>36 | <br>in-lb<br>J<br>in-lb<br>J | |
| **Thermal** | | | |
| Heat Deflection Temp.<br>@ 66 psi<br><br>@ 264 psi | <br>255<br>124<br>245<br>118 | <br>°F<br>°C<br>°F<br>°C | D648 |
| CLTE ($10^{-5}$) | 3.4<br>6.1 | in/in/°F<br>m/m/°C | D696 |
| UL Flammability | — | | UL 94 |

# Research Polymers International

## Contact Address:

Research Polymers International
1205 Avenue H East
Grand Prairie, TX 75050

Phone: 1 (214) 641-7152

## RPI 507EP

### Product Information:

RPI 507EP is a polycarbonate alloy designed for high heat resistance, excellent impact strength, and superior chemical resistance. RPI 507EP is unique relative to other polycarbonate alloys in resistance to attack by cleaning compounds and organic solvents.

RPI 507EP applications include injection molded parts where heat and chemical resistance are necessary, specifically appliance and automotive components.

### Processing:

Material should be dried prior to molding for 5 to 6 hours at 200°F (90°C). Target melt temperature range from 450–480°F (230–250°C), depending on part geometry, cycle times and machine size. Material should be purged at end of run or for extended shutdown.

### Product Data:

**RPI 507EP**

| Property | Typical Value | Units | ASTM Method |
|---|---|---|---|
| **Physical** | | | |
| Specific Gravity | 1.16 | | D792 |
| Water Absorption (24 hr.) | 0.5 | % | D570 |
| Mold Shrinkage | 0.007–0.008 | in/in<br>mm/mm | D955 |
| Melt Flow Rate | — | g/10 min | D1238 |
| **Mechanical** | | | |
| Tensile Strength | 8,600<br>60 | psi<br>MPa | D638 |
| Elongation @ Break | 160 | % | D638 |
| Flexural Strength | 12,000<br>83 | psi<br>MPa | D790 |

*(continued)*

NOTE – Information provided in this book has been condensed. The publisher recommends that the product manufacturer be contacted, at the address or phone number provided, for more complete information on product safety, processing parameters, and applications not provided in this information sheet.

RPI 507EP *(continued)*

| Property | Typical Value | Units | ASTM Method |
|---|---|---|---|
| **Mechanical** | | | |
| Flexural Modulus | 340,000<br>2,350 | psi<br>MPa | D790 |
| Izod Impact Strength<br>Notched 1/8″<br><br>Notched 1/4″ | <br>20<br>1.1<br>12<br>0.6 | <br>ft lb/in<br>KJ/m<br>ft lb/in<br>KJ/m | D256 |
| Gardner Impact 1/8″<br>23°C<br><br>−30°C | <br>>320<br>36<br>>320<br>36 | <br>in-lb<br>J<br>in-lb<br>J | |
| **Thermal** | | | |
| Heat Deflection Temp.<br>@ 66 psi<br><br>@ 264 psi | <br>220<br>104<br>210<br>99 | <br>°F<br>°C<br>°F<br>°C | D648 |
| CLTE ($10^{-5}$) | 4.0<br>7.2 | in/in/°F<br>m/m/°C | D696 |
| UL Flammability | — | | UL 94 |

# Research Polymers International

## Contact Address:

Research Polymers International
1205 Avenue H East
Grand Prairie, TX 75050

Phone: 1 (214) 641-7152

## RPI 602EP

### Product Information:

RPI 602EP is a high impact 6,6 nylon alloy designed to provide good tensile and stiffness properties while delivering the excellent chemical resistance associated with nylons.

RPI 602EP applications include injection molded parts where environmental conditions require excellent toughness and chemical resistance.

### Processing:

Material should be dried prior to molding for 3 to 4 hours at 170°F (75°C). Target melt temperature range from 530–560°F depending on part geometry, cycle times and machine size. Material should be purged at end of run or for extended shutdown.

### Product Data:

**RPI 602EP**

| Property | Typical Value | Units | ASTM Method |
|---|---|---|---|
| **Physical** | | | |
| Specific Gravity | 1.13 | | D792 |
| Water Absorption (24 hr.) | 1.1 | % | D570 |
| Mold Shrinkage | 0.015 | in/in<br>mm/mm | D955 |
| Melt Flow Rate | — | g/10 min | D1238 |
| **Mechanical** | | | |
| Tensile Strength | 7,800<br>54 | psi<br>MPa | D638 |
| Elongation @ Break | 110 | % | D638 |
| Flexural Strength | 12,000<br>83 | psi<br>MPa | D790 |
| Flexural Modulus | 300,000<br>2,070 | psi<br>MPa | D790 |

*(continued)*

NOTE – Information provided in this book has been condensed. The publisher recommends that the product manufacturer be contacted, at the address or phone number provided, for more complete information on product safety, processing parameters, and applications not provided in this information sheet.

**RPI 602EP *(continued)***

| Property | Typical Value | Units | ASTM Method |
|---|---|---|---|
| **Mechanical** | | | |
| Izod Impact Strength | | | D256 |
| Notched 1/8″ | 18 | ft lb/in | |
| | 0.96 | KJ/m | |
| Notched 1/4″ | — | ft lb/in | |
| | — | KJ/m | |
| Gardner Impact 1/8″ | | | |
| 23°C | >320 | in-lb | |
| | 36 | J | |
| −30°C | >320 | in-lb | |
| | 36 | J | |
| **Thermal** | | | |
| Heat Deflection Temp. | | | D648 |
| @ 66 psi | 421 | °F | |
| | 216 | °C | |
| @ 264 psi | 162 | °F | |
| | 72 | °C | |
| CLTE ($10^{-5}$) | — | in/in/°F | D696 |
| | — | m/m/°C | |
| UL Flammability | — | | UL 94 |

# Research Polymers International

Contact Address:

Research Polymers International
1205 Avenue H East
Grand Prairie, TX 75050

Phone: 1 (214) 641-7152

## RPI 609EP

Product Information:

RPI 609EP is a low modulus non-plasticized nylon alloy developed to balance heat and chemical resistance with processing ease.

RPI 609EP applications include extrusion operations and specifically tubing components where plasticized nylon II systems fail due to extraction at elevated temperatures.

Processing:

Material should be dried prior to run for 3 to 4 hours at 170°F (75°C). Target melt temperature range from 380–440°F (195–225°C), depending on geometry and machine size. Material should be purged at end of run or for extended shutdown.

Product Data:

**RPI 609EP**

| Property | Typical Value | Units | ASTM Method |
|---|---|---|---|
| **Physical** | | | |
| Specific Gravity | 1.00 | | D792 |
| Water Absorption (24 hr.) | 0.3 | % | D570 |
| Mold Shrinkage | 0.007 | in/in<br>mm/mm | D955 |
| Melt Flow Rate | — | g/10 min | D1238 |
| **Mechanical** | | | |
| Tensile Strength | 2,200<br>15 | psi<br>MPa | D638 |
| Elongation @ Break | 350 | % | D638 |
| Flexural Strength | —<br>— | psi<br>MPa | D790 |
| Flexural Modulus | 42,000<br>290 | psi<br>MPa | D790 |

*(continued)*

NOTE—Information provided in this book has been condensed. The publisher recommends that the product manufacturer be contacted, at the address or phone number provided, for more complete information on product safety, processing parameters, and applications not provided in this information sheet.

RPI 609EP *(continued)*

| Property | Typical Value | Units | ASTM Method |
|---|---|---|---|
| **Mechanical** | | | |
| Izod Impact Strength | | | D256 |
| Notched 1/8″ | NB | ft lb/in | |
| | NB | KJ/m | |
| Notched 1/4″ | NB | ft lb/in | |
| | NB | KJ/m | |
| Gardner Impact 1/8″ | | | |
| 23°C | 210 | in-lb | |
| | 24 | J | |
| −30°C | >320 | in-lb | |
| | 36 | J | |
| **Thermal** | | | |
| Heat Deflection Temp. | | | D648 |
| @ 66 psi | — | °F | |
| | — | °C | |
| @ 264 psi | — | °F | |
| | — | °C | |
| CLTE ($10^{-5}$) | — | in/in/°F | D696 |
| | — | m/m/°C | |
| UL Flammability | — | | UL 94 |

# Research Polymers International

## Contact Address:

Research Polymers International
1205 Avenue H East
Grand Prairie, TX 75050

Phone: 1 (214) 641-7152

## RPI 632EP

### Product Information:

RPI 632EP is a glass reinforced 6,6 nylon designed to reduce inherent moisture absorption. The reduction in moisture absorption imparts dimensional stability over a wide range of environmental conditions.

RPI 632EP applications include injection molded parts where environmental conditions require stringent dimensional tolerance.

### Processing:

Material should be dried prior to molding for 3 to 4 hours at 200°F (95°C). Target melt temperature range from 530–580°F (275–300°C), depending on part geometry, cycle times and machine size. Material should be purged at end of run or for extended shutdown.

### Product Data:

**RPI 632EP**

| Property | Typical Value | Units | ASTM Method |
|---|---|---|---|
| **Physical** | | | |
| Specific Gravity | 1.35 | | D792 |
| Water Absorption (24 hr.) | 0.7 | % | D570 |
| Mold Shrinkage | 0.002 | in/in<br>mm/mm | D955 |
| Melt Flow Rate | — | g/10 min | D1238 |
| **Mechanical** | | | |
| Tensile Strength | 24,000<br>166 | psi<br>MPa | D638 |
| Elongation @ Break | 3 | % | D638 |
| Flexural Strength | 38,000<br>262 | psi<br>MPa | D790 |
| Flexural Modulus | 1,000,000<br>6,900 | psi<br>MPa | D790 |

*(continued)*

NOTE – Information provided in this book has been condensed. The publisher recommends that the product manufacturer be contacted, at the address or phone number provided, for more complete information on product safety, processing parameters, and applications not provided in this information sheet.

**RPI 632EP *(continued)***

| Property | Typical Value | Units | ASTM Method |
|---|---|---|---|
| **Mechanical** | | | |
| Izod Impact Strength | | | D256 |
| Notched 1/8″ | 1.9 | ft lb/in | |
| | 0.10 | KJ/m | |
| Notched 1/4″ | — | ft lb/in | |
| | — | KJ/m | |
| Gardner Impact 1/8″ | | | |
| 23°C | 100 | in-lb | |
| | 11.3 | J | |
| −30°C | — | in-lb | |
| | — | J | |
| **Thermal** | | | |
| Heat Deflection Temp. | | | D648 |
| @ 66 psi | 490 | °F | |
| | 254 | °C | |
| @ 264 psi | 480 | °F | |
| | 249 | °C | |
| CLTE ($10^{-5}$) | 1.4 | in/in/°F | D696 |
| | 2.5 | m/m/°C | |
| UL Flammability | — | | UL 94 |

# Shell Chemical Company

Contact Address:

Shell Chemical Company
1415 West 22nd Street
Oak Brook, IL 60522-9088

Phone: 1 (800) 323-3405

## KRATON RUBBER

Product Information:

Kraton rubber is a unique high performance thermoplastic elastomer engineered for a wide variety of end uses. There are over 65 grades of Kraton polymers and compounds having performance characteristics developed for specific end uses.

Having an unusual combination of high strength and low viscosity, Kraton combines the characteristics of vulcanized rubber in a product that is easy to process. Krayton does not require vulcanization.

Kraton elastomers are available in a Shore A hardness range from 28 to 95, and a tensile strength range from 150 to 5000 psi. It remains flexible at temperatures as low as −80°C. Many Kraton products meet FDA standards for contact with regulated products. Many are listed to meet UL specifications, and find end uses as electrical insulation and molded appliance components. Normally clear and colorless, some products can be molded which retain these characteristics. Kraton can be extended and/or reinforced using fillers, and can be colored with excellent control, due to the clarity of the base polymer.

Many grades, not listed in the molding products data pages, are used in compounding other resin systems to improve physical properties and facilitate processing. As an additive, it can improve the toughness of plastics. It improves the shrink characteristics of polyester thermosets. At higher concentrations, it improves flexibility, softness, elasticity, and surface characteristics of certain plastics. It has been used to formulate pressure sensitive adhesives, sealants, asphalts, oil gels, and even motor oil.

It can be processed using a wide variety of thermoplastic processing methods—including injection molding, blow molding, compression molding, extrusion, hot melt crating, and even solution coating processes.

Directly molded or extruded Kraton have the following end uses:

| Medical/Food (Series numbers noted) | Automotive | Sporting Goods | Molded Items |
|---|---|---|---|
| D-2100 | | D-2100 | D-2100 |
| | | D-3000 | D-3000 |
| | | D-5000 | D-5000 |
| G-2700 | G-2700 | G-2700 | G-2700 |
| | G-7000 | G-7000 | G-7000 |
| | G-8000 | G-8000 | G-8000 |

The D-5000 series also is used to mold products for the shoe industry. The G-8000 series finds extensive use as insulation for wire and cable.

*(continued)*

NOTE—Information provided in this book has been condensed. The publisher recommends that the product manufacturer be contacted, at the address or phone number provided, for more complete information on product safety, processing parameters, and applications not provided in this information sheet.

## Product Data:

### Kraton D-2103

| Property | |
|---|---|
| Hardness<br>Shore A (D-2240)<br>Shore D | <br>70<br>— |
| Tensile Properties (D-412)<br>Tensile Strength, psi | <br>4300 |
| 300% Modulus, psi | 400 |
| Elongation, % | 880 |
| Tear Resistance<br>Die C<br>pli (D-624) | <br><br>— |
| Yerzley Resilience<br>% (D-945) | <br>— |

| Property | |
|---|---|
| Specific Gravity | 0.94 |
| Melt Index<br>Condition E, gms/10 min<br>Condition G, gms/10 min | <br>—<br>15 |
| Physical Form | Pellet |
| Standard Color | Natural (Clear) |
| Ozone Resistance<br>—Stressed<br>—Unstressed | <br>Poor<br>Poor |
| Comments | FDA |

## Product Data:

### Kraton D-2104

| Property | |
|---|---|
| Hardness<br>Shore A (D-2240)<br>Shore D | <br>43<br>— |
| Tensile Properties (D-412)<br>Tensile Strength, psi | <br>1700 |
| 300% Modulus, psi | 200 |
| Elongation, % | 1350 |
| Tear Resistance<br>Die C<br>pli (D-624) | <br><br>180 |
| Yerzley Resilience<br>% (D-945) | <br>80 |

| Property | |
|---|---|
| Specific Gravity | 0.93 |
| Melt Index<br>Condition E, gms/10 min<br>Condition G, gms/10 min | <br>—<br>22 |
| Physical Form | Pellet |
| Standard Color | Natural (Clear) |
| Ozone Resistance<br>—Stressed<br>—Unstressed | <br>Poor<br>Poor |
| Comments | FDA |

*(continued)*

Product Data:

**Kraton D-2109**

| **Property** | |
|---|---|
| Hardness<br>Shore A (D-2240)<br>Shore D | <br>44<br>— |
| Tensile Properties (D-412)<br>Tensile Strength, psi | <br>950 |
| 300% Modulus, psi | 300 |
| Elongation, % | 800 |
| Tear Resistance<br>Die C<br>pli (D-624) | <br><br>160 |
| Yerzley Resilience<br>% (D-945) | <br>80 |

| **Property** | |
|---|---|
| Specific Gravity | 0.94 |
| Melt Index<br>Condition E, gms/10 min<br>Condition G, gms/10 min | <br>—<br>15 |
| Physical Form | Pellet |
| Standard Color | White |
| Ozone Resistance<br>—Stressed<br>—Unstressed | <br>Poor<br>Fair |
| Comments | FDA |

Product Data:

**Kraton D-3202**

| **Property** | |
|---|---|
| Hardness<br>Shore A (D-2240)<br>Shore D | <br>47<br>— |
| Tensile Properties (D-412)<br>Tensile Strength, psi | <br>850 |
| 300% Modulus, psi | 600 |
| Elongation, % | 500 |
| Tear Resistance<br>Die C<br>pli (D-624) | <br><br>320 |
| Yerzley Resilience<br>% (D-945) | <br>72 |

| **Property** | |
|---|---|
| Specific Gravity | 1.00 |
| Melt Index<br>Condition E, gms/10 min<br>Condition G, gms/10 min | <br>14<br>— |
| Physical Form | Pellet |
| Standard Color | Natural |
| Ozone Resistance<br>—Stressed<br>—Unstressed | <br>Poor<br>Fair |
| Comments | — |

*(continued)*

## Product Data:

### Kraton D-3204

| Property | |
|---|---|
| Hardness<br>Shore A (D-2240)<br>Shore D | <br>75<br>— |
| Tensile Properties (D-412)<br>Tensile Strength, psi | <br>1350 |
| 300% Modulus, psi | 970 |
| Elongation, % | 550 |
| Tear Resistance<br>Die C<br>pli (D-624) | <br><br>270 |
| Yerzley Resilience<br>% (D-945) | <br>67 |

| Property | |
|---|---|
| Specific Gravity | 1.01 |
| Melt Index<br>Condition E, gms/10 min<br>Condition G, gms/10 min | <br>—<br>35 |
| Physical Form | Pellet |
| Standard Color | Natural |
| Ozone Resistance<br>—Stressed<br>—Unstressed | <br>Poor<br>Fair |
| Comments | — |

## Product Data:

### Kraton D-3226

| Property | |
|---|---|
| Hardness<br>Shore A (D-2240)<br>Shore D | <br>34<br>— |
| Tensile Properties (D-412)<br>Tensile Strength, psi | <br>625 |
| 300% Modulus, psi | 275 |
| Elongation, % | 640 |
| Tear Resistance<br>Die C<br>pli (D-624) | <br><br>115 |
| Yerzley Resilience<br>% (D-945) | <br>84 |

| Property | |
|---|---|
| Specific Gravity | 1.00 |
| Melt Index<br>Condition E, gms/10 min<br>Condition G, gms/10 min | <br>20<br>— |
| Physical Form | Pellet |
| Standard Color | Natural |
| Ozone Resistance<br>—Stressed<br>—Unstressed | <br>Poor<br>Fair |
| Comments | — |

*(continued)*

## Product Data:

### Kraton D-5119

| Property | |
|---|---|
| Hardness<br>Shore A (D-2240)<br>Shore D | <br>47<br>— |
| Tensile Properties (D-412)<br>Tensile Strength, psi | <br>560 |
| 300% Modulus, psi | 565 |
| Elongation, % | 565 |
| Tear Resistance<br>Die C<br>pli (D-624) | <br><br>105 |
| Yerzley Resilience<br>% (D-945) | <br>— |

| Property | |
|---|---|
| Specific Gravity | 1.09 |
| Melt Index<br>Condition E, gms/10 min<br>Condition G, gms/10 min | <br>27<br>— |
| Physical Form | Pellet |
| Standard Color | Natural<br>White or Tan |
| Ozone Resistance<br>—Stressed<br>—Unstressed | <br>Poor<br>Good |
| Comments | — |

## Product Data:

### Kraton D-5122

| Property | |
|---|---|
| Hardness<br>Shore A (D-2240)<br>Shore D | <br>60<br>— |
| Tensile Properties (D-412)<br>Tensile Strength, psi | <br>895 |
| 300% Modulus, psi | 880 |
| Elongation, % | 325 |
| Tear Resistance<br>Die C<br>pli (D-624) | <br><br>220 |
| Yerzley Resilience<br>% (D-945) | <br>— |

| Property | |
|---|---|
| Specific Gravity | 1.08 |
| Melt Index<br>Condition E, gms/10 min<br>Condition G, gms/10 min | <br>25<br>— |
| Physical Form | Pellet |
| Standard Color | Natural<br>or White |
| Ozone Resistance<br>—Stressed<br>—Unstressed | <br>Poor<br>Good |
| Comments | — |

*(continued)*

## Product Data:

**Kraton D-5152**

| Property | |
|---|---|
| Hardness<br>Shore A (D-2240)<br>Shore D | <br>46<br>— |
| Tensile Properties (D-412)<br>Tensile Strength, psi | <br>525 |
| 300% Modulus, psi | 325 |
| Elongation, % | 450 |
| Tear Resistance<br>Die C<br>pli (D-624) | <br><br>80 |
| Yerzley Resilience<br>% (D-945) | <br>— |

| Property | |
|---|---|
| Specific Gravity | 1.00 |
| Melt Index<br>Condition E, gms/10 min<br>Condition G, gms/10 min | <br>26<br>— |
| Physical Form | Pellet |
| Standard Color | Natural or<br>Dark Crepe |
| Ozone Resistance<br>—Stressed<br>—Unstressed | <br>Poor<br>Good |
| Comments | — |

## Product Data:

**Kraton D-5239**

| Property | |
|---|---|
| Hardness<br>Shore A (D-2240)<br>Shore D | <br>36<br>— |
| Tensile Properties (D-412)<br>Tensile Strength, psi | <br>545 |
| 300% Modulus, psi | 190 |
| Elongation, % | 785 |
| Tear Resistance<br>Die C<br>pli (D-624) | <br><br>85 |
| Yerzley Resilience<br>% (D-945) | <br>— |

| Property | |
|---|---|
| Specific Gravity | 0.98 |
| Melt Index<br>Condition E, gms/10 min<br>Condition G, gms/10 min | <br>21<br>— |
| Physical Form | Pellet |
| Standard Color | Natural |
| Ozone Resistance<br>—Stressed<br>—Unstressed | <br>Poor<br>Good |
| Comments | — |

*(continued)*

## Product Data:

### Kraton D-5250

| Property | |
|---|---|
| Hardness<br>Shore A (D-2240)<br>Shore D | <br>80<br>— |
| Tensile Properties (D-412)<br>Tensile Strength, psi | <br>685 |
| 300% Modulus, psi | 550 |
| Elongation, % | 465 |
| Tear Resistance<br>Die C<br>pli (D-624) | <br><br>175 |
| Yerzley Resilience<br>% (D-945) | <br>— |

| Property | |
|---|---|
| Specific Gravity | 1.10 |
| Melt Index<br>Condition E, gms/10 min<br>Condition G, gms/10 min | <br>23<br>— |
| Physical Form | Pellet |
| Standard Color | Natural |
| Ozone Resistance<br>—Stressed<br>—Unstressed | <br>Poor<br>Good |
| Comments | — |

## Product Data:

### Kraton D-5310

| Property | |
|---|---|
| Hardness<br>Shore A (D-2240)<br>Shore D | <br>50<br>— |
| Tensile Properties (D-412)<br>Tensile Strength, psi | <br>980 |
| 300% Modulus, psi | 220 |
| Elongation, % | 960 |
| Tear Resistance<br>Die C<br>pli (D-624) | <br><br>140 |
| Yerzley Resilience<br>% (D-945) | <br>— |

| Property | |
|---|---|
| Specific Gravity | 0.94 |
| Melt Index<br>Condition E, gms/10 min<br>Condition G, gms/10 min | <br>23<br>— |
| Physical Form | Pellet |
| Standard Color | Natural |
| Ozone Resistance<br>—Stressed<br>—Unstressed | <br>Poor<br>Good |
| Comments | — |

*(continued)*

## Product Data:

### Kraton D-7340

| Property | |
|---|---|
| Hardness<br>Shore A (D-2240)<br>Shore D | <br>—<br>40 |
| Tensile Properties (D-412)<br>Tensile Strength, psi | <br>1300 |
| 300% Modulus, psi | — |
| Elongation, % | 350 |
| Tear Resistance<br>Die C<br>pli (D-624) | <br><br>400 |
| Yerzley Resilience<br>% (D-945) | <br>— |

| Property | |
|---|---|
| Specific Gravity | 1.05 |
| Melt Index<br>Condition E, gms/10 min<br>Condition G, gms/10 min | <br>—<br>— |
| Physical Form | Pellet |
| Standard Color | Black |
| Ozone Resistance<br>—Stressed<br>—Unstressed | <br>Good<br>Good |
| Comments | — |

## Product Data:

### Kraton G-2701

| Property | |
|---|---|
| Hardness<br>Shore A (D-2240)<br>Shore D | <br>67<br>— |
| Tensile Properties (D-412)<br>Tensile Strength, psi | <br>1600 |
| 300% Modulus, psi | 480 |
| Elongation, % | 800 |
| Tear Resistance<br>Die C<br>pli (D-624) | <br><br>260 |
| Yerzley Resilience<br>% (D-945) | <br>— |

| Property | |
|---|---|
| Specific Gravity | 0.90 |
| Melt Index<br>Condition E, gms/10 min<br>Condition G, gms/10 min | <br>—<br>— |
| Physical Form | Pellet |
| Standard Color | Natural<br>(Clear) |
| Ozone Resistance<br>—Stressed<br>—Unstressed | <br>Excellent<br>Excellent |
| Comments | FDA |

*(continued)*

## Product Data:

### Kraton G-2703

| Property | |
|---|---|
| Hardness<br>Shore A (D-2240)<br>Shore D | <br>63<br>— |
| Tensile Properties (D-412)<br>Tensile Strength, psi | <br>1200 |
| 300% Modulus, psi | 470 |
| Elongation, % | 670 |
| Tear Resistance<br>Die C<br>pli (D-624) | <br><br>— |
| Yerzley Resilience<br>% (D-945) | <br>— |

| Property | |
|---|---|
| Specific Gravity | 0.90 |
| Melt Index<br>Condition E, gms/10 min<br>Condition G, gms/10 min | <br>—<br>— |
| Physical Form | Pellet |
| Standard Color | Natural<br>(Clear) |
| Ozone Resistance<br>—Stressed<br>—Unstressed | <br>Excellent<br>Excellent |
| Comments | FDA |

## Product Data:

### Kraton G-2705

| Property | |
|---|---|
| Hardness<br>Shore A (D-2240)<br>Shore D | <br>55<br>— |
| Tensile Properties (D-412)<br>Tensile Strength, psi | <br>1200 |
| 300% Modulus, psi | 350 |
| Elongation, % | 700 |
| Tear Resistance<br>Die C<br>pli (D-624) | <br><br>120 |
| Yerzley Resilience<br>% (D-945) | <br>75 |

| Property | |
|---|---|
| Specific Gravity | 0.90 |
| Melt Index<br>Condition E, gms/10 min<br>Condition G, gms/10 min | <br>—<br>— |
| Physical Form | Pellet |
| Standard Color | Natural<br>(Clear) |
| Ozone Resistance<br>—Stressed<br>—Unstressed | <br>Excellent<br>Excellent |
| Comments | FDA |

*(continued)*

## Product Data:

### Kraton G-2706

| Property | |
|---|---|
| Hardness<br>Shore A (D-2240)<br>Shore D | <br>28<br>— |
| Tensile Properties (D-412)<br>Tensile Strength, psi | <br>850 |
| 300% Modulus, psi | 130 |
| Elongation, % | 950 |
| Tear Resistance<br>Die C<br>pli (D-624) | <br><br>— |
| Yerzley Resilience<br>% (D-945) | <br>— |

| Property | |
|---|---|
| Specific Gravity | 0.90 |
| Melt Index<br>Condition E, gms/10 min<br>Condition G, gms/10 min | <br>—<br>— |
| Physical Form | Pellet |
| Standard Color | Natural<br>(Clear) |
| Ozone Resistance<br>—Stressed<br>—Unstressed | <br>Excellent<br>Excellent |
| Comments | FDA |

## Product Data:

### Kraton G-7150

| Property | |
|---|---|
| Hardness<br>Shore A (D-2240)<br>Shore D | <br>90<br>35 |
| Tensile Properties (D-412)<br>Tensile Strength, psi | <br>700 |
| 300% Modulus, psi | — |
| Elongation, % | 250 |
| Tear Resistance<br>Die C<br>pli (D-624) | <br><br>200 |
| Yerzley Resilience<br>% (D-945) | <br>— |

| Property | |
|---|---|
| Specific Gravity | 1.95 |
| Melt Index<br>Condition E, gms/10 min<br>Condition G, gms/10 min | <br>—<br>— |
| Physical Form | Pellet |
| Standard Color | Black |
| Ozone Resistance<br>—Stressed<br>—Unstressed | <br>Excellent<br>Excellent |
| Comments | — |

*(continued)*

## Product Data:

### Kraton G-7155

| Property | |
|---|---|
| Hardness<br>Shore A (D-2240)<br>Shore D | <br>—<br>45 |
| Tensile Properties (D-412)<br>Tensile Strength, psi | <br>900 |
| 300% Modulus, psi | — |
| Elongation, % | 250 |
| Tear Resistance<br>Die C<br>pli (D-624) | <br><br>225 |
| Yerzley Resilience<br>% (D-945) | <br>— |

| Property | |
|---|---|
| Specific Gravity | 1.85 |
| Melt Index<br>Condition E, gms/10 min<br>Condition G, gms/10 min | <br>—<br>— |
| Physical Form | Pellet |
| Standard Color | Black |
| Ozone Resistance<br>—Stressed<br>—Unstressed | <br>Excellent<br>Excellent |
| Comments | — |

## Product Data:

### Kraton G-7430

| Property | |
|---|---|
| Hardness<br>Shore A (D-2240)<br>Shore D | <br>65<br>— |
| Tensile Properties (D-412)<br>Tensile Strength, psi | <br>1400 |
| 300% Modulus, psi | — |
| Elongation, % | 550 |
| Tear Resistance<br>Die C<br>pli (D-624) | <br><br>170 |
| Yerzley Resilience<br>% (D-945) | <br>— |

| Property | |
|---|---|
| Specific Gravity | 0.93 |
| Melt Index<br>Condition E, gms/10 min<br>Condition G, gms/10 min | <br>—<br>— |
| Physical Form | Pellet |
| Standard Color | Black |
| Ozone Resistance<br>—Stressed<br>—Unstressed | <br>Excellent<br>Excellent |
| Comments | — |

*(continued)*

## Product Data:

### Kraton G-7440

| Property | |
|---|---|
| Hardness<br>Shore A (D-2240)<br>Shore D | <br>70<br>— |
| Tensile Properties (D-412)<br>Tensile Strength, psi | <br>700 |
| 300% Modulus, psi | — |
| Elongation, % | 400 |
| Tear Resistance<br>Die C<br>pli (D-624) | <br><br>200 |
| Yerzley Resilience<br>% (D-945) | <br>— |

| Property | |
|---|---|
| Specific Gravity | 1.18 |
| Melt Index<br>Condition E, gms/10 min<br>Condition G, gms/10 min | <br>—<br>— |
| Physical Form | Pellet |
| Standard Color | Natural or Black |
| Ozone Resistance<br>—Stressed<br>—Unstressed | <br>Excellent<br>Excellent |
| Comments | Fire Retardant LOI 25 |

## Product Data:

### Kraton G-7450

| Property | |
|---|---|
| Hardness<br>Shore A (D-2240)<br>Shore D | <br>70<br>— |
| Tensile Properties (D-412)<br>Tensile Strength, psi | <br>1200 |
| 300% Modulus, psi | — |
| Elongation, % | 400 |
| Tear Resistance<br>Die C<br>pli (D-624) | <br><br>200 |
| Yerzley Resilience<br>% (D-945) | <br>— |

| Property | |
|---|---|
| Specific Gravity | 1.00 |
| Melt Index<br>Condition E, gms/10 min<br>Condition G, gms/10 min | <br>—<br>— |
| Physical Form | Pellet |
| Standard Color | Black |
| Ozone Resistance<br>—Stressed<br>—Unstressed | <br>Excellent<br>Excellent |
| Comments | — |

*(continued)*

## Product Data:

**Kraton G-7455**

| Property | |
|---|---|
| Hardness<br>Shore A (D-2240)<br>Shore D | <br>85<br>— |
| Tensile Properties (D-412)<br>Tensile Strength, psi | <br>1300 |
| 300% Modulus, psi | — |
| Elongation, % | 400 |
| Tear Resistance<br>Die C<br>pli (D-624) | <br><br>280 |
| Yerzley Resilience<br>% (D-945) | <br>— |

| Property | |
|---|---|
| Specific Gravity | 1.08 |
| Melt Index<br>Condition E, gms/10 min<br>Condition G, gms/10 min | <br>—<br>— |
| Physical Form | Pellet |
| Standard Color | Natural or Black |
| Ozone Resistance<br>—Stressed<br>—Unstressed | <br>Excellent<br>Excellent |
| Comments | Fire Retardant LOI 25 |

## Product Data:

**Kraton G-7680**

| Property | |
|---|---|
| Hardness<br>Shore A (D-2240)<br>Shore D | <br>—<br>55 |
| Tensile Properties (D-412)<br>Tensile Strength, psi | <br>3000 |
| 300% Modulus, psi | — |
| Elongation, % | 400 |
| Tear Resistance<br>Die C<br>pli (D-624) | <br><br>700 |
| Yerzley Resilience<br>% (D-945) | <br>— |

| Property | |
|---|---|
| Specific Gravity | 1.03 |
| Melt Index<br>Condition E, gms/10 min<br>Condition G, gms/10 min | <br>—<br>— |
| Physical Form | Pellet |
| Standard Color | Black |
| Ozone Resistance<br>—Stressed<br>—Unstressed | <br>Excellent<br>Excellent |
| Comments | — |

*(continued)*

Product Data:

**Kraton G-7705**

| Property | |
|---|---|
| Hardness<br>Shore A (D-2240)<br>Shore D | <br>45<br>— |
| Tensile Properties (D-412)<br>Tensile Strength, psi | <br>500 |
| 300% Modulus, psi | 300 |
| Elongation, % | 600 |
| Tear Resistance<br>Die C<br>pli (D-624) | <br><br>85 |
| Yerzley Resilience<br>% (D-945) | <br>— |

| Property | |
|---|---|
| Specific Gravity | 1.19 |
| Melt Index<br>Condition E, gms/10 min<br>Condition G, gms/10 min | <br>—<br>— |
| Physical Form | Pellet |
| Standard Color | Natural<br>or Black |
| Ozone Resistance<br>—Stressed<br>—Unstressed | <br>Excellent<br>Excellent |
| Comments | — |

Product Data:

**Kraton G-7715**

| Property | |
|---|---|
| Hardness<br>Shore A (D-2240)<br>Shore D | <br>55<br>— |
| Tensile Properties (D-412)<br>Tensile Strength, psi | <br>1200 |
| 300% Modulus, psi | 300 |
| Elongation, % | 500 |
| Tear Resistance<br>Die C<br>pli (D-624) | <br><br>120 |
| Yerzley Resilience<br>% (D-945) | <br>— |

| Property | |
|---|---|
| Specific Gravity | 0.91 |
| Melt Index<br>Condition E, gms/10 min<br>Condition G, gms/10 min | <br>—<br>— |
| Physical Form | Pellet |
| Standard Color | Black |
| Ozone Resistance<br>—Stressed<br>—Unstressed | <br>Excellent<br>Excellent |
| Comments | — |

*(continued)*

## Product Data:

### Kraton G-7720

| Property | |
|---|---|
| Hardness<br>Shore A (D-2240)<br>Shore D | <br>60<br>— |
| Tensile Properties (D-412)<br>Tensile Strength, psi | <br>900 |
| 300% Modulus, psi | 350 |
| Elongation, % | 600 |
| Tear Resistance<br>Die C<br>pli (D-624) | <br><br>140 |
| Yerzley Resilience<br>% (D-945) | <br>— |

| Property | |
|---|---|
| Specific Gravity | 1.20 |
| Melt Index<br>Condition E, gms/10 min<br>Condition G, gms/10 min | <br>—<br>— |
| Physical Form | Pellet |
| Standard Color | Natural<br>or Black |
| Ozone Resistance<br>—Stressed<br>—Unstressed | <br>Excellent<br>Excellent |
| Comments | — |

## Product Data:

### Kraton G-7820

| Property | |
|---|---|
| Hardness<br>Shore A (D-2240)<br>Shore D | <br>90<br>35 |
| Tensile Properties (D-412)<br>Tensile Strength, psi | <br>1800 |
| 300% Modulus, psi | 1000 |
| Elongation, % | 500 |
| Tear Resistance<br>Die C<br>pli (D-624) | <br><br>340 |
| Yerzley Resilience<br>% (D-945) | <br>— |

| Property | |
|---|---|
| Specific Gravity | 1.14 |
| Melt Index<br>Condition E, gms/10 min<br>Condition G, gms/10 min | <br>—<br>— |
| Physical Form | Pellet |
| Standard Color | Natural<br>or Black |
| Ozone Resistance<br>—Stressed<br>—Unstressed | <br>Excellent<br>Excellent |
| Comments | — |

*(continued)*

## Product Data:

### Kraton G-7821X

| Property | |
|---|---|
| Hardness<br>Shore A (D-2240)<br>Shore D | <br>90<br>— |
| Tensile Properties (D-412)<br>Tensile Strength, psi | <br>1400 |
| 300% Modulus, psi | 1000 |
| Elongation, % | 400 |
| Tear Resistance<br>Die C<br>pli (D-624) | <br><br>— |
| Yerzley Resilience<br>% (D-945) | <br>— |

| Property | |
|---|---|
| Specific Gravity | 1.20 |
| Melt Index<br>Condition E, gms/10 min<br>Condition G, gms/10 min | <br>—<br>— |
| Physical Form | Pellet |
| Standard Color | Black |
| Ozone Resistance<br>—Stressed<br>—Unstressed | <br>Excellent<br>Excellent |
| Comments | High Service Temperature |

## Product Data:

### Kraton G-7827

| Property | |
|---|---|
| Hardness<br>Shore A (D-2240)<br>Shore D | <br>—<br>40 |
| Tensile Properties (D-412)<br>Tensile Strength, psi | <br>2400 |
| 300% Modulus, psi | 1400 |
| Elongation, % | 500 |
| Tear Resistance<br>Die C<br>pli (D-624) | <br><br>540 |
| Yerzley Resilience<br>% (D-945) | <br>— |

| Property | |
|---|---|
| Specific Gravity | 1.14 |
| Melt Index<br>Condition E, gms/10 min<br>Condition G, gms/10 min | <br>—<br>— |
| Physical Form | Pellet |
| Standard Color | Grey |
| Ozone Resistance<br>—Stressed<br>—Unstressed | <br>Excellent<br>Excellent |
| Comments | — |

*(continued)*

## Product Data:

### Kraton G-7890X

| Property | |
|---|---|
| Hardness<br>Shore A (D-2240)<br>Shore D | <br>—<br>55 |
| Tensile Properties (D-412)<br>Tensile Strength, psi | <br>2600 |
| 300% Modulus, psi | — |
| Elongation, % | 450 |
| Tear Resistance<br>Die C<br>pli (D-624) | <br><br>750 |
| Yerzley Resilience<br>% (D-945) | <br>— |

| Property | |
|---|---|
| Specific Gravity | 0.90 |
| Melt Index<br>Condition E, gms/10 min<br>Condition G, gms/10 min | <br>—<br>— |
| Physical Form | Pellet |
| Standard Color | Black |
| Ozone Resistance<br>—Stressed<br>—Unstressed | <br>Excellent<br>Excellent |
| Comments | — |

## Product Data:

### Kraton G-7940

| Property | |
|---|---|
| Hardness<br>Shore A (D-2240)<br>Shore D | <br>—<br>50 |
| Tensile Properties (D-412)<br>Tensile Strength, psi | <br>1400 |
| 300% Modulus, psi | — |
| Elongation, % | 450 |
| Tear Resistance<br>Die C<br>pli (D-624) | <br><br>420 |
| Yerzley Resilience<br>% (D-945) | <br>— |

| Property | |
|---|---|
| Specific Gravity | 0.96 |
| Melt Index<br>Condition E, gms/10 min<br>Condition G, gms/10 min | <br>—<br>— |
| Physical Form | Pellet |
| Standard Color | Black |
| Ozone Resistance<br>—Stressed<br>—Unstressed | <br>Excellent<br>Excellent |
| Comments | — |

*(continued)*

## Product Data:

### Shell Elexar 8230

| Property | |
|---|---|
| Hardness<br>Shore A (D-2240)<br>Shore D | <br>70<br>— |
| Tensile Properties (D-412)<br>Tensile Strength, psi | <br>1800 |
| 300% Modulus, psi | — |
| Elongation, % | 830 |
| Tear Resistance<br>Die C<br>pli (D-624) | <br><br>180 |
| Yerzley Resilience<br>% (D-945) | <br>— |

| Property | |
|---|---|
| Specific Gravity | 0.90 |
| Melt Index<br>Condition E, gms/10 min<br>Condition G, gms/10 min | <br>—<br>— |
| Physical Form | Pellet |
| Standard Color | Natural<br>or Black |
| Ozone Resistance<br>—Stressed<br>—Unstressed | <br>Excellent<br>Excellent |
| Comments | — |

## Product Data:

### Shell Elexar 8313

| Property | |
|---|---|
| Hardness<br>Shore A (D-2240)<br>Shore D | <br>—<br>62 |
| Tensile Properties (D-412)<br>Tensile Strength, psi | <br>2900 |
| 300% Modulus, psi | — |
| Elongation, % | 450 |
| Tear Resistance<br>Die C<br>pli (D-624) | <br><br>570 |
| Yerzley Resilience<br>% (D-945) | <br>— |

| Property | |
|---|---|
| Specific Gravity | 1.29 |
| Melt Index<br>Condition E, gms/10 min<br>Condition G, gms/10 min | <br>—<br>— |
| Physical Form | Pellet |
| Standard Color | Natural |
| Ozone Resistance<br>—Stressed<br>—Unstressed | <br>Excellent<br>Excellent |
| Comments | Fire<br>Retardant<br>LOI 25 |

*(continued)*

## Product Data:

### Shell Elexar 8421

| Property | |
|---|---|
| Hardness<br>Shore A (D-2240)<br>Shore D | <br>89<br>— |
| Tensile Properties (D-412)<br>Tensile Strength, psi | <br>2100 |
| 300% Modulus, psi | — |
| Elongation, % | 600 |
| Tear Resistance<br>Die C<br>pli (D-624) | <br><br>470 |
| Yerzley Resilience<br>% (D-945) | <br>— |

| Property | |
|---|---|
| Specific Gravity | 0.99 |
| Melt Index<br>Condition E, gms/10 min<br>Condition G, gms/10 min | <br>—<br>— |
| Physical Form | Pellet |
| Standard Color | Natural |
| Ozone Resistance<br>—Stressed<br>—Unstressed | <br>Excellent<br>Excellent |
| Comments | UL Listed |

## Product Data:

### Shell Elexar 8431

| Property | |
|---|---|
| Hardness<br>Shore A (D-2240)<br>Shore D | <br>67<br>— |
| Tensile Properties (D-412)<br>Tensile Strength, psi | <br>2300 |
| 300% Modulus, psi | — |
| Elongation, % | 675 |
| Tear Resistance<br>Die C<br>pli (D-624) | <br><br>275 |
| Yerzley Resilience<br>% (D-945) | <br>— |

| Property | |
|---|---|
| Specific Gravity | 0.93 |
| Melt Index<br>Condition E, gms/10 min<br>Condition G, gms/10 min | <br>—<br>— |
| Physical Form | Pellet |
| Standard Color | Natural |
| Ozone Resistance<br>—Stressed<br>—Unstressed | <br>Excellent<br>Excellent |
| Comments | UL Listed |

*(continued)*

Product Data:

**Shell Elexar 8451**

| Property | |
|---|---|
| Hardness<br>Shore A (D-2240)<br>Shore D | <br>82<br>— |
| Tensile Properties (D-412)<br>Tensile Strength, psi | <br>2000 |
| 300% Modulus, psi | — |
| Elongation, % | 670 |
| Tear Resistance<br>Die C<br>pli (D-624) | <br><br>360 |
| Yerzley Resilience<br>% (D-945) | <br>— |

| Property | |
|---|---|
| Specific Gravity | 1.00 |
| Melt Index<br>Condition E, gms/10 min<br>Condition G, gms/10 min | <br>—<br>— |
| Physical Form | Pellet |
| Standard Color | Natural |
| Ozone Resistance<br>—Stressed<br>—Unstressed | <br>Excellent<br>Excellent |
| Comments | UL Listed |

Product Data:

**Shell Elexar 8614**

| Property | |
|---|---|
| Hardness<br>Shore A (D-2240)<br>Shore D | <br>95<br>45 |
| Tensile Properties (D-412)<br>Tensile Strength, psi | <br>2200 |
| 300% Modulus, psi | — |
| Elongation, % | 500 |
| Tear Resistance<br>Die C<br>pli (D-624) | <br><br>380 |
| Yerzley Resilience<br>% (D-945) | <br>— |

| Property | |
|---|---|
| Specific Gravity | 1.17 |
| Melt Index<br>Condition E, gms/10 min<br>Condition G, gms/10 min | <br>—<br>— |
| Physical Form | Pellet |
| Standard Color | Natural |
| Ozone Resistance<br>—Stressed<br>—Unstressed | <br>Excellent<br>Excellent |
| Comments | UL Listed<br>Fire Retard-<br>ant LOI 29 |

*(continued)*

## Product Data:

### Shell Elexar 8623

| Property | |
|---|---|
| Hardness<br>Shore A (D-2240)<br>Shore D | <br>95<br>42 |
| Tensile Properties (D-412)<br>Tensile Strength, psi | <br>2300 |
| 300% Modulus, psi | — |
| Elongation, % | 530 |
| Tear Resistance<br>Die C<br>pli (D-624) | <br><br>450 |
| Yerzley Resilience<br>% (D-945) | <br>— |

| Property | |
|---|---|
| Specific Gravity | 1.16 |
| Melt Index<br>Condition E, gms/10 min<br>Condition G, gms/10 min | <br>—<br>— |
| Physical Form | Pellet |
| Standard Color | Natural |
| Ozone Resistance<br>—Stressed<br>—Unstressed | <br>Excellent<br>Excellent |
| Comments | UL Listed Fire Retard-ant LOI 23 |

## Product Data:

### Shell Elexar 8712

| Property | |
|---|---|
| Hardness<br>Shore A (D-2240)<br>Shore D | <br>65<br>— |
| Tensile Properties (D-412)<br>Tensile Strength, psi | <br>1400 |
| 300% Modulus, psi | 650 |
| Elongation, % | 190 |
| Tear Resistance<br>Die C<br>pli (D-624) | <br><br>190 |
| Yerzley Resilience<br>% (D-945) | <br>— |

| Property | |
|---|---|
| Specific Gravity | 1.10 |
| Melt Index<br>Condition E, gms/10 min<br>Condition G, gms/10 min | <br>—<br>— |
| Physical Form | Pellet |
| Standard Color | Natural |
| Ozone Resistance<br>—Stressed<br>—Unstressed | <br>Excellent<br>Excellent |
| Comments | Fire Retardant LOI 21 |

*(continued)*

## Product Data:

### Shell Elexar 8725

| Property | |
|---|---|
| Hardness<br>Shore A (D-2240)<br>Shore D | <br>84<br>— |
| Tensile Properties (D-412)<br>Tensile Strength, psi | <br>1700 |
| 300% Modulus, psi | — |
| Elongation, % | 540 |
| Tear Resistance<br>Die C<br>pli (D-624) | <br><br>250 |
| Yerzley Resilience<br>% (D-945) | <br>— |

| Property | |
|---|---|
| Specific Gravity | 1.08 |
| Melt Index<br>Condition E, gms/10 min<br>Condition G, gms/10 min | <br>—<br>— |
| Physical Form | Pellet |
| Standard Color | Natural |
| Ozone Resistance<br>—Stressed<br>—Unstressed | <br>Excellent<br>Excellent |
| Comments | UL Listed Fire Retardant LOI 24 |

## Product Data:

### Shell Elexar 8921

| Property | |
|---|---|
| Hardness<br>Shore A (D-2240)<br>Shore D | <br>86<br>— |
| Tensile Properties (D-412)<br>Tensile Strength, psi | <br>2000 |
| 300% Modulus, psi | — |
| Elongation, % | 510 |
| Tear Resistance<br>Die C<br>pli (D-624) | <br><br>300 |
| Yerzley Resilience<br>% (D-945) | <br>— |

| Property | |
|---|---|
| Specific Gravity | 1.00 |
| Melt Index<br>Condition E, gms/10 min<br>Condition G, gms/10 min | <br>—<br>— |
| Physical Form | Pellet |
| Standard Color | Natural |
| Ozone Resistance<br>—Stressed<br>—Unstressed | <br>Excellent<br>Excellent |
| Comments | — |

# Thermofil Inc.

Contact Address:

Thermofil Inc.
P.O. Box 489
Brighton, MI 48116-0489

Phone: 1 (313) 227-3500
1 (800) 282-1482
Facsimile #: 1 (313) 227-3824

## THERMOFIL E1-10FG-0100

Product Data:

**10% Glass Fiber Reinforced PBT/PET Alloy**

| Property | Test Methods ASTM | English System | Metric System | International System |
|---|---|---|---|---|
| **General** | | | | |
| Reinforcement Content | D2584 | 10% | 10% | 10% |
| Specific Gravity/Density | D792 | 89.23 lb/ft$^3$ | 1.43 g/cm$^3$ | 1430 kg/m$^3$ |
| Mold Shrinkage | D955 | .006 in/in | .006 cm/cm | .006 mm/mm |
| Hardness (Test) | D785 | R-120 | R-120 | R-120 |
| Water Absorption, 24 hrs. | D570 | .09% | .09% | .09% |
| **Mechanical** | | | | |
| Tensile Strength | D638 | 11,000 psi | 773.3 kg/cm$^2$ | 75.9 MPa |
| Tensile Elongation | D638 | 9% | 9% | 9% |
| Tensile Modulus | D638 | 7.8 psi $\times$ 10$^5$ | 5.5 kg/cm$^2$ $\times$ 10$^4$ | 5382.0 MPa |
| Flexural Strength | D790 | 18,000 psi | 1265 kg/cm$^2$ | 124 MPa |
| Flexural Modulus | D790 | 5.6 psi $\times$ 10$^5$ | 3.9 kg/cm$^2$ $\times$ 10$^4$ | 3864.0 MPa |
| Izod Impact (Notched) | D256 | 1.00 ft lb/in | 5.44 kg cm/cm | 53.3 J/m |
| Izod Impact (Unnotched) | D256 | 8.00 ft lb/in | 43.52 kg cm/cm | 426.4 J/m |
| Compressive Strength | D695 | 12,200 psi | 858 kg/cm$^2$ | 84 MPa |

*(continued)*

NOTE – Information provided in this book has been condensed. The publisher recommends that the product manufacturer be contacted, at the address or phone number provided, for more complete information on product safety, processing parameters, and applications not provided in this information sheet.

10% Glass Fiber Reinforced PBT/PET Alloy *(continued)*

| Property | Test Methods ASTM | English System | Metric System | International System |
|---|---|---|---|---|
| **Thermal** | | | | |
| Coefficient of Expansion | D696 | 2.0 $10^{-5}$in/in°F | 3.6 $10^{-5}$cm/cm°C | 3.6 $10^{-5}$m/m°K |
| Heat Deflection Temp. | D648 | 340°F @ 264 psi | 308°C @ 18.5 kg/cm² | 581°K @ 1.82 MPa |
| Heat Deflection Temp. | D648 | 350°F @ 66 psi | 177°C @ 4.6 kg/cm² | 450°K @ 46 MPa |
| Max. Temp. Continuous Use | | 280°F | 138°C | 411°K |
| **Electrical** | | | | |
| Dielectric Strength (ST) | D149 | 500 V/mil | 18 kV/mm | 18 MV/m |
| Volume Resistivity | D257 | 16.00 Log 10 Ohm in²/in | 17.00 Log 10 Ohm cm²/cm | 15.00 Log 10 Ohm m²/m |
| Dielectric Constant @ $10^3$Hz | D150 | 3.20 | 3.20 | 3.20 |
| Dissipation Factor @ $10^3$Hz | D150 | .015 | .015 | .015 |
| Arc Resistance | D495 | 150 sec. | 150 sec. | 150 sec. |
| **Flammability** | | | | |
| UL 94 Class, 1/16″ | | HB | HB | HB |
| Oxygen Index | D2863 | <26 Vol% | <26 Vol% | <26 Vol% |

# Thermofil Inc.

## Contact Address:

Thermofil Inc.
P.O. Box 489
Brighton, MI 48116-0489

Phone: 1 (313) 227-3500
1 (800) 282-1482
Facsimile #: 1 (313) 227-3824

## THERMOFIL E1-20FG-0100

### Product Data:

**20% Glass Reinforced PBT/PET Alloy**

| Property | Test Methods ASTM | English System | Metric System | International System |
|---|---|---|---|---|
| **General** | | | | |
| Reinforcement Content | D2584 | 20% | 20% | 20% |
| Specific Gravity/Density | D792 | 93.60 $lb/ft^3$ | 1.50 $g/cm^3$ | 1500 $kg/m^3$ |
| Mold Shrinkage | D955 | .004 in/in | .004 cm/cm | .004 mm/mm |
| Hardness (Test) | D785 | R-120 | R-120 | R-120 |
| Water Absorption, 24 hrs. | D570 | .12% | .12% | .12% |
| **Mechanical** | | | | |
| Tensile Strength | D638 | 14,500 psi | 1019.4 $kg/cm^2$ | 100.1 MPa |
| Tensile Elongation | D638 | 8% | 8% | 8% |
| Tensile Modulus | D638 | 10.8 psi $\times 10^5$ | 7.6 $kg/cm^2 \times 10^4$ | 7452.0 MPa |
| Flexural Strength | D790 | 23,500 psi | 1652 $kg/cm^2$ | 162 MPa |
| Flexural Modulus | D790 | 8.4 psi $\times 10^5$ | 5.9 $kg/cm^2 \times 10^4$ | 5796.0 MPa |
| Izod Impact (Notched) | D256 | 1.70 ft lb/in | 9.25 kg cm/cm | 90.6 J/m |
| Izod Impact (Unnotched) | D256 | 13.00 ft lb/in | 70.72 kg cm/cm | 692.9 J/m |
| Compressive Strength | D695 | 16,400 psi | 1153 $kg/cm^2$ | 113 MPa |

*(continued)*

NOTE – Information provided in this book has been condensed. The publisher recommends that the product manufacturer be contacted, at the address or phone number provided, for more complete information on product safety, processing parameters, and applications not provided in this information sheet.

**20% Glass Reinforced PBT/PET Alloy *(continued)***

| Property | Test Methods ASTM | English System | Metric System | International System |
|---|---|---|---|---|
| **Thermal** | | | | |
| Coefficient of Expansion | D696 | 1.8 $10^{-5}$in/in°F | 3.2 $10^{-5}$cm/cm°C | 3.2 $10^{-5}$m/m°K |
| Heat Deflection Temp. | D648 | 380°F @ 264 psi | 348°C @ 18.5 kg/cm² | 621°K @ 1.82 MPa |
| Heat Deflection Temp. | D648 | 390°F @ 66 psi | 199°C @ 4.6 kg/cm² | 472°K @ 46 MPa |
| Max. Temp. Continuous Use | | 266°F | 130°C | 403°K |
| **Electrical** | | | | |
| Dielectric Strength (ST) | D149 | 500 V/mil | 18 kV/mm | 18 MV/m |
| Volume Resistivity | D257 | 16.00 Log 10 Ohm in²/in | 17.00 Log 10 Ohm cm²/cm | 15.00 Log 10 Ohm m²/m |
| Dielectric Constant @ $10^3$Hz | D150 | 3.40 | 3.40 | 3.40 |
| Dissipation Factor @ $10^3$Hz | D150 | .015 | .015 | .015 |
| Arc Resistance | D495 | 140 sec. | 140 sec. | 140 sec. |
| **Flammability** | | | | |
| UL 94 Class, 1/16″ | | HB | HB | HB |
| Oxygen Index | D2863 | Vol% | Vol% | Vol% |

# Thermofil Inc.

## Contact Address:

Thermofil Inc.
P.O. Box 489
Brighton, MI 48116-0489

Phone: 1 (313) 227-3500
1 (800) 282-1482
Facsimile #: 1 (313) 227-3824

## THERMOFIL E1-30FG-0100

### Product Data:

**30% Glass Fiber Reinforced PBT/PET Alloy**

| Property | Test Methods ASTM | English System | Metric System | International System |
|---|---|---|---|---|
| **General** | | | | |
| Reinforcement Content | D2584 | 30% | 30% | 30% |
| Specific Gravity/Density | D792 | 99.22 $lb/ft^3$ | 1.59 $g/cm^3$ | 1590 $kg/m^3$ |
| Mold Shrinkage | D955 | .003 in/in | .003 cm/cm | .003 mm/mm |
| Hardness (Test) | D785 | R-121 | R-121 | R-121 |
| Water Absorption, 24 hrs. | D570 | .11% | .11% | .11% |
| **Mechanical** | | | | |
| Tensile Strength | D638 | 18,000 psi | 1265.4 $kg/cm^2$ | 124.2 MPa |
| Tensile Elongation | D638 | 3% | 3% | 3% |
| Tensile Modulus | D638 | 13.8 psi $\times 10^5$ | 9.7 $kg/cm^2 \times 10^4$ | 9522.0 MPa |
| Flexural Strength | D790 | 28,000 psi | 1968 $kg/cm^2$ | 193 MPa |
| Flexural Modulus | D790 | 11.9 psi $\times 10^5$ | 8.4 $kg/cm^2 \times 10^4$ | 8211.0 MPa |
| Izod Impact (Notched) | D256 | 1.70 ft lb/in | 9.25 kg cm/cm | 90.6 J/m |
| Izod Impact (Unnotched) | D256 | 16.00 ft lb/in | 87.04 kg cm/cm | 852.8 J/m |
| Compressive Strength | D695 | 20,700 psi | 1455 $kg/cm^2$ | 143 MPa |

*(continued)*

NOTE – Information provided in this book has been condensed. The publisher recommends that the product manufacturer be contacted, at the address or phone number provided, for more complete information on product safety, processing parameters, and applications not provided in this information sheet.

### 30% Glass Fiber Reinforced PBT/PET Alloy *(continued)*

| Property | Test Methods ASTM | English System | Metric System | International System |
|---|---|---|---|---|
| **Thermal** | | | | |
| Coefficient of Expansion | D696 | 1.4 $10^{-5}$in/in °F | 2.5 $10^{-5}$cm/cm °C | 2.5 $10^{-5}$m/m °K |
| Heat Deflection Temp. | D648 | 400 °F @ 264 psi | 368 °C @ 18.5 kg/cm² | 641 °K @ 1.82 MPa |
| Heat Deflection Temp. | D648 | 405 °F @ 66 psi | 207 °C @ 4.6 kg/cm² | 480 °K @ 46 MPa |
| Max. Temp. Continuous Use | | 266 °F | 130 °C | 403 °K |
| **Electrical** | | | | |
| Dielectric Strength (ST) | D149 | 500 V/mil | 18 kV/mm | 18 MV/m |
| Volume Resistivity | D257 | 16.00 Log 10 Ohm in²/in | 17.00 Log 10 Ohm cm²/cm | 15.00 Log 10 Ohm m²/m |
| Dielectric Constant @ $10^3$Hz | D150 | 3.60 | 3.60 | 3.60 |
| Dissipation Factor @ $10^3$Hz | D150 | .015 | .015 | .015 |
| Arc Resistance | D495 | 130 sec. | 130 sec. | 130 sec. |
| **Flammability** | | | | |
| UL 94 Class, 1/16″ | | HB | HB | HB |
| Oxygen Index | D2863 | <26 Vol% | <26 Vol% | <26 Vol% |

# Thermofil Inc.

## Contact Address:

Thermofil Inc.
P.O. Box 489
Brighton, MI 48116-0489

Phone: 1 (313) 227-3500
1 (800) 282-1482
Facsimile #: 1 (313) 227-3824

## THERMOFIL E1-40FG-0100

### Product Data:

**40% Glass Reinforced Modified PBT/PET Alloy**

| Property | Test Methods ASTM | English System | Metric System | International System |
|---|---|---|---|---|
| **General** | | | | |
| Reinforcement Content | D2584 | 40% | 40% | 40% |
| Specific Gravity/Density | D792 | 104.83 lb/ft$^3$ | 1.68 g/cm$^3$ | 1680 kg/m$^3$ |
| Mold Shrinkage | D955 | .002 in/in | .002 cm/cm | .002 mm/mm |
| Hardness (Test) | D785 | R-121 | R-121 | R-121 |
| Water Absorption, 24 hrs. | D570 | .10% | .10% | .10% |
| **Mechanical** | | | | |
| Tensile Strength | D638 | 19,500 psi | 1370.9 kg/cm$^2$ | 134.6 MPa |
| Tensile Elongation | D638 | 6% | 6% | 6% |
| Tensile Modulus | D638 | 16.8 psi × 10$^5$ | 11.8 kg/cm$^2$ × 10$^4$ | 11,592.0 MPa |
| Flexural Strength | D790 | 29,000 psi | 2039 kg/cm$^2$ | 200 MPa |
| Flexural Modulus | D790 | 15.0 psi × 10$^5$ | 10.5 kg/cm$^2$ × 10$^4$ | 10,350.0 MPa |
| Izod Impact (Notched) | D256 | 2.60 ft lb/in | 14.14 kg cm/cm | 138.6 J/m |
| Izod Impact (Unnotched) | D256 | 16.00 ft lb/in | 87.04 kg cm/cm | 852.8 J/m |
| Compressive Strength | D695 | 24,900 psi | 1750 kg/cm$^2$ | 172 MPa |

*(continued)*

NOTE—Information provided in this book has been condensed. The publisher recommends that the product manufacturer be contacted, at the address or phone number provided, for more complete information on product safety, processing parameters, and applications not provided in this information sheet.

### 40% Glass Reinforced Modified PBT/PET Alloy *(continued)*

| Property | Test Methods ASTM | English System | Metric System | International System |
|---|---|---|---|---|
| **Thermal** | | | | |
| Coefficient of Expansion | D696 | 1.3 $10^{-5}$in/in°F | 2.3 $10^{-5}$cm/cm°C | 2.3 $10^{-5}$m/m°K |
| Heat Deflection Temp. | D648 | 410°F @ 264 psi | 378°C @ 18.5 kg/cm$^2$ | 651°K @ 1.82 MPa |
| Heat Deflection Temp. | D648 | 415°F @ 66 psi | 213°C @ 4.6 kg/cm$^2$ | 486°K @ 46 MPa |
| Max. Temp. Continuous Use | | 266°F | 130°C | 403°K |
| **Electrical** | | | | |
| Dielectric Strength (ST) | D149 | 500 V/mil | 18 kV/mm | 18 MV/m |
| Volume Resistivity | D257 | 16.00 Log 10 Ohm in$^2$/in | 17.00 Log 10 Ohm cm$^2$/cm | 15.00 Log 10 Ohm m$^2$/m |
| Dielectric Constant @ $10^3$Hz | D150 | 3.80 | 3.80 | 3.80 |
| Dissipation Factor @ $10^3$Hz | D150 | .010 | .010 | .010 |
| Arc Resistance | D495 | 120 sec. | 120 sec. | 120 sec. |
| **Flammability** | | | | |
| UL 94 Class, 1/16″ | | HB | HB | HB |
| Oxygen Index | D2863 | Vol% | Vol% | Vol% |

# Thermofil Inc.

## Contact Address:

Thermofil Inc.
P.O. Box 489
Brighton, MI 48116-0489

Phone: 1 (313) 227-3500
1 (800) 282-1482
Facsimile #: 1 (313) 227-3824

## THERMOFIL E1-45FG-0100

### Product Data:

**45% Glass Reinforced PBT/PET Alloy**

| Property | Test Methods ASTM | English System | Metric System | International System |
|---|---|---|---|---|
| **General** | | | | |
| Reinforcement Content | D2584 | 45% | 45% | 45% |
| Specific Gravity/Density | D792 | 107.95 lb/ft$^3$ | 1.73 g/cm$^3$ | 1730 kg/m$^3$ |
| Mold Shrinkage | D955 | .001 in/in | .001 cm/cm | .001 mm/mm |
| Hardness (Test) | D785 | R-121 | R-121 | R-121 |
| Water Absorption, 24 hrs. | D570 | .10% | .10% | .10% |
| **Mechanical** | | | | |
| Tensile Strength | D638 | 20,500 psi | 1441.2 kg/cm$^2$ | 141.5 MPa |
| Tensile Elongation | D638 | 5% | 5% | 5% |
| Tensile Modulus | D638 | 18.0 psi $\times$ 10$^5$ | 12.7 kg/cm$^2$ $\times$ 10$^4$ | 12,420.0 MPa |
| Flexural Strength | D790 | 31,500 psi | 2214 kg/cm$^2$ | 217 MPa |
| Flexural Modulus | D790 | 17.5 psi $\times$ 10$^5$ | 12.3 kg/cm$^2$ $\times$ 10$^4$ | 12,075.0 MPa |
| Izod Impact (Notched) | D256 | 2.50 ft lb/in | 13.60 kg cm/cm | 133.3 J/m |
| Izod Impact (Unnotched) | D256 | 15.00 ft lb/in | 81.60 kg cm/cm | 799.5 J/m |
| Compressive Strength | D695 | 26,000 psi | 1828 kg/cm$^2$ | 179 MPa |

*(continued)*

NOTE—Information provided in this book has been condensed. The publisher recommends that the product manufacturer be contacted, at the address or phone number provided, for more complete information on product safety, processing parameters, and applications not provided in this information sheet.

## 45% Glass Reinforced PBT/PET Alloy *(continued)*

| Property | Test Methods ASTM | English System | Metric System | International System |
|---|---|---|---|---|
| **Thermal** | | | | |
| Coefficient of Expansion | D696 | 1.2 $10^{-5}$in/in°F | 2.2 $10^{-5}$cm/cm°C | 2.2 $10^{-5}$m/m°K |
| Heat Deflection Temp. | D648 | 415°F @ 264 psi | 383°C @ 18.5 kg/cm² | 656°K @ 1.82 MPa |
| Heat Deflection Temp. | D648 | 420°F @ 66 psi | 216°C @ 4.6 kg/cm² | 489°K @ 46 MPa |
| Max. Temp. Continuous Use | | 266°F | 130°C | 403°K |
| **Electrical** | | | | |
| Dielectric Strength (ST) | D149 | 500 V/mil | 18 kV/mm | 18 MV/m |
| Volume Resistivity | D257 | 16.00 Log 10 Ohm in²/in | 17.00 Log 10 Ohm cm²/cm | 15.00 Log 10 Ohm m²/m |
| Dielectric Constant @ $10^3$Hz | D150 | 3.90 | 3.90 | 3.90 |
| Dissipation Factor @ $10^3$Hz | D150 | .010 | .010 | .010 |
| Arc Resistance | D495 | 120 sec. | 120 sec. | 120 sec. |
| **Flammability** | | | | |
| UL 94 Class, 1/16″ | | HB | HB | HB |
| Oxygen Index | D2863 | <27 Vol% | <27 Vol% | <27 Vol% |

# Thermofil Inc.

## Contact Address:

Thermofil Inc.
P.O. Box 489
Brighton, MI 48116-0489

Phone: 1 (313) 227-3500
1 (800) 282-1482
Facsimile #: 1 (313) 227-3824

## THERMOFIL E1-50FG-0100

### Product Data:

**50% Glass Reinforced PBT/PET Alloy**

| Property | Test Methods ASTM | English System | Metric System | International System |
|---|---|---|---|---|
| **General** | | | | |
| Reinforcement Content | D2584 | 50% | 50% | 50% |
| Specific Gravity/Density | D792 | 111.70 lb/ft³ | 1.79 g/cm³ | 1790 kg/m³ |
| Mold Shrinkage | D955 | .001 in/in | .001 cm/cm | .001 mm/mm |
| Hardness (Test) | D785 | R-122 | R-122 | R-122 |
| Water Absorption, 24 hrs. | D570 | .10% | .10% | .10% |
| **Mechanical** | | | | |
| Tensile Strength | D638 | 21,500 psi | 1511.5 kg/cm² | 148.4 MPa |
| Tensile Elongation | D638 | 5% | 5% | 5% |
| Tensile Modulus | D638 | 19.8 psi × $10^5$ | 13.9 kg/cm² × $10^4$ | 13,662.0 MPa |
| Flexural Strength | D790 | 32,000 psi | 2250 kg/cm² | 221 MPa |
| Flexural Modulus | D790 | 20.0 psi × $10^5$ | 14.1 kg/cm² × $10^4$ | 13,800.0 MPa |
| Izod Impact (Notched) | D256 | 2.40 ft lb/in | 13.06 kg cm/cm | 127.9 J/m |
| Izod Impact (Unnotched) | D256 | 15.00 ft lb/in | 81.60 kg cm/cm | 799.5 J/m |
| Compressive Strength | D695 | 29,100 psi | 2046 kg/cm² | 201 MPa |

*(continued)*

NOTE – Information provided in this book has been condensed. The publisher recommends that the product manufacturer be contacted, at the address or phone number provided, for more complete information on product safety, processing parameters, and applications not provided in this information sheet.

**50% Glass Reinforced PBT/PET Alloy *(continued)***

| Property | Test Methods ASTM | English System | Metric System | International System |
|---|---|---|---|---|
| **Thermal** | | | | |
| Coefficient of Expansion | D696 | 1.2 $10^{-5}$in/in °F | 2.2 $10^{-5}$cm/cm °C | 2.2 $10^{-5}$m/m °K |
| Heat Deflection Temp. | D648 | 415 °F @ 264 psi | 383 °C @ 18.5 kg/cm² | 656 °K @ 1.82 MPa |
| Heat Deflection Temp. | D648 | 420 °F @ 66 psi | 216 °C @ 4.6 kg/cm² | 489 °K @ 46 MPa |
| Max. Temp. Continuous Use | | 266 °F | 130 °C | 403 °K |
| **Electrical** | | | | |
| Dielectric Strength (ST) | D149 | 500 V/mil | 18 kV/mm | 18 MV/m |
| Volume Resistivity | D257 | 16.00 Log 10 Ohm in²/in | 17.00 Log 10 Ohm cm²/cm | 15.00 Log 10 Ohm m²/m |
| Dielectric Constant @ $10^3$Hz | D150 | 4.00 | 4.00 | 4.00 |
| Dissipation Factor @ $10^3$Hz | D150 | .010 | .010 | .010 |
| Arc Resistance | D495 | 120 sec. | 120 sec. | 120 sec. |
| **Flammability** | | | | |
| UL 94 Class, 1/16″ | | HB | HB | HB |
| Oxygen Index | D2863 | Vol% | Vol% | Vol% |

# Thermofil Inc.

## Contact Address:

Thermofil Inc.
P.O. Box 489
Brighton, MI 48116-0489

Phone: 1 (313) 227-3500
1 (800) 282-1482
Facsimile #: 1 (313) 227-3824

## THERMOFIL N7-40AF-0100

### Product Data:

**Aluminum Flake Reinforced High Impact Nylon 6/6**

| Property | Test Methods ASTM | English System | Metric System | International System |
|---|---|---|---|---|
| **General** | | | | |
| Reinforcement Content | D2584 | 40% | 40% | 40% |
| Specific Gravity/Density | D792 | 88.61 lb/ft$^3$ | 1.42 g/cm$^3$ | 1420 kg/m$^3$ |
| Mold Shrinkage | D955 | .005 in/in | .005 cm/cm | .005 mm/mm |
| Hardness (Test) | D785 | R-114 | R-114 | R-114 |
| Water Absorption, 24 hrs. | D570 | 1.00% | 1.00% | 1.00% |
| **Mechanical** | | | | |
| Tensile Strength | D638 | 8100 psi | 569.4 kg/cm$^2$ | 55.9 MPa |
| Tensile Elongation | D638 | 3% | 3% | 3% |
| Tensile Modulus | D638 | 6.0 psi × 10$^5$ | 4.2 kg/cm$^2$ × 10$^4$ | 4140.0 MPa |
| Flexural Strength | D790 | 14,000 psi | 984 kg/cm$^2$ | 97 MPa |
| Flexural Modulus | D790 | 6.2 psi × 10$^5$ | 4.4 kg/cm$^2$ × 10$^4$ | 4278.0 MPa |
| Izod Impact (Notched) | D256 | 6.00 ft lb/in | 32.64 kg cm/cm | 319.8 J/m |
| Izod Impact (Unnotched) | D256 | 15.00 ft lb/in | 81.60 kg cm/cm | 799.5 J/m |
| Compressive Strength | D695 | 10,000 psi | 703 kg/cm$^2$ | 69 MPa |

*(continued)*

NOTE – Information provided in this book has been condensed. The publisher recommends that the product manufacturer be contacted, at the address or phone number provided, for more complete information on product safety, processing parameters, and applications not provided in this information sheet.

**Aluminum Flake Reinforced High Impact Nylon 6/6 *(continued)***

| Property | Test Methods ASTM | English System | Metric System | International System |
|---|---|---|---|---|
| **Thermal** | | | | |
| Coefficient of Expansion | D696 | 2.2 $10^{-5}$in/in°F | 4.0 $10^{-5}$cm/cm°C | 4.0 $10^{-5}$m/m°K |
| Heat Deflection Temp. | D648 | 350°F @ 264 psi | 318°C @ 18.5 kg/cm² | 591°K @ 1.82 MPa |
| Heat Deflection Temp. | D648 | 390°F @ 66 psi | 199°C @ 4.6 kg/cm² | 472°K @ 46 MPa |
| Max. Temp. Continuous Use | | 248°F | 120°C | 393°K |
| **Electrical** | | | | |
| Dielectric Strength (ST) | D149 | 0 V/mil | 0 kV/mm | 0 MV/m |
| Volume Resistivity | D257 | 1.50 Log 10 Ohm in²/in | 2.50 Log 10 Ohm cm²/cm | .50 Log 10 Ohm m²/m |
| Dielectric Constant @ $10^3$Hz | D150 | | | |
| Dissipation Factor @ $10^3$Hz | D150 | | | |
| Arc Resistance | D495 | | | |
| **Flammability** | | | | |
| UL 94 Class, 1/16″ | | HB | HB | HB |
| Oxygen Index | D2863 | 22 Vol% | 22 Vol% | 22 Vol% |

# Thermofil Inc.

## Contact Address:

Thermofil Inc.
P.O. Box 489
Brighton, MI 48116-0489

Phone: 1 (313) 227-3500
1 (800) 282-1482
Facsimile #: 1 (313) 227-3824

## THERMOFIL N5-15FG-0100

### Product Data:

**15% Glass Fiber Reinforced, Dimensionally Stable Nylon 6/6 Alloy**

| Property | Test Methods ASTM | English System | Metric System | International System |
|---|---|---|---|---|
| **General** | | | | |
| Reinforcement Content | D2584 | 15% | 15% | 15% |
| Specific Gravity/Density | D792 | 76.13 lb/ft$^3$ | 1.22 g/cm$^3$ | 1220 kg/m$^3$ |
| Mold Shrinkage | D955 | .004 in/in | .004 cm/cm | .004 mm/mm |
| Hardness (Test) | D785 | R-115 | R-115 | R-115 |
| Water Absorption, 24 hrs. | D570 | .40% | .40% | .40% |
| **Mechanical** | | | | |
| Tensile Strength | D638 | 16,000 psi | 1124.8 kg/cm$^2$ | 110.4 MPa |
| Tensile Elongation | D638 | 3% | 3% | 3% |
| Tensile Modulus | D638 | 9.0 psi × 10$^5$ | 6.3 kg/cm$^2$ × 10$^4$ | 6210.0 MPa |
| Flexural Strength | D790 | 26,000 psi | 1828 kg/cm$^2$ | 179 MPa |
| Flexural Modulus | D790 | 9.0 psi × 10$^5$ | 6.3 kg/cm$^2$ × 10$^4$ | 6210.0 MPa |
| Izod Impact (Notched) | D256 | 1.10 ft lb/in | 5.98 kg cm/cm | 58.6 J/m |
| Izod Impact (Unnotched) | D256 | 10.00 ft lb/in | 54.40 kg cm/cm | 533.0 J/m |
| Compressive Strength | D695 | 18,000 psi | 1265 kg/cm$^2$ | 124 MPa |

*(continued)*

NOTE—Information provided in this book has been condensed. The publisher recommends that the product manufacturer be contacted, at the address or phone number provided, for more complete information on product safety, processing parameters, and applications not provided in this information sheet.

**15% Glass Fiber Reinforced, Dimensionally Stable Nylon 6/6 Alloy *(continued)***

| Property | Test Methods ASTM | English System | Metric System | International System |
|---|---|---|---|---|
| **Thermal** | | | | |
| Coefficient of Expansion | D696 | 2.4 $10^{-5}$in/in°F | 4.3 $10^{-5}$cm/cm°C | 4.3 $10^{-5}$m/m°K |
| Heat Deflection Temp. | D648 | 430°F @ 264 psi | 398°C @ 18.5 kg/cm² | 671°K @ 1.82 MPa |
| Heat Deflection Temp. | D648 | 475°F @ 66 psi | 246°C @ 4.6 kg/cm² | 519°K @ 46 MPa |
| Max. Temp. Continuous Use | | 221°F | 105°C | 378°K |
| **Electrical** | | | | |
| Dielectric Strength (ST) | D149 | 410 V/mil | 15 kV/mm | 15 MV/m |
| Volume Resistivity | D257 | 15.00 Log 10 Ohm in²/in | 16.00 Log 10 Ohm cm²/cm | 14.00 Log 10 Ohm m²/m |
| Dielectric Constant @ $10^3$Hz | D150 | 3.60 | 3.60 | 3.60 |
| Dissipation Factor @ $10^3$Hz | D150 | .014 | .014 | .014 |
| Arc Resistance | D495 | 110 sec. | 110 sec. | 110 sec. |
| **Flammability** | | | | |
| UL 94 Class, 1/16″ | | HB | HB | HB |
| Oxygen Index | D2863 | <25 Vol% | <25 Vol% | <25 Vol% |

# Thermofil Inc.

## Contact Address:

Thermofil Inc.
P.O. Box 489
Brighton, MI 48116-0489

Phone: 1 (313) 227-3500
1 (800) 282-1482
Facsimile #: 1 (313) 227-3824

## THERMOFIL N5-30FG-0100

### Product Data:

**30% Glass Fiber Reinforced, Dimensionally Stable Nylon 6/6 Alloy**

| Property | Test Methods ASTM | English System | Metric System | International System |
|---|---|---|---|---|
| **General** | | | | |
| Reinforcement Content | D2584 | 30% | 30% | 30% |
| Specific Gravity/Density | D792 | 84.24 lb/ft$^3$ | 1.35 g/cm$^3$ | 1350 kg/m$^3$ |
| Mold Shrinkage | D955 | .002 in/in | .002 cm/cm | .002 mm/mm |
| Hardness (Test) | D785 | R-118 | R-118 | R-118 |
| Water Absorption, 24 hrs. | D570 | .30% | .30% | .30% |
| **Mechanical** | | | | |
| Tensile Strength | D638 | 21,000 psi | 1476.3 kg/cm$^2$ | 144.9 MPa |
| Tensile Elongation | D638 | 2% | 2% | 2% |
| Tensile Modulus | D638 | 14.0 psi $\times$ 10$^5$ | 9.8 kg/cm$^2$ $\times$ 10$^4$ | 9660.0 MPa |
| Flexural Strength | D790 | 31,000 psi | 2179 kg/cm$^2$ | 214 MPa |
| Flexural Modulus | D790 | 14.0 psi $\times$ 10$^5$ | 9.8 kg/cm$^2$ $\times$ 10$^4$ | 9660.0 MPa |
| Izod Impact (Notched) | D256 | 1.30 ft lb/in | 7.07 kg cm/cm | 69.3 J/m |
| Izod Impact (Unnotched) | D256 | 12.00 ft lb/in | 65.28 kg cm/cm | 639.6 J/m |
| Compressive Strength | D695 | 26,000 psi | 1828 kg/cm$^2$ | 179 MPa |

*(continued)*

NOTE—Information provided in this book has been condensed. The publisher recommends that the product manufacturer be contacted, at the address or phone number provided, for more complete information on product safety, processing parameters, and applications not provided in this information sheet.

**30% Glass Fiber Reinforced, Dimensionally Stable Nylon 6/6 Alloy *(continued)***

| Property | Test Methods ASTM | English System | Metric System | International System |
|---|---|---|---|---|
| **Thermal** | | | | |
| Coefficient of Expansion | D696 | 2.0 $10^{-5}$in/in°F | 3.6 $10^{-5}$cm/cm°C | 3.6 $10^{-5}$m/m°K |
| Heat Deflection Temp. | D648 | 465°F @ 264 psi | 433°C @ 18.5 $kg/cm^2$ | 706°K @ 1.82 MPa |
| Heat Deflection Temp. | D648 | 495°F @ 66 psi | 257°C @ 4.6 $kg/cm^2$ | 530°K @ 46 MPa |
| Max. Temp. Continuous Use | | 221°F | 105°C | 378°K |
| **Electrical** | | | | |
| Dielectric Strength (ST) | D149 | 400 V/mil | 14 kV/mm | 14 MV/m |
| Volume Resistivity | D257 | 15.00 Log 10 Ohm $in^2$/in | 16.00 Log 10 Ohm $cm^2$/cm | 14.00 Log 10 Ohm $m^2$/m |
| Dielectric Constant @ $10^3$Hz | D150 | 3.70 | 3.70 | 3.70 |
| Dissipation Factor @ $10^3$Hz | D150 | .015 | .015 | .015 |
| Arc Resistance | D495 | 90 sec. | 90 sec. | 90 sec. |
| **Flammability** | | | | |
| UL 94 Class, 1/16″ | | HB | HB | HB |
| Oxygen Index | D2863 | <27 Vol% | <27 Vol% | <27 Vol% |

# Thermofil Inc.

## Contact Address:

Thermofil Inc.
P.O. Box 489
Brighton, MI 48116-0489

Phone: 1 (313) 227-3500
1 (800) 282-1482
Facsimile #: 1 (313) 227-3824

## THERMOFIL N7-13FG-0100

### Product Data:

**Glass Fiber Reinforced, High Impact Nylon 6/6**

| Property | Test Methods ASTM | English System | Metric System | International System |
|---|---|---|---|---|
| **General** | | | | |
| Reinforcement Content | D2584 | 13% | 13% | 13% |
| Specific Gravity/Density | D792 | 69.89 lb/ft$^3$ | 1.12 g/cm$^3$ | 1120 kg/m$^3$ |
| Mold Shrinkage | D955 | .004 in/in | .004 cm/cm | .004 mm/mm |
| Hardness (Test) | D785 | M-90 | M-90 | M-90 |
| Water Absorption, 24 hrs. | D570 | .50% | .50% | .50% |
| **Mechanical** | | | | |
| Tensile Strength | D638 | 11,600 psi | 815.5 kg/cm$^2$ | 80.0 MPa |
| Tensile Elongation | D638 | 5% | 5% | 5% |
| Tensile Modulus | D638 | 5.0 psi $\times$ 10$^5$ | 3.5 kg/cm$^2$ $\times$ 10$^4$ | 3450.0 MPa |
| Flexural Strength | D790 | 18,200 psi | 1279 kg/cm$^2$ | 126 MPa |
| Flexural Modulus | D790 | 5.1 psi $\times$ 10$^5$ | 3.6 kg/cm$^2$ $\times$ 10$^4$ | 3519.0 MPa |
| Izod Impact (Notched) | D256 | 3.50 ft lb/in | 19.04 kg cm/cm | 186.6 J/m |
| Izod Impact (Unnotched) | D256 | 14.00 ft lb/in | 76.16 kg cm/cm | 746.2 J/m |
| Compressive Strength | D695 | 12,000 psi | 844 kg/cm$^2$ | 83 MPa |

*(continued)*

NOTE–Information provided in this book has been condensed. The publisher recommends that the product manufacturer be contacted, at the address or phone number provided, for more complete information on product safety, processing parameters, and applications not provided in this information sheet.

**Glass Fiber Reinforced, High Impact Nylon 6/6 *(continued)***

| Property | Test Methods ASTM | English System | Metric System | International System |
|---|---|---|---|---|
| **Thermal** | | | | |
| Coefficient of Expansion | D696 | 2.0 $10^{-5}$in/in°F | 3.6 $10^{-5}$cm/cm°C | 3.6 $10^{-5}$m/m°K |
| Heat Deflection Temp. | D648 | 385°F @ 264 psi | 353°C @ 18.5 kg/cm² | 626°K @ 1.82 MPa |
| Heat Deflection Temp. | D648 | 400°F @ 66 psi | 204°C @ 4.6 kg/cm² | 477°K @ 46 MPa |
| Max. Temp. Continuous Use | | 240°F | 116°C | 389°K |
| **Electrical** | | | | |
| Dielectric Strength (ST) | D149 | 400 V/mil | 14 kV/mm | 14 MV/m |
| Volume Resistivity | D257 | 15.00 Log 10 Ohm in²/in | 16.00 Log 10 Ohm cm²/cm | 14.00 Log 10 Ohm m²/m |
| Dielectric Constant @ $10^3$Hz | D150 | 3.50 | 3.50 | 3.50 |
| Dissipation Factor @ $10^3$Hz | D150 | .015 | .015 | .015 |
| Arc Resistance | D495 | 120 sec. | 120 sec. | 120 sec. |
| **Flammability** | | | | |
| UL 94 Class, 1/16″ | | HB | HB | HB |
| Oxygen Index | D2863 | <28 Vol% | <28 Vol% | <28 Vol% |

# Thermofil Inc.

## Contact Address:

Thermofil Inc.
P.O. Box 489
Brighton, MI 48116-0489

Phone: 1 (313) 227-3500
1 (800) 282-1482
Facsimile #: 1 (313) 227-3824

## THERMOFIL N7-15FG-0100

### Product Data:

**15% Fiberglass Reinforced Super Impact Type 6/6 Nylon**

| Property | Test Methods ASTM | English System | Metric System | International System |
|---|---|---|---|---|
| **General** | | | | |
| Reinforcement Content | D2584 | 15% | 15% | 15% |
| Specific Gravity/Density | D792 | 76.13 lb/ft$^3$ | 1.22 g/cm$^3$ | 1220 kg/m$^3$ |
| Mold Shrinkage | D955 | .004 in/in | .004 cm/cm | .004 mm/mm |
| Hardness (Test) | D785 | M-80 | M-80 | M-80 |
| Water Absorption, 24 hrs. | D570 | .80% | .80% | .80% |
| **Mechanical** | | | | |
| Tensile Strength | D638 | 12,500 psi | 878.8 kg/cm$^2$ | 86.3 MPa |
| Tensile Elongation | D638 | 9% | 9% | 9% |
| Tensile Modulus | D638 | 6.0 psi × 10$^5$ | 4.2 kg/cm$^2$ × 10$^4$ | 4140.0 MPa |
| Flexural Strength | D790 | 20,500 psi | 1441 kg/cm$^2$ | 141 MPa |
| Flexural Modulus | D790 | 5.4 psi × 10$^5$ | 3.8 kg/cm$^2$ × 10$^4$ | 3726.0 MPa |
| Izod Impact (Notched) | D256 | 4.00 ft lb/in | 21.76 kg cm/cm | 213.2 J/m |
| Izod Impact (Unnotched) | D256 | 15.00 ft lb/in | 81.60 kg cm/cm | 799.5 J/m |
| Compressive Strength | D695 | 13,000 psi | 914 kg/cm$^2$ | 90 MPa |

*(continued)*

NOTE—Information provided in this book has been condensed. The publisher recommends that the product manufacturer be contacted, at the address or phone number provided, for more complete information on product safety, processing parameters, and applications not provided in this information sheet.

15% Fiberglass Reinforced Super Impact 6/6 Nylon *(continued)*

| Property | Test Methods ASTM | English System | Metric System | International System |
|---|---|---|---|---|
| **Thermal** | | | | |
| Coefficient of Expansion | D696 | 2.8 $10^{-5}$in/in °F | 5.0 $10^{-5}$cm/cm °C | 5.0 $10^{-5}$m/m °K |
| Heat Deflection Temp. | D648 | 445 °F @ 264 psi | 413 °C @ 18.5 kg/cm$^2$ | 686 °K @ 1.82 MPa |
| Heat Deflection Temp. | D648 | 460 °F @ 66 psi | 238 °C @ 4.6 kg/cm$^2$ | 511 °K @ 46 MPa |
| Max. Temp. Continuous Use | | 220 °F | 104 °C | 377 °K |
| **Electrical** | | | | |
| Dielectric Strength (ST) | D149 | 500 V/mil | 18 kV/mm | 18 MV/m |
| Volume Resistivity | D257 | 15.00 Log 10 Ohm in$^2$/in | 16.00 Log 10 Ohm cm$^2$/cm | 14.00 Log 10 Ohm m$^2$/m |
| Dielectric Constant @ $10^3$Hz | D150 | 3.40 | 3.40 | 3.40 |
| Dissipation Factor @ $10^3$Hz | D150 | .015 | .015 | .015 |
| Arc Resistance | D495 | 120 sec. | 120 sec. | 120 sec. |
| **Flammability** | | | | |
| UL 94 Class, 1/16″ | | HB | HB | HB |
| Oxygen Index | D2863 | <28 Vol% | <28 Vol% | <28 Vol% |

# Thermofil Inc.

## Contact Address:

Thermofil Inc.
P.O. Box 489
Brighton, MI 48116-0489

Phone: 1 (313) 227-3500
1 (800) 282-1482
Facsimile #: 1 (313) 227-3824

## THERMOFIL N7-30FG-0100

### Product Data:

**30% Glass Fiber Reinforced Nylon 6/6 (High Impact)**

| Property | Test Methods ASTM | English System | Metric System | International System |
|---|---|---|---|---|
| **General** | | | | |
| Reinforcement Content | D2584 | 30% | 30% | 30% |
| Specific Gravity/Density | D792 | 82.37 lb/ft$^3$ | 1.32 g/cm$^3$ | 1320 kg/m$^3$ |
| Mold Shrinkage | D955 | .004 in/in | .004 cm/cm | .004 mm/mm |
| Hardness (Test) | D785 | R-114 | R-114 | R-114 |
| Water Absorption, 24 hrs. | D570 | .60% | .60% | .60% |
| **Mechanical** | | | | |
| Tensile Strength | D638 | 19,000 psi | 1335.7 kg/cm$^2$ | 131.1 MPa |
| Tensile Elongation | D638 | 9% | 9% | 9% |
| Tensile Modulus | D638 | 10.0 psi $\times$ 10$^5$ | 7.0 kg/cm$^2$ $\times$ 10$^4$ | 6900.0 MPa |
| Flexural Strength | D790 | 27,500 psi | 1933 kg/cm$^2$ | 190 MPa |
| Flexural Modulus | D790 | 9.7 psi $\times$ 10$^5$ | 6.8 kg/cm$^2$ $\times$ 10$^4$ | 6693.0 MPa |
| Izod Impact (Notched) | D256 | 5.00 ft lb/in | 27.20 kg cm/cm | 266.5 J/m |
| Izod Impact (Unnotched) | D256 | 17.50 ft lb/in | 95.20 kg cm/cm | 932.8 J/m |
| Compressive Strength | D695 | 20,000 psi | 1406 kg/cm$^2$ | 138 MPa |

*(continued)*

NOTE – Information provided in this book has been condensed. The publisher recommends that the product manufacturer be contacted, at the address or phone number provided, for more complete information on product safety, processing parameters, and applications not provided in this information sheet.

## 30% Glass Fiber Reinforced Nylon 6/6 (High Impact) *(continued)*

| Property | Test Methods ASTM | English System | Metric System | International System |
|---|---|---|---|---|
| **Thermal** | | | | |
| Coefficient of Expansion | D696 | 1.8 $10^{-5}$in/in°F | 3.2 $10^{-5}$cm/cm°C | 3.2 $10^{-5}$m/m°K |
| Heat Deflection Temp. | D648 | 480°F @ 264 psi | 448°C @ 18.5 kg/cm$^2$ | 721°K @ 1.82 MPa |
| Heat Deflection Temp. | D648 | 490°F @ 66 psi | 254°C @ 4.6 kg/cm$^2$ | 527°K @ 46 MPa |
| Max. Temp. Continuous Use | | 248°F | 120°C | 393°K |
| **Electrical** | | | | |
| Dielectric Strength (ST) | D149 | 500 V/mil | 18 kV/mm | 18 MV/m |
| Volume Resistivity | D257 | 14.00 Log 10 Ohm in$^2$/in | 15.00 Log 10 Ohm cm$^2$/cm | 13.00 Log 10 Ohm m$^2$/m |
| Dielectric Constant @ $10^3$Hz | D150 | 3.70 | 3.70 | 3.70 |
| Dissipation Factor @ $10^3$Hz | D150 | .010 | .010 | .010 |
| Arc Resistance | D495 | 120 sec. | 120 sec. | 120 sec. |
| **Flammability** | | | | |
| UL 94 Class, 1/16″ | | HB | HB | HB |
| Oxygen Index | D2863 | <25 Vol% | <25 Vol% | <25 Vol% |

# Thermofil Inc.

## Contact Address:

Thermofil Inc.
P.O. Box 489
Brighton, MI 48116-0489

Phone: 1 (313) 227-3500
1 (800) 282-1482
Facsimile #: 1 (313) 227-3824

## THERMOFIL N7-33FG-0100

### Product Data:

**Glass Fiber Reinforced, Highly Impact Modified Nylon 6/6**

| Property | Test Methods ASTM | English System | Metric System | International System |
|---|---|---|---|---|
| **General** | | | | |
| Reinforcement Content | D2584 | 33% | 33% | 33% |
| Specific Gravity/Density | D792 | 82.37 $lb/ft^3$ | 1.32 $g/cm^3$ | 1320 $kg/m^3$ |
| Mold Shrinkage | D955 | .003 in/in | .003 cm/cm | .003 mm/mm |
| Hardness (Test) | D785 | M-95 | M-95 | M-95 |
| Water Absorption, 24 hrs. | D570 | .60% | .60% | .60% |
| **Mechanical** | | | | |
| Tensile Strength | D638 | 19,000 psi | 1335.7 $kg/cm^2$ | 131.1 MPa |
| Tensile Elongation | D638 | 8% | 8% | 8% |
| Tensile Modulus | D638 | 10.2 psi $\times 10^5$ | 7.2 $kg/cm^2 \times 10^4$ | 7038.0 MPa |
| Flexural Strength | D790 | 27,500 psi | 1933 $kg/cm^2$ | 190 MPa |
| Flexural Modulus | D790 | 9.8 psi $\times 10^5$ | 6.9 $kg/cm^2 \times 10^4$ | 6762.0 MPa |
| Izod Impact (Notched) | D256 | 5.00 ft lb/in | 27.20 kg cm/cm | 266.5 J/m |
| Izod Impact (Unnotched) | D256 | 17.50 ft lb/in | 95.20 kg cm/cm | 932.8 J/m |
| Compressive Strength | D695 | 20,000 psi | 1406 $kg/cm^2$ | 138 MPa |

*(continued)*

NOTE – Information provided in this book has been condensed. The publisher recommends that the product manufacturer be contacted, at the address or phone number provided, for more complete information on product safety, processing parameters, and applications not provided in this information sheet.

### Glass Fiber Reinforced, Highly Impact Modified Nylon 6/6 *(continued)*

| Property | Test Methods ASTM | English System | Metric System | International System |
|---|---|---|---|---|
| **Thermal** | | | | |
| Coefficient of Expansion | D696 | 1.8 $10^{-5}$in/in°F | 3.2 $10^{-5}$cm/cm°C | 3.2 $10^{-5}$m/m°K |
| Heat Deflection Temp. | D648 | 480°F @ 264 psi | 448°C @ 18.5 kg/cm² | 721°K @ 1.82 MPa |
| Heat Deflection Temp. | D648 | 490°F @ 66 psi | 254°C @ 4.6 kg/cm² | 527°K @ 46 MPa |
| Max. Temp. Continuous Use | | 248°F | 120°C | 393°K |
| **Electrical** | | | | |
| Dielectric Strength (ST) | D149 | 500 V/mil | 18 kV/mm | 18 MV/m |
| Volume Resistivity | D257 | 15.00 Log 10 Ohm in²/in | 16.00 Log 10 Ohm cm²/cm | 14.00 Log 10 Ohm m²/m |
| Dielectric Constant @ $10^3$Hz | D150 | 3.40 | 3.40 | 3.40 |
| Dissipation Factor @ $10^3$Hz | D150 | .030 | .030 | .030 |
| Arc Resistance | D495 | 120 sec. | 120 sec. | 120 sec. |
| **Flammability** | | | | |
| UL 94 Class, 1/16″ | | HB | HB | HB |
| Oxygen Index | D2863 | <25 Vol% | <25 Vol% | <25 Vol% |

# Thermofil Inc.

## Contact Address:

Thermofil Inc.
P.O. Box 489
Brighton, MI 48116-0489

Phone: 1 (313) 227-3500
1 (800) 282-1482
Facsimile #: 1 (313) 227-3824

## THERMOFIL N7-40FG-0100

### Product Data:

**Impact Modified Type 6/6 Nylon**

| Property | Test Methods ASTM | English System | Metric System | International System |
|---|---|---|---|---|
| **General** | | | | |
| Reinforcement Content | D2584 | 40% | 40% | 40% |
| Specific Gravity/Density | D792 | 87.98 $lb/ft^3$ | 1.41 $g/cm^3$ | 1410 $kg/m^3$ |
| Mold Shrinkage | D955 | .003 in/in | .003 cm/cm | .003 mm/mm |
| Hardness (Test) | D785 | R-114 | R-114 | R-114 |
| Water Absorption, 24 hrs. | D570 | .70% | .70% | .70% |
| **Mechanical** | | | | |
| Tensile Strength | D638 | 22,000 psi | 1546.6 $kg/cm^2$ | 151.8 MPa |
| Tensile Elongation | D638 | 7% | 7% | 7% |
| Tensile Modulus | D638 | 13.4 psi $\times 10^5$ | 9.4 $kg/cm^2 \times 10^4$ | 9246.0 MPa |
| Flexural Strength | D790 | 33,000 psi | 2320 $kg/cm^2$ | 228 MPa |
| Flexural Modulus | D790 | 13.0 psi $\times 10^5$ | 9.1 $kg/cm^2 \times 10^4$ | 8970.0 MPa |
| Izod Impact (Notched) | D256 | 4.00 ft lb/in | 21.76 kg cm/cm | 213.2 J/m |
| Izod Impact (Unnotched) | D256 | 20.00 ft lb/in | 108.80 kg cm/cm | 1066.0 J/m |
| Compressive Strength | D695 | 0 psi | 0 $kg/cm^2$ | 0 MPa |

*(continued)*

NOTE – Information provided in this book has been condensed. The publisher recommends that the product manufacturer be contacted, at the address or phone number provided, for more complete information on product safety, processing parameters, and applications not provided in this information sheet.

## Impact Modified Type 6/6 Nylon *(continued)*

| Property | Test Methods ASTM | English System | Metric System | International System |
|---|---|---|---|---|
| **Thermal** | | | | |
| Coefficient of Expansion | D696 | 1.7 $10^{-5}$in/in°F | 3.1 $10^{-5}$cm/cm°C | 3.1 $10^{-5}$m/m°K |
| Heat Deflection Temp. | D648 | 480°F @ 264 psi | 448°C @ 18.5 kg/cm² | 721°K @ 1.82 MPa |
| Heat Deflection Temp. | D648 | 490°F @ 66 psi | 254°C @ 4.6 kg/cm² | 527°K @ 46 MPa |
| Max. Temp. Continuous Use | | 248°F | 120°C | 393°K |
| **Electrical** | | | | |
| Dielectric Strength (ST) | D149 | 500 V/mil | 18 kV/mm | 18 MV/m |
| Volume Resistivity | D257 | 14.00 Log 10 Ohm in²/in | 15.00 Log 10 Ohm cm²/cm | 13.00 Log 10 Ohm m²/m |
| Dielectric Constant @ $10^3$Hz | D150 | 3.40 | 3.40 | 3.40 |
| Dissipation Factor @ $10^3$Hz | D150 | .030 | .030 | .030 |
| Arc Resistance | D495 | 120 sec. | 120 sec. | 120 sec. |
| **Flammability** | | | | |
| UL 94 Class, 1/16″ | | | | |
| Oxygen Index | D2863 | Vol% | Vol% | Vol% |

# Thermofil Inc.

**Contact Address:**

Thermofil Inc.
P.O. Box 489
Brighton, MI 48116-0489

Phone: 1 (313) 227-3500
1 (800) 282-1482
Facsimile #: 1 (313) 227-3824

## THERMOFIL N7-20NF-0100

**Product Data:**

**20% Graphite Fiber Reinforced Nylon 6/6 (High Impact)**

| Property | Test Methods ASTM | English System | Metric System | International System |
|---|---|---|---|---|
| **General** | | | | |
| Reinforcement Content | D2584 | 20% | 20% | 20% |
| Specific Gravity/Density | D792 | 74.88 lb/ft$^3$ | 1.20 g/cm$^3$ | 1200 kg/m$_3$ |
| Mold Shrinkage | D955 | .001 in/in | .001 cm/cm | .001 mm/mm |
| Hardness (Test) | D785 | R-113 | R-113 | R-113 |
| Water Absorption, 24 hrs. | D570 | 1.10% | 1.10% | 1.10% |
| **Mechanical** | | | | |
| Tensile Strength | D638 | 19,000 psi | 1335.7 kg/cm$^2$ | 131.1 MPa |
| Tensile Elongation | D638 | 6% | 6% | 6% |
| Tensile Modulus | D638 | 16.0 psi $\times$ 10$^5$ | 11.2 kg/cm$^2$ $\times$ 10$^4$ | 11,040.0 MPa |
| Flexural Strength | D790 | 29,000 psi | 2039 kg/cm$^2$ | 200 MPa |
| Flexural Modulus | D790 | 14.5 psi $\times$ 10$^5$ | 10.2 kg/cm$^2$ $\times$ 10$^4$ | 10,005.0 MPa |
| Izod Impact (Notched) | D256 | 3.00 ft lb/in | 16.32 kg cm/cm | 159.9 J/m |
| Izod Impact (Unnotched) | D256 | 15.00 ft lb/in | 81.60 kg cm/cm | 799.5 J/m |
| Compressive Strength | D695 | 18,000 psi | 1265 kg/cm$^2$ | 124 MPa |

*(continued)*

NOTE – Information provided in this book has been condensed. The publisher recommends that the product manufacturer be contacted, at the address or phone number provided, for more complete information on product safety, processing parameters, and applications not provided in this information sheet.

**20% Graphite Fiber Reinforced Nylon 6/6 (High Impact) *(continued)***

| Property | Test Methods ASTM | English System | Metric System | International System |
|---|---|---|---|---|
| **Thermal** | | | | |
| Coefficient of Expansion | D696 | 1.4 $10^{-5}$in/in°F | 2.5 $10^{-5}$cm/cm°C | 2.5 $10^{-5}$m/m°K |
| Heat Deflection Temp. | D648 | 380°F @ 264 psi | 348°C @ 18.5 kg/cm² | 621°K @ 1.82 MPa |
| Heat Deflection Temp. | D648 | 425°F @ 66 psi | 218°C @ 4.6 kg/cm² | 491°K @ 46 MPa |
| Max. Temp. Continuous Use | | 248°F | 120°C | 393°K |
| **Electrical** | | | | |
| Dielectric Strength (ST) | D149 | 0 V/mil | 0 kV/mm | 0 MV/m |
| Volume Resistivity | D257 | 3.30 Log 10 Ohm in²/in | 4.30 Log 10 Ohm cm²/cm | 2.30 Log 10 Ohm m²/m |
| Dielectric Constant @ $10^3$Hz | D150 | | | |
| Dissipation Factor @ $10^3$Hz | D150 | | | |
| Arc Resistance | D495 | | | |
| **Flammability** | | | | |
| UL 94 Class, 1/16″ | | HB | HB | HB |
| Oxygen Index | D2863 | <24 Vol% | <24 Vol% | <24 Vol% |

# Thermofil Inc.

## Contact Address:

Thermofil Inc.
P.O. Box 489
Brighton, MI 48116-0489

Phone: 1 (313) 227-3500
1 (800) 282-1482
Facsimile #: 1 (313) 227-3824

## THERMOFIL N7-30NF-0100

### Product Data:

**30% Graphite Fiber Reinforced High Impact Nylon 6/6**

| Property | Test Methods ASTM | English System | Metric System | International System |
|---|---|---|---|---|
| **General** | | | | |
| Reinforcement Content | D2584 | 30% | 30% | 30% |
| Specific Gravity/Density | D792 | 77.38 $lb/ft^3$ | 1.24 $g/cm^3$ | 1240 $kg/m^3$ |
| Mold Shrinkage | D955 | .001 in/in | .001 cm/cm | .001 mm/mm |
| Hardness (Test) | D785 | R-114 | R-114 | R-114 |
| Water Absorption, 24 hrs. | D570 | 1.10% | 1.10% | 1.10% |
| **Mechanical** | | | | |
| Tensile Strength | D638 | 25,000 psi | 1757.5 $kg/cm^2$ | 172.5 MPa |
| Tensile Elongation | D638 | 4% | 4% | 4% |
| Tensile Modulus | D638 | 21.0 psi $\times 10^5$ | 14.8 $kg/cm^2 \times 10^4$ | 14,490.0 MPa |
| Flexural Strength | D790 | 35,000 psi | 2461 $kg/cm^2$ | 242 MPa |
| Flexural Modulus | D790 | 22.0 psi $\times 10^5$ | 15.5 $kg/cm^2 \times 10^4$ | 15,180.0 MPa |
| Izod Impact (Notched) | D256 | 3.20 ft lb/in | 17.41 kg cm/cm | 170.6 J/m |
| Izod Impact (Unnotched) | D256 | 16.00 ft lb/in | 87.04 kg cm/cm | 852.8 J/m |
| Compressive Strength | D695 | 25,000 psi | 1758 $kg/cm^2$ | 173 MPa |

*(continued)*

NOTE – Information provided in this book has been condensed. The publisher recommends that the product manufacturer be contacted, at the address or phone number provided, for more complete information on product safety, processing parameters, and applications not provided in this information sheet.

**30% Graphite Fiber Reinforced High Impact Nylon 6/6 *(continued)***

| Property | Test Methods ASTM | English System | Metric System | International System |
|---|---|---|---|---|
| **Thermal** | | | | |
| Coefficient of Expansion | D696 | 1.2 $10^{-5}$in/in°F | 2.2 $10^{-5}$cm/cm°C | 2.2 $10^{-5}$m/m°K |
| Heat Deflection Temp. | D648 | 470°F @ 264 psi | 438°C @ 18.5 kg/cm² | 711°K @ 1.82 MPa |
| Heat Deflection Temp. | D648 | 490°F @ 66 psi | 254°C @ 4.6 kg/cm² | 527°K @ 46 MPa |
| Max. Temp. Continuous Use | | 248°F | 120°C | 393°K |
| **Electrical** | | | | |
| Dielectric Strength (ST) | D149 | 0 V/mil | 0 kV/mm | 0 MV/m |
| Volume Resistivity | D257 | | 1.00 Log 10 Ohm cm²/cm | 1.00 Log 10 Ohm m²/m |
| Dielectric Constant @ $10^3$Hz | D150 | | | |
| Dissipation Factor @ $10^3$Hz | D150 | | | |
| Arc Resistance | D495 | | | |
| **Flammability** | | | | |
| UL 94 Class, 1/16″ | | HB | HB | HB |
| Oxygen Index | D2863 | Vol% | Vol% | Vol% |

# Thermofil Inc.

## Contact Address:

Thermofil Inc.
P.O. Box 489
Brighton, MI 48116-0489

Phone: 1 (313) 227-3500
1 (800) 282-1482
Facsimile #: 1 (313) 227-3824

## THERMOFIL N7-40NF-0100

### Product Data:

**40% Graphite Fiber Reinforced High Impact Nylon 6/6**

| Property | Test Methods ASTM | English System | Metric System | International System |
|---|---|---|---|---|
| **General** | | | | |
| Reinforcement Content | D2584 | 40% | 40% | 40% |
| Specific Gravity/Density | D792 | 80.50 lb/ft$^3$ | 1.29 g/cm$^3$ | 1290 kg/m$^3$ |
| Mold Shrinkage | D955 | .001 in/in | .001 cm/cm | .001 mm/mm |
| Hardness (Test) | D785 | R-114 | R-114 | R-114 |
| Water Absorption, 24 hrs. | D570 | 1.10% | 1.10% | 1.10% |
| **Mechanical** | | | | |
| Tensile Strength | D638 | 30,000 psi | 2109.0 kg/cm$^2$ | 207.0 MPa |
| Tensile Elongation | D638 | 3% | 3% | 3% |
| Tensile Modulus | D638 | 23.0 psi $\times$ 10$^5$ | 16.2 kg/cm$^2$ $\times$ 10$^4$ | 15,870.0 MPa |
| Flexural Strength | D790 | 39,000 psi | 2742 kg/cm$^2$ | 269 MPa |
| Flexural Modulus | D790 | 26.0 psi $\times$ 10$^5$ | 18.3 kg/cm$^2$ $\times$ 10$^4$ | 17,940.0 MPa |
| Izod Impact (Notched) | D256 | 3.50 ft lb/in | 19.04 kg cm/cm | 186.6 J/m |
| Izod Impact (Unnotched) | D256 | 16.50 ft lb/in | 89.76 kg cm/cm | 879.5 J/m |
| Compressive Strength | D695 | 30,000 psi | 2109 kg/cm$^2$ | 207 MPa |

*(continued)*

NOTE – Information provided in this book has been condensed. The publisher recommends that the product manufacturer be contacted, at the address or phone number provided, for more complete information on product safety, processing parameters, and applications not provided in this information sheet.

**40% Graphite Fiber Reinforced High Impact Nylon 6/6 *(continued)***

| Property | Test Methods ASTM | English System | Metric System | International System |
|---|---|---|---|---|
| **Thermal** | | | | |
| Coefficient of Expansion | D696 | 1.1 $10^{-5}$in/in°F | 2.0 $10^{-5}$cm/cm°C | 2.0 $10^{-5}$m/m°K |
| Heat Deflection Temp. | D648 | 480°F @ 264 psi | 448°C @ 18.5 kg/cm$^2$ | 721°K @ 1.82 MPa |
| Heat Deflection Temp. | D648 | 490°F @ 66 psi | 254°C @ 4.6 kg/cm$^2$ | 527°K @ 46 MPa |
| Max. Temp. Continuous Use | | 248°F | 120°C | 393°K |
| **Electrical** | | | | |
| Dielectric Strength (ST) | D149 | 0 V/mil | 0 kV/mm | 0 MV/m |
| Volume Resistivity | D257 | | 1.00 Log 10 Ohm cm$^2$/cm | 1.00 Log 10 Ohm m$^2$/m |
| Dielectric Constant @ $10^3$Hz | D150 | | | |
| Dissipation Factor @ $10^3$Hz | D150 | | | |
| Arc Resistance | D495 | | | |
| **Flammability** | | | | |
| UL 94 Class, 1/16″ | | HB | HB | HB |
| Oxygen Index | D2863 | Vol% | Vol% | Vol% |

# Thermofil Inc.

## Contact Address:

Thermofil Inc.
P.O. Box 489
Brighton, MI 48116-0489

Phone: 1 (313) 227-3500
1 (800) 282-1482
Facsimile #: 1 (313) 227-3824

## THERMOFIL R2-10FG-0100

### Product Data:

**10% Glass Fiber Reinforced P/C Alloy**

| Property | Test Methods ASTM | English System | Metric System | International System |
|---|---|---|---|---|
| **General** | | | | |
| Reinforcement Content | D2584 | 10% | 10% | 10% |
| Specific Gravity/Density | D792 | 81.74 lb/ft$^3$ | 1.31 g/cm$^3$ | 1310 kg/m$_3$ |
| Mold Shrinkage | D955 | .003 in/in | .003 cm/cm | .003 mm/mm |
| Hardness (Test) | D785 | R-118 | R-118 | R-118 |
| Water Absorption, 24 hrs. | D570 | .08% | .08% | .08% |
| **Mechanical** | | | | |
| Tensile Strength | D638 | 10,500 psi | 738.2 kg/cm$^2$ | 72.5 MPa |
| Tensile Elongation | D638 | 3% | 3% | 3% |
| Tensile Modulus | D638 | 4.8 psi $\times$ 10$^5$ | 3.4 kg/cm$^2$ $\times$ 10$^4$ | 3312.0 MPa |
| Flexural Strength | D790 | 19,000 psi | 1336 kg/cm$^2$ | 131 MPa |
| Flexural Modulus | D790 | 5.2 psi $\times$ 10$^5$ | 3.7 kg/cm$^2$ $\times$ 10$^4$ | 3588.0 MPa |
| Izod Impact (Notched) | D256 | 1.80 ft lb/in | 9.79 kg cm/cm | 95.9 J/m |
| Izod Impact (Unnotched) | D256 | 15.00 ft lb/in | 81.60 kg cm/cm | 799.5 J/m |
| Compressive Strength | D695 | 12,000 psi | 844 kg/cm$^2$ | 83 MPa |

*(continued)*

NOTE—Information provided in this book has been condensed. The publisher recommends that the product manufacturer be contacted, at the address or phone number provided, for more complete information on product safety, processing parameters, and applications not provided in this information sheet.

**10% Glass Fiber Reinforced P/C Alloy *(continued)***

| Property | Test Methods ASTM | English System | Metric System | International System |
|---|---|---|---|---|
| **Thermal** | | | | |
| Coefficient of Expansion | D696 | 1.8 $10^{-5}$in/in°F | 3.2 $10^{-5}$cm/cm°C | 3.2 $10^{-5}$m/m°K |
| Heat Deflection Temp. | D648 | 270°F @ 264 psi | 238°C @ 18.5 kg/cm$^2$ | 511°K @ 1.82 MPa |
| Heat Deflection Temp. | D648 | 280°F @ 66 psi | 138°C @ 4.6 kg/cm$^2$ | 416°K @ 46 MPa |
| Max. Temp. Continuous Use | | 240°F | 116°C | 389°K |
| **Electrical** | | | | |
| Dielectric Strength (ST) | D149 | 450 V/mil | 16 kV/mm | 16 MV/m |
| Volume Resistivity | D257 | 15.00 Log 10 Ohm in$^2$/in | 16.00 Log 10 Ohm cm$^2$/cm | 14.00 Log 10 Ohm m$^2$/m |
| Dielectric Constant @ $10^3$Hz | D150 | 3.20 | 3.20 | 3.20 |
| Dissipation Factor @ $10^3$Hz | D150 | .010 | .010 | .010 |
| Arc Resistance | D495 | 125 sec. | 125 sec. | 125 sec. |
| **Flammability** | | | | |
| UL 94 Class, 1/16″ | | HB | HB | HB |
| Oxygen Index | D2863 | <28 Vol% | <28 Vol% | <28 Vol% |

# Thermofil Inc.

## Contact Address:

Thermofil Inc.
P.O. Box 489
Brighton, MI 48116-0489

Phone: 1 (313) 227-3500
1 (800) 282-1482
Facsimile #: 1 (313) 227-3824

## THERMOFIL R2-20FG-0100

### Product Data:

**Glass Fiber Reinforced Polycarbonate Alloy**

| Property | Test Methods ASTM | English System | Metric System | International System |
|---|---|---|---|---|
| **General** | | | | |
| Reinforcement Content | D2584 | 20% | 20% | 20% |
| Specific Gravity/Density | D792 | 87.36 lb/ft$^3$ | 1.40 g/cm$^3$ | 1400 kg/m$^3$ |
| Mold Shrinkage | D955 | .003 in/in | .003 cm/cm | .003 mm/mm |
| Hardness (Test) | D785 | R-118 | R-118 | R-118 |
| Water Absorption, 24 hrs. | D570 | .09% | .09% | .09% |
| **Mechanical** | | | | |
| Tensile Strength | D638 | 14,000 psi | 984.2 kg/cm$^2$ | 96.6 MPa |
| Tensile Elongation | D638 | 10% | 10% | 10% |
| Tensile Modulus | D638 | 8.0 psi $\times$ 10$^5$ | 5.6 kg/cm$^2$ $\times$ 10$^4$ | 5520.0 MPa |
| Flexural Strength | D790 | 20,000 psi | 1406 kg/cm$^2$ | 138 MPa |
| Flexural Modulus | D790 | 8.2 psi $\times$ 10$^5$ | 5.8 kg/cm$^2$ $\times$ 10$^4$ | 5658.0 MPa |
| Izod Impact (Notched) | D256 | 3.00 ft lb/in | 16.32 kg cm/cm | 159.9 J/m |
| Izod Impact (Unnotched) | D256 | 16.00 ft lb/in | 87.04 kg cm/cm | 852.8 J/m |
| Compressive Strength | D695 | 17,000 psi | 1195 kg/cm$^2$ | 117 MPa |

*(continued)*

NOTE – Information provided in this book has been condensed. The publisher recommends that the product manufacturer be contacted, at the address or phone number provided, for more complete information on product safety, processing parameters, and applications not provided in this information sheet.

### Glass Fiber Reinforced Polycarbonate Alloy *(continued)*

| Property | Test Methods ASTM | English System | Metric System | International System |
|---|---|---|---|---|
| **Thermal** | | | | |
| Coefficient of Expansion | D696 | 1.0 $10^{-5}$in/in°F | 1.8 $10^{-5}$cm/cm°C | 1.8 $10^{-5}$m/m°K |
| Heat Deflection Temp. | D648 | 270°F @ 264 psi | 238°C @ 18.5 kg/cm$^2$ | 527°K @ 46 MPa |
| Heat Deflection Temp. | D648 | 300°F @ 66 psi | 149°C @ 4.6 kg/cm$^2$ | 422°K @ 46 MPa |
| Max. Temp. Continuous Use | | 240°F | 116°C | 389°K |
| **Electrical** | | | | |
| Dielectric Strength (ST) | D149 | 450 V/mil | 16 kV/mm | 16 MV/m |
| Volume Resistivity | D257 | 15.00 Log 10 Ohm in$^2$/in | 16.00 Log 10 Ohm cm$^2$/cm | 14.00 Log 10 Ohm m$^2$/m |
| Dielectric Constant @ $10^3$Hz | D150 | 3.20 | 3.20 | 3.20 |
| Dissipation Factor @ $10^3$Hz | D150 | .010 | .010 | .010 |
| Arc Resistance | D495 | 125 sec. | 125 sec. | 125 sec. |
| **Flammability** | | | | |
| UL 94 Class, 1/16″ | | HB | HB | HB |
| Oxygen Index | D2863 | <25 Vol% | <25 Vol% | <25 Vol% |

# Thermofil Inc.

## Contact Address:

Thermofil Inc.
P.O. Box 489
Brighton, MI 48116-0489

Phone: 1 (313) 227-3500
1 (800) 282-1482
Facsimile #: 1 (313) 227-3824

## THERMOFIL R2-30FG-0100

### Product Data:

**30% Glass Fiber Reinforced P/C Alloy**

| Property | Test Methods ASTM | English System | Metric System | International System |
|---|---|---|---|---|
| **General** | | | | |
| Reinforcement Content | D2584 | 30% | 30% | 30% |
| Specific Gravity/Density | D792 | 92.35 $lb/ft^3$ | 1.48 $g/cm^3$ | 1480 $kg/m^3$ |
| Mold Shrinkage | D955 | .001 in/in | .001 cm/cm | .001 mm/mm |
| Hardness (Test) | D785 | R-118 | R-118 | R-118 |
| Water Absorption, 24 hrs. | D570 | .10% | .10% | .10% |
| **Mechanical** | | | | |
| Tensile Strength | D638 | 15,000 psi | 1054.5 $kg/cm^2$ | 103.5 MPa |
| Tensile Elongation | D638 | 2% | 2% | 2% |
| Tensile Modulus | D638 | 11.0 psi $\times$ $10^5$ | 7.7 $kg/cm^2$ $\times$ $10^4$ | 7590.0 MPa |
| Flexural Strength | D790 | 24,000 psi | 1687 $kg/cm^2$ | 166 MPa |
| Flexural Modulus | D790 | 10.0 psi $\times$ $10^5$ | 7.0 $kg/cm^2$ $\times$ $10^4$ | 6900.0 MPa |
| Izod Impact (Notched) | D256 | 3.00 ft lb/in | 16.32 kg cm/cm | 159.9 J/m |
| Izod Impact (Unnotched) | D256 | 15.00 ft lb/in | 81.60 kg cm/cm | 799.5 J/m |
| Compressive Strength | D695 | 14,500 psi | 1019 $kg/cm^2$ | 100 MPa |

*(continued)*

NOTE – Information provided in this book has been condensed. The publisher recommends that the product manufacturer be contacted, at the address or phone number provided, for more complete information on product safety, processing parameters, and applications not provided in this information sheet.

## 30% Glass Fiber Reinforced P/C Alloy *(continued)*

| Property | Test Methods ASTM | English System | Metric System | International System |
|---|---|---|---|---|
| **Thermal** | | | | |
| Coefficient of Expansion | D696 | 1.3 $10^{-5}$in/in °F | 2.3 $10^{-5}$cm/cm °C | 2.3 $10^{-5}$m/m °K |
| Heat Deflection Temp. | D648 | 270 °F @ 264 psi | 238 °C @ 18.5 kg/cm² | 511 °K @ 1.82 MPa |
| Heat Deflection Temp. | D648 | 280 °F @ 66 psi | 138 °C @ 4.6 kg/cm² | 411 °K @ 46 MPa |
| Max. Temp. Continuous Use | | 240 °F | 116 °C | 389 °K |
| **Electrical** | | | | |
| Dielectric Strength (ST) | D149 | 400 V/mil | 14 kV/mm | 14 MV/m |
| Volume Resistivity | D257 | 15.00 Log 10 Ohm in²/in | 16.00 Log 10 Ohm cm²/cm | 14.00 Log 10 Ohm m²/m |
| Dielectric Constant @ $10^3$Hz | D150 | 3.30 | 3.30 | 3.30 |
| Dissipation Factor @ $10^3$Hz | D150 | .009 | .009 | .009 |
| Arc Resistance | D495 | 125 sec. | 125 sec. | 125 sec. |
| **Flammability** | | | | |
| UL 94 Class, 1/16″ | | HB | HB | HB |
| Oxygen Index | D2863 | <28 Vol% | <28 Vol% | <28 Vol% |

# Thermofil Inc.

## Contact Address:

Thermofil Inc.
P.O. Box 489
Brighton, MI 48116-0489

Phone: 1 (313) 227-3500
1 (800) 282-1482
Facsimile #: 1 (313) 227-3824

## THERMOFIL R2-40FG-0100

### Product Data:

**Glass Fiber Reinforced Polycarbonate Alloy**

| Property | Test Methods ASTM | English System | Metric System | International System |
|---|---|---|---|---|
| **General** | | | | |
| Reinforcement Content | D2584 | 40% | 40% | 40% |
| Specific Gravity/Density | D792 | 98.59 lb/ft$^3$ | 1.58 g/cm$^3$ | 1580 kg/m$^3$ |
| Mold Shrinkage | D955 | .002 in/in | .002 cm/cm | .002 mm/mm |
| Hardness (Test) | D785 | R-118 | R-118 | R-118 |
| Water Absorption, 24 hrs. | D570 | .09% | .09% | .09% |
| **Mechanical** | | | | |
| Tensile Strength | D638 | 19,700 psi | 1384.9 kg/cm$^2$ | 135.9 MPa |
| Tensile Elongation | D638 | 6% | 6% | 6% |
| Tensile Modulus | D638 | 14.5 psi × 10$^5$ | 10.2 kg/cm$^2$ × 10$^4$ | 10,005.0 MPa |
| Flexural Strength | D790 | 32,600 psi | 2292 kg/cm$^2$ | 225 MPa |
| Flexural Modulus | D790 | 14.3 psi × 10$^5$ | 10.1 kg/cm$^2$ × 10$^4$ | 9867.0 MPa |
| Izod Impact (Notched) | D256 | 2.00 ft lb/in | 10.88 kg cm/cm | 106.6 J/m |
| Izod Impact (Unnotched) | D256 | 13.50 ft lb/in | 73.44 kg cm/cm | 719.6 J/m |
| Compressive Strength | D695 | 21,000 psi | 1476 kg/cm$^2$ | 145 MPa |

*(continued)*

NOTE—Information provided in this book has been condensed. The publisher recommends that the product manufacturer be contacted, at the address or phone number provided, for more complete information on product safety, processing parameters, and applications not provided in this information sheet.

**Glass Fiber Reinforced Polycarbonate Alloy *(continued)***

| Property | Test Methods ASTM | English System | Metric System | International System |
|---|---|---|---|---|
| **Thermal** | | | | |
| Coefficient of Expansion | D696 | 1.0 $10^{-5}$in/in°F | 1.8 $10^{-5}$cm/cm°C | 1.8 $10^{-5}$m/m°K |
| Heat Deflection Temp. | D648 | 280°F @ 264 psi | 248°C @ 18.5 kg/cm$^2$ | 521°K @ 1.82 MPa |
| Heat Deflection Temp. | D648 | 300°F @ 66 psi | 149°C @ 4.6 kg/cm$^2$ | 422°K @ 46 MPa |
| Max. Temp. Continuous Use | | 240°F | 116°C | 389°K |
| **Electrical** | | | | |
| Dielectric Strength (ST) | D149 | 450 V/mil | 16 kV/mm | 16 MV/m |
| Volume Resistivity | D257 | 15.00 Log 10 Ohm in$^2$/in | 16.00 Log 10 Ohm cm$^2$/cm | 14.00 Log 10 Ohm m$^2$/m |
| Dielectric Constant @ $10^3$Hz | D150 | 3.00 | 3.00 | 3.00 |
| Dissipation Factor @ $10^3$Hz | D150 | .010 | .010 | .010 |
| Arc Resistance | D495 | 125 sec. | 125 sec. | 125 sec. |
| **Flammability** | | | | |
| UL 94 Class, 1/16″ | | HB | HB | HB |
| Oxygen Index | D2863 | <25 Vol% | <25 Vol% | <25 Vol% |

# Uniroyal Chemical Company Inc.

Contact Address:

Uniroyal Chemical Company, Inc.
Middlebury, CT 06749

Phone: 1 (800) 243-5098—Technical Service
Phone: 1 (800) 243-3024—Customer Service

## ROYALTHERM

### Description:

Uniroyal Chemical Company, Inc. is a worldwide producer of specialty chemicals and polymers. A pioneer in the development of ethylene-propylene elastomers, Uniroyal Chemical is one of the largest suppliers of chemicals and polymers to the rubber and plastics industries.

ROYALTHERM elastomers are silicone-modified ethylene-propylene (EPDM) elastomers. Available in four basic types that can be blended and vulcanized with different curatives, ROYALTHERM elastomers offer a wide range of properties targeted to specific end uses. Superior heat resistance, excellent electrical properties, and outstanding physical properties over a wide range of temperatures and moisture conditions make ROYALTHERM elastomers excellent candidates for radiator and steam hose, gaskets and seals, spark plug boots, cable insulation and jacketing, and sponge weatherstripping.

### Processing Information:

ROYALTHERM elastomers are blended with selected vulcanizers and accelerators, and formulated to specific processes and product end uses. Equipment used to process EPDM or silicone can be used.

**Mill Mixing**

| | | |
|---|---|---|
| Heat both mill rolls | 175°F–190°F | ROYALTHERM will band on front roll after initial breakdown on a tighter nip setting. Add other ingredients, such as fillers, antioxidants, coagents and curatives. |
| Front roll | 175°F–180°F | |
| Back roll | 185°F–190°F | |

**Banbury Mixing**

Batch size important; calculate to a full or slightly oversized batch.

Cooling water on full; control speed for temperature not to exceed 265°F–270°F.

| | |
|---|---|
| 0″ | Add ROYALTHERM and other ingredients except curatives. |
| 2″ | Brush down. If temperature is below 240°F–245°F, add cure system. |
| 4″–5″ | Dump. Temperature of batch at discharge should be 240°F for best results. If batch temperature on discharge exceeds 250°F, then add cure system on mill. |

**Calendering Conditions**

Compound feed should be at least 140°F–160°F for uniform plasticity and to maintain a uniform bank on the calender.

Top roll—Hot (180°F–200°F); Center roll—Hot (170°F–190°F); Bottom roll—Warm (140°F–160°F).

NOTE—Information provided in this book has been condensed. The publisher recommends that the product manufacturer be contacted, at the address or phone number provided, for more complete information on product safety, processing parameters, and applications not provided in this information sheet.

### Extrusion Processing

It is desirable to warm compound prior to feeding extruder. However, this may not be necessary on high L/D ratio extruders.

Barrel—Warm (130°F–150°F); Screw—Cooling water usually ON; Head—180°F–230°F; Die—230°F–270°F.

Screw speed is more important than temperature for extrusion rate of ROYALTHERM compounds. However, temperature conditions on the extruder are important to produce a smooth extrudate.

**Typical Physical Properties:**

**ROYALTHERM Elastomers**

| Properties | 1411 | 1421 | 1711 | 1721 |
|---|---|---|---|---|
| **Typical** | | | | |
| Color | ----BEIGE---- | | | |
| Specific Gravity @ 25° | 1.00 | 1.03 | 1.13 | 1.15 |
| Mooney Viscosity, ML 1+4 @ 100°C | 35 | 30 | 80 | 85 |
| **Formulation, Parts by Weight** | | | | |
| ROYALTHERM | 100 | 100 | 100 | 100 |
| T-Butyl Cumyl Peroxide (100% active) | 1.4 | 1.4 | 1.4 | 1.4 |
| **Press Cured 10 min. @ 170°C** | | | | |
| Tensile Strength, MPa | 7.8 | 6.6 | 17.9 | 10.8 |
| Elongation, % | 955 | 880 | 690 | 683 |
| Hardness, Shore A Durometer | 40 | 40 | 66 | 65 |
| Linear Shrinkage, % | 2.8 | 2.8 | 2.0 | 2.1 |
| Resiliency, % | 65 | 58 | 47 | 48 |
| Compression set, 22 hrs. @ 100°C | 25 | 28 | 33 | 30 |

*(continued)*

ROYALTHERM Elastomers

| Properties | 1411 | 1421 | 1711 | 1721 |
|---|---|---|---|---|
| **Post Cured 2 hrs. @ 150°C** | | | | |
| 100% Modulus, MPa | 1.1 | 1.9 | 2.2 | 2.2 |
| 300% Modulus, MPa | 2.2 | 1.4 | 6.9 | 7.3 |
| Tensile Strength, MPa | 8.6 | 6.9 | 18.6 | 11.9 |
| Elongation, % | 910 | 890 | 652 | 633 |
| Hardness, Shore A Durometer | 41 | 42 | 68 | 68 |
| Linear Shrinkage, % | 3.5 | 3.5 | 2.5 | 2.7 |
| Resiliency, % | 62 | 62 | 50 | 51 |
| Compression Set | | | | |
| 22 hrs @ 150°C | 31 | 37 | 38 | 35 |
| 22 hrs @ 100°C | 18 | 22 | 22 | 19 |
| **After Heat Aging, 72 hrs @ 200°C** | | | | |
| Tensile Strength, MPa | | 5.3 | 12.4 | 11.7 |
| Elongation, % | | 445 | 339 | 342 |
| Hardness, Shore A Durometer | | 50 | 77 | 77 |

# Wilson-Fiberfil International

## Contact Address:

Wilson-Fiberfil International
P.O. Box 3333
2267 West Mill Road
Evansville, IN 47732

Phone: 1 (800) 457-3764

## NYLAFIL

### Product Data:

The J-7/XX series of compounds are an extension of our new Fiberfil TN Toughened Nylons. These products exhibit notched and unnotched Izod impact values not formerly achievable in a resin system with modulus and heat distortion values in the ranges shown below. The combination of strength, stiffness, toughness, chemical and thermal resistance make the J-7 series compounds an excellent choice for tool housings and other industrial components.

Advantages of the J-7/XX series include lower cost per cubic inch, improved material flow than our J-8 Toughened 6/6 Nylon.

**Glass Reinforced Fiberfil® TN Toughened Type 6 Nylon Fiberfil "J-7/XX" Series**

| Properties | Unit | ASTM Test | NY-7 | J-7/13 | J-7/33 | J-7/43 |
|---|---|---|---|---|---|---|
| **Mechanical** | | | | | | |
| Tensile Strength | psi | D638 | 6500 | 12,000 | 17,800 | 20,900 |
| Elongation | % | D638 | 65 | 5.0 | 4.0 | 4.0 |
| Flexural Strength | psi | D790 | 9100 | 18,500 | 25,800 | 31,000 |
| Flexural Modulus | psi × $10^5$ | D790 | 2.5 | 5.6 | 11.1 | 14.5 |
| Izod Impact Strength | | | | | | |
| Notched (1/4″ bars) | ft-lb/in | D256 | 16.4 | 2.5 | 3.5 | 3.8 |
| Unnotched (1/4″ bars) | ft-lb/in | D256 | NB* | 15 | 16.5 | 20.0 |

*NB = 1/4″ thick bar, did not break with 8 pound hammer.

*(continued)*

NOTE – Information provided in this book has been condensed. The publisher recommends that the product manufacturer be contacted, at the address or phone number provided, for more complete information on product safety, processing parameters, and applications not provided in this information sheet.

**Glass Reinforced Fiberfil® TN Toughened Type 6 Nylon Fiberfil "J-7/XX" Series *(continued)***

| Properties | Unit | ASTM Test | NY-7 | J-7/13 | J-7/33 | J-7/43 |
|---|---|---|---|---|---|---|
| **Physical** | | | | | | |
| Specific Gravity | | D792 | 1.07 | 1.19 | 1.33 | 1.43 |
| Water Absorption, 24 hours | % | D570 | | 1.15 | .86 | .71 |
| Linear Mold Shrinkage | | | | | | |
| 1/8″ | in/in | | .0170 | .0036 | .001 | .0007 |
| 1/4″ | in/in | | .0204 | .0054 | .0028 | .0022 |
| **Thermal** | | | | | | |
| Deflection Temp. Under Load | | | | | | |
| @ 264 psi | °F | D648 | 130 | 390 | 398 | 407 |
| @ 66 psi | °F | D648 | | 424 | 430 | 430 |
| **Flammability** | | | | | | |
| UL Rating, 1/32″ | UL94 | HB | HB | HB | HB | HB |

# Wilson-Fiberfil International

## Contact Address:

Wilson-Fiberfil International
P.O. Box 3333
2267 West Mill Road
Evansville, IN 47732

Phone: 1 (800) 457-3764

## NYLAFIL

### Product Data:

NY-8 unreinforced nylon and J-8 reinforced compounds are a new family of nylons exhibiting notched and unnotched Izod values not formerly achievable in a nylon 6/6. The combination of strength, stiffness, toughness, chemical and thermal resistance, make the Fiberfil TN compounds an excellent choice for products which require impact resistance.

**Fiberfil® TN Toughened Type 6/6 Nylon**

| Properties | Unit | ASTM Test | J-8/13 | J-8/33 |
|---|---|---|---|---|
| **Mechanical** | | | | |
| Tensile Strength | psi | D638 | 12,700 | 20,300 |
| Elongation | % | D638 | 5–6 | 4–6 |
| Flexural Strength | psi | D790 | 19,000 | 29,900 |
| Flexural Modulus | psi × $10^5$ | D790 | 5.5 | 11.0 |
| Izod Impact Strength @ 73° | | | | |
| Notched (1/4″ bars) | ft-lb/in | D256 | 2.9 | 4.5 |
| Unnotched (1/4″ bars) | ft-lb/in | D256 | 19 | 28 |
| Gardner Impact @ 73° | | | 30 | 17 |
| Izod Impact Strength @ 40° | | | | |
| Notched (1/4″ bars) | ft-lb/in | D256 | 2.0 | 3.1 |
| Unnotched (1/4″ bars) | ft-lb/in | D256 | 19 | 30 |
| Gardner Impact @ 40° | | | 22 | 19 |

*(continued)*

NOTE – Information provided in this book has been condensed. The publisher recommends that the product manufacturer be contacted, at the address or phone number provided, for more complete information on product safety, processing parameters, and applications not provided in this information sheet.

**Fiberfil® TN Toughened Type 6/6 Nylon *(continued)***

| Properties | Unit | ASTM Test | J-8/13 | J-8/33 |
|---|---|---|---|---|
| **Physical** | | | | |
| Rockwell Hardness | | D785 | R-105 | R-107 |
| Water Absorption, 24 hours | % | D570 | 0.9 | 0.7 |
| Linear Mold Shrinkage | | | | |
| 1/8″ Avg. Section | in/in | | .005 | .0025 |
| 1/4″ Avg. Section | in/in | | .008 | .0045 |
| Specific Gravity | | D792 | 1.19 | 1.34 |
| **Thermal** | | | | |
| Deflection Temp. Under Load | | | | |
| @ 264 psi | °F | D648 | 450 | 470 |
| @ 66 psi | °F | D648 | | 495 |
| **Flammability** | | | | |
| UL Rating, 1/16″ | | | | |

All Mechanical tests conducted at 73° unless noted otherwise.

*NB = 32 ft-lb/in.

# Wilson-Fiberfil International

Contact Address:

Wilson-Fiberfil International
P.O. Box 3333
2267 West Mill Road
Evansville, IN 47732

Phone: 1 (800) 457-3764

## NYLAFIL

Product Data:

NY-8 unreinforced nylon and J-8 reinforced compounds are a new family of nylons exhibiting notched and unnotched Izod values not formerly achievable in a nylon 6/6. The combination of strength, stiffness, toughness, chemical and thermal resistance, make the Fiberfil TN compounds an excellent choice for products which require impact resistance.

**Fiberfil® TN Toughened 6/6 Nylon**

| Properties | Unit | ASTM Test | J-8/43 | NY-8 | NY-8/ MF/36 |
|---|---|---|---|---|---|
| **Mechanical** | | | | | |
| Tensile Strength | psi | D638 | 23,000 | 7000 | 10,300 |
| Elongation | % | D638 | 4–5 | 125 | 26 |
| Flexural Strength | psi | D790 | 33,000 | 8500 | 16,700 |
| Flexural Modulus | psi × $10^5$ | D790 | 13.5 | 2.4 | 5.6 |
| Izod Impact Strength @ 73° | | | | | |
| Notched (1/4″ bars) | ft-lb/in | D256 | 4.5 | 17.0 | 2.1 |
| Unnotched (1/4″ bars) | ft-lb/in | D256 | 25 | NB* | |
| Gardner Impact @ 73° | | | | >320 | |
| Izod Impact Strength @ 40° | | | | | |
| Notched (1/4″ bars) | ft-lb/in | D256 | | 5.3 | |
| Unnotched (1/4″ bars) | ft-lb/in | D256 | | >32 | |
| Gardner Impact @ 40° | | | | >320 | |

*NB = 32 ft-lb/in.

(continued)

NOTE—Information provided in this book has been condensed. The publisher recommends that the product manufacturer be contacted, at the address or phone number provided, for more complete information on product safety, processing parameters, and applications not provided in this information sheet.

**Fiberfil® TN Toughened 6/6 Nylon *(continued)***

| Properties | Unit | ASTM Test | J-8/43 | NY-8 | NY-8/ MF/36 |
|---|---|---|---|---|---|
| **Physical** | | | | | |
| Rockwell Hardness | | D785 | R-111 | R-100 | |
| Water Absorption, 24 hours | % | D570 | 0.6 | 1.0 | 0.7 |
| Linear Mold Shrinkage | | | | | |
| 1/8″ Avg. Section | in/in | | .0015 | 0.025 | 0.017 |
| 1/4″ Avg. Section | in/in | | .0035 | 0.040 | 0.022 |
| Specific Gravity | | D792 | 1.42 | 1.08 | 1.41 |
| **Thermal** | | | | | |
| Deflection Temp. Under Load | | | | | |
| @ 264 psi | °F | D648 | 475 | 150 | 262 |
| @ 66 psi | °F | D648 | | | |
| **Flammability** | | | | | |
| UL Rating, 1/16″ | | | | | |

All Mechanical tests conducted at 73° unless noted otherwise.

# Wilson-Fiberfil International

## Contact Address:

Wilson-Fiberfil International
P.O. Box 3333
2267 West Mill Road
Evansville, IN 47732

Phone: 1 (800) 457-3764

## NYLAFIL®/FR

## Product Data:

**Fiberglass Reinforced Nylon With Flame Retardant Additive**

| Properties | Unit | ASTM Test | J-1/15/FR | J-1/20/FR | J-1/30/FR |
|---|---|---|---|---|---|
| **Mechanical** | | | | | |
| Tensile Strength @ 73°F | psi | D638 | 11,000 | 12,500 | 16,000 |
| Elongation @ 73°F | % | D638 | 2.6 | 2.3 | 2.1 |
| Flexural Strength @ 73°F | psi | D790 | 19,000 | 23,000 | 27,000 |
| Izod Impact Strength @ 73° (1/4″ bars) | ft-lb/in | D256 | 1.0 | 1.1 | 1.6 |
| **Physical** | | | | | |
| Water Absorption, 24 hours | % | D570 | 0.7 | 0.7 | 0.6 |
| Linear Mold Shrinkage 1/8″ to 1/4″ Avg. Section | in/in | | .005–.006 | .004–.005 | .003–.006 |
| Specific Gravity | | D792 | 1.47 | 1.51 | 1.60 |
| UL Subject 94 Vertical Test (1/16″) | | | V-0 | V-0 | V-0 |
| UL Subject 94 Vertical Test (1/32″) | | | V-0 | V-0 | V-0 |
| **Thermal** | | | | | |
| Deflection Temp. Under Load @ 264 psi | °F | D648 | 406 | 412 | 420 |

*(continued)*

NOTE – Information provided in this book has been condensed. The publisher recommends that the product manufacturer be contacted, at the address or phone number provided, for more complete information on product safety, processing parameters, and applications not provided in this information sheet.

## Fiberglass Reinforced Nylon With Flame Retardant Additive *(continued)*

| Properties | Unit | ASTM Test | J-1/15/FR | J-1/20/FR | J-1/30/FR |
|---|---|---|---|---|---|
| **Electrical*** | | | | | |
| Dielectric Strength | volts/mil | D149 | | | |
| Short Time | | | | 430 | 400 |
| Step by Step | | | | 400 | 380 |
| Dielectric Constant | | D150 | | | |
| 60 Hz | | | | 3.95 | 4.58 |
| $10^3$ Hz | | | | 3.78 | 4.26 |
| $10^6$ Hz | | | | 3.39 | 3.72 |
| Dissipation Factor | | D150 | | | |
| 60 Hz | | | | .029 | .053 |
| $10^3$ Hz | | | | .0252 | .0375 |
| $10^6$ Hz | | | | .0178 | .0198 |
| Volume Resistivity | ohm/cm | D257 | | $8.1 \times 10^{14}$ | |
| Surface Resistivity | ohm/sq | D257 | | $1.73 \times 10^{13}$ | |
| Arc Resistance | sec | D495 | | 123 | |

*All samples conditioned for 48 hrs. @ 50% RH prior to testing.

# Wilson-Fiberfil International

## Contact Address:

Wilson-Fiberfil International
P.O. Box 3333
2267 West Mill Road
Evansville, IN 47732

Phone: 1 (800) 457-3764

## NYLAFIL/VO

### Product Data:

NY-7/VO is a toughened flame retardant nylon composition developed to meet UL 94V-O flammability requirements in thin sections. Based upon Wilson-Fiberfil International's toughening technology, it offers the highest impact resistance of any commercially available flame retardant nylon.

**NY-7/VO**

| Properties | Unit | ASTM Test | NY-7/VO |
|---|---|---|---|
| **General** | | | |
| Specific Gravity | | D792 | 1.26 |
| Water Absorption, 24 Hours | % | D570 | 0.55 |
| Linear Mold Shrinkage 1/8″ Section | in/in | | 0.014 |
| **Mechanical** | | | |
| Tensile Strength | psi | D638 | 5300 |
| Elongation | % | D638 | 40 |
| Flexural Strength | psi | D638 | 7800 |
| Flexural Modulus | psi | D790 | 220,000 |
| Izod Impact Strength 1/8″ Section | ft-lb/in | D256 | 10 |
| Unnotched Impact Strength 1/8″ Section | ft-lb/in | D256 | >32 |
| Gardner Impact Strength | in-lb | D1709 | >320 |
| Rockwell Hardness | | D785 | R97 |

*(continued)*

NOTE – Information provided in this book has been condensed. The publisher recommends that the product manufacturer be contacted, at the address or phone number provided, for more complete information on product safety, processing parameters, and applications not provided in this information sheet.

NY-7/VO *(continued)*

| Properties | Unit | ASTM Test | NY-7/VO |
|---|---|---|---|
| **Thermal** | | | |
| DTUL @ 264 psi | °F | D648 | 130 |
| DTUL @ 66 psi | °F | D648 | 330 |
| **Flammability** | | | |
| UL Subject 94, 1/16″ Section | | | 94V-O |

# Wilson-Fiberfil International

Contact Address:

Wilson-Fiberfil International
P.O. Box 3333
2267 West Mill Road
Evansville, IN 47732

Phone: 1 (800) 457-3764

## PLASTALLOY®

Product Data:

**Fiberglass Reinforced Polysulfone Alloy**

| Properties | Unit | ASTM Test | J-1505/10 | J-1505/20 | J-1505/30 | J-1505/40 |
|---|---|---|---|---|---|---|
| **Mechanical** | | | | | | |
| Tensile Strength @ 73°F | psi | D638 | 14,200 | 17,000 | 19,000 | 21,000 |
| Elongation @ 73°F | % | D638 | 3.8 | 3.5 | 3.0 | 2.5 |
| Modulus of Elasticity @ 73°F | psi × $10^5$ | D638 | 6.0 | 7.2 | 8.3 | 9.5 |
| Flexural Strength @ 73°F | psi | D790 | 21,500 | 24,000 | 25,500 | 26,500 |
| Flexural Modulus @ 73°F | psi × $10^5$ | D790 | 6.4 | 9.2 | 12.4 | 15.5 |
| Izod Impact Strength @ 73°F (1/4″) | ft-lb/in | D256 | 1.0 | 1.6 | 2.0 | 2.0 |
| **Physical** | | | | | | |
| Water Absorption, 24 hours | % | D570 | 0.3 | 0.25 | 0.20 | 0.15 |
| Linear Mold Shrinkage | | | | | | |
| 1/8″ Average Section | in/in | | 0.004 | 0.002 | 0.001 | 0.001 |
| 1/4″ Average Section | in/in | | 0.004 | 0.002 | 0.002 | 0.001 |
| 1/2″ Average Section | in/in | | 0.006 | 0.005 | 0.003 | 0.002 |
| Specific Gravity | | D792 | 1.31 | 1.38 | 1.47 | 1.56 |
| **Thermal** | | | | | | |
| Deflection Temp. Under Load @ 264 psi | °F | D648 | 315 | 320 | 320 | 320 |
| Flammability @ 1/16″ | * | UL94 | V-1 | V-1 | V-1 | V-1 |

*Values obtained using UL test procedures at Fiberfil. Information relating to Temperature Index data available on request.

NOTE – Information provided in this book has been condensed. The publisher recommends that the product manufacturer be contacted, at the address or phone number provided, for more complete information on product safety, processing parameters, and applications not provided in this information sheet.

# Wilson-Fiberfil International

Contact Address:

Wilson-Fiberfil International
P.O. Box 3333
2267 West Mill Road
Evansville, IN 47732

Phone: 1 (800) 457-3764

## PROFIL

Product Data:

**Extrusion Grade Calcium Carbonate Reinforced Polypropylene**

| Properties | Unit | ASTM Test | PF-6000 | Machine Direction | Transverse Direction |
|---|---|---|---|---|---|
| **Mechanical** | | | | | |
| Tensile Strength | psi | D638 | 2200 | 1600 | 1400 |
| Elongation | % | D638 | 200 | >110 | <260 |
| Tensile Modulus | psi × $10^5$ | D638 | 2.0 | | |
| Flexural Strength | psi | D790 | 2800 | 2500 | 3000 |
| Flexural Modulus | psi × $10^5$ | D790 | 1.75 | 1.33 | 1.76 |
| Izod Impact Strength (1/4″ bars) | ft-lb/in | D256 | non-breaking | | |
| Gardner Impact Strength | in/lb | | | | |
| @ 73°F | | | non-breaking | | non-breaking |
| @ −20°F | | | non-breaking | | |
| @ −40°F | | | 190 | | |
| **Physical** | | | | | |
| Water Absorption in 24 hours | % | D570 | 0.05 | | |
| Specific Gravity | | D792 | 1.23 | | 1.24 |
| **Thermal** | | | | | |
| Deflection Temp. Under Load | °F | D648 | | | |
| @ 264 psi | | | 135 | | |
| @ 66 psi | | | 200 | | |
| Melt Flow, Condition L | GM/10 min | | <1 | | |

*(continued)*

NOTE–Information provided in this book has been condensed. The publisher recommends that the product manufacturer be contacted, at the address or phone number provided, for more complete information on product safety, processing parameters, and applications not provided in this information sheet.

# Wilson-Fiberfil International

## Contact Address:

Wilson-Fiberfil International
P.O. Box 3333
2267 West Mill Road
Evansville, IN 47732

Phone: 1 (800) 457-3764

## PROFIL

### Product Data:

**Extrusion Grade Calcium Carbonate Reinforced Polypropylene**

| Properties | Unit | ASTM Test | PF-6020 | PF-6010[1] Impact Modified |
|---|---|---|---|---|
| **Mechanical** | | | | |
| Tensile Strength | psi | D638 | 3500 | 3100 |
| Elongation | % | D638 | >50 | >50 |
| Tensile Modulus | psi × $10^5$ | D638 | 4.2 | 3.5 |
| Flexural Strength | psi | D790 | 5000 | 4100 |
| Flexural Modulus | psi × $10^5$ | D790 | 3.5 | 3.0 |
| Izod Impact Strength (1/4″ bars) | ft-lb/in | D256 | 1.0 | non-breaking |
| Hardness, Rockwell | | D785 | R-88 | R-82 |
| Gardner Impact Strength | in/lb | | | |
| @ 73°F | | | 200 | non-breaking |
| @ −20°F | | | 20 | 160 |
| @ −40°F | | | 7 | 25 |
| **Physical** | | | | |
| Water Absorption in 24 hours | % | D570 | 0.02 | 0.01 |
| Specific Gravity | | D792 | 1.25 | 1.25 |
| **Thermal** | | | | |
| Deflection Temp. Under Load | °F | D648 | | |
| @ 264 psi | | | 155 | 138 |
| @ 66 psi | | | 240 | 240 |
| Melt Flow, Condition L | GM/10 min | | <1 | <1 |

[1]Data obtained from injection molded specimens.

NOTE—Information provided in this book has been condensed. The publisher recommends that the product manufacturer be contacted, at the address or phone number provided, for more complete information on product safety, processing parameters, and applications not provided in this information sheet.

# Wilson-Fiberfil International

## Contact Address:

Wilson-Fiberfil International
P.O. Box 3333
2267 West Mill Road
Evansville, IN 47732

Phone: 1 (800) 457-3764

# PROFIL

## Product Data:

**Extrusion Grade Talc Reinforced Polypropylene**

| Properties | Unit | ASTM Test | PF-6001 Impact Modified | PF-6024 |
|---|---|---|---|---|
| **Mechanical** | | | | |
| Tensile Strength @ 73°F | psi | D638 | 2500 | 4100 |
| Elongation @ 73°F | % | D638 | 50 | 22.3 |
| Modulus of Elasticity @ 73°F | psi × $10^5$ | D638 | 3400 | |
| Flexural Modulus @ 73°F | psi × $10^5$ | D790 | 2.5 | 4.7 |
| Izod Impact Strength @ 73°F (1/4″ bars) | ft-lb/in | D256 | 4.2 | 0.6 |
| Unnotched Impact @ 73°F (1/8″ bars) | ft-lb/in | | | 7.5 |
| Gardner Impact Strength | | | | |
| @ 73°F | in/lbs | | 230 | 12 |
| @ −40°F | in/lbs | | 18 | |
| **Physical** | | | | |
| Water Absorption in 24 hours | % | D570 | 0.06 | |
| Specific Gravity | | D792 | 1.23 | 1.23 |
| **Thermal** | | | | |
| Deflection Temp. Under Load | °F | D648 | | |
| @ 264 psi | | | 150 | 182 |
| @ 66 psi | | | 240 | 276 |
| Melt Flow, Condition L | GM/10 min | | <1 | <1 |

NOTE—Information provided in this book has been condensed. The publisher recommends that the product manufacturer be contacted, at the address or phone number provided, for more complete information on product safety, processing parameters, and applications not provided in this information sheet.

# Wilson-Fiberfil International

Contact Address:

Wilson-Fiberfil International
P.O. Box 3333
2267 West Mill Road
Evansville, IN 47732

Phone: 1 (800) 457-3764

## PROFIL

Product Data:

**Extrusion Grade Mica Reinforced Polypropylene**

| Properties | Unit | ASTM Test | PF-6022 | PF-6012 Impact Modified |
|---|---|---|---|---|
| **Mechanical** | | | | |
| Tensile Strength @ 73°F | psi | D638 | 4500 | 4000 |
| Elongation @ 73°F | % | D638 | 3.0 | 6.0 |
| Modulus of Elasticity @ 73°F | psi × $10^5$ | D638 | 7.0 | 6.2 |
| Flexural Strength @ 73°F | psi | D790 | 7000 | 5500 |
| Flexural Modulus @ 73°F | psi × $10^5$ | D790 | 7.5 | 5.5 |
| Izod Impact Strength @ 73°F (1/4″ bars) | ft-lb/in | D256 | 0.6 | 1.0 |
| **Physical** | | | | |
| Water Absorption in 24 hours | % | D570 | 0.03 | 0.03 |
| Linear Mold Shrinkage | | | | |
| 1/8″ Average Section | in/in | | .007 | .007 |
| 1/4″ Average Section | in/in | | .010 | .009 |
| 1/2″ Average Section | in/in | | .011 | .011 |
| Specific Gravity | | D792 | 1.26 | 1.22 |
| **Thermal** | | | | |
| Deflection Temp. Under Load @ 264 psi | °F | D648 | 210 | 190 |
| Melt Flow, Condition L | GM/10 min | | <1 | <1 |

NOTE – Information provided in this book has been condensed. The publisher recommends that the product manufacturer be contacted, at the address or phone number provided, for more complete information on product safety, processing parameters, and applications not provided in this information sheet.

# Wilson-Fiberfil International

## Contact Address:

Wilson-Fiberfil International
P.O. Box 3333
2267 West Mill Road
Evansville, IN 47732

Phone: 1 (800) 457-3764

## PROFIL

### Product Data:

**Extrusion Grade Mica Reinforced Polypropylene**

| Properties | Unit | ASTM Test | PF-6023 | PF-6013 Impact Modified |
|---|---|---|---|---|
| **Mechanical** | | | | |
| Tensile Strength @ 73°F | psi | D638 | 5400 | 4200 |
| Elongation @ 73°F | % | D638 | 3.1 | 4.0 |
| Modulus of Elasticity @ 73°F | psi × $10^5$ | D638 | 6.1 | 6.3 |
| Flexural Strength @ 73°F | psi | D790 | 8100 | 6000 |
| Flexural Modulus @ 73°F | psi × $10^5$ | D790 | 7.5 | 6.0 |
| Izod Impact Strength @ 73°F (1/4″ bars) | ft-lb/in | D256 | 0.7 | 1.0 |
| **Physical** | | | | |
| Water Absorption in 24 hours | % | D570 | 0.03 | 0.03 |
| Linear Mold Shrinkage | | | | |
| 1/8″ Average Section | in/in | | .007 | .006 |
| 1/4″ Average Section | in/in | | .010 | .008 |
| 1/2″ Average Section | in/in | | .011 | .011 |
| Specific Gravity | | D792 | 1.25 | 1.22 |
| **Thermal** | | | | |
| Deflection Temp. Under Load @ 264 psi | °F | D648 | 124 | 195 |
| Melt Flow, Condition L | GM/10 min | | <1 | <1 |

NOTE—Information provided in this book has been condensed. The publisher recommends that the product manufacturer be contacted, at the address or phone number provided, for more complete information on product safety, processing parameters, and applications not provided in this information sheet.

# Wilson-Fiberfil International

## Contact Address:

Wilson-Fiberfil International
P.O. Box 3333
2267 West Mill Road
Evansville, IN 47732

Phone: 1 (800) 457-3764

## PROFIL

### Product Data:

**Extrusion Grade Calcium Carbonate Reinforced Polypropylene with Electrically Conductive Carbon Black**

| Properties | Unit | ASTM Test | PF-6000/EC[1] |
|---|---|---|---|
| **Mechanical** | | | |
| Tensile Strength @ 73°F | psi | D638 | 2500 |
| Elongation @73°F | % | D638 | 42.0 |
| Modulus of Elasticity @ 73°F | psi × $10^5$ | D638 | 2.4 |
| Flexural Strength @ 73°F | psi | D790 | 3000 |
| Flexural Modulus @ 73°F | psi × $10^5$ | D790 | 1.75 |
| Izod Impact Strength @ 73°F (1/4″ bars) | ft-lb/in | D256 | 5.5 |
| **Physical** | | | |
| Water Absorption in 24 hours | % | D570 | 0.02 |
| Linear Mold Shrinkage | | | |
| 1/8″ Average Section | in/in | | .012 |
| 1/4″ Average Section | in/in | | .014 |
| 1/2″ Average Section | in/in | | .020 |
| Specific Gravity | | D792 | 1.21 |
| **Thermal** | | | |
| Deflection Temp. Under Load @ 264 psi | °F | D648 | 135 |
| Melt Flow, Condition L | GM/10 min | | <1 |
| **Electrical** | | | |
| Surface Resistivity | ohms/sq. | Fiberfil | 2500 |

[1]Data obtained from injection molded specimens.

NOTE – Information provided in this book has been condensed. The publisher recommends that the product manufacturer be contacted, at the address or phone number provided, for more complete information on product safety, processing parameters, and applications not provided in this information sheet.

# Wilson-Fiberfil International

Contact Address:

Wilson-Fiberfil International
P.O. Box 3333
2267 West Mill Road
Evansville, IN 47732

Phone: 1 (800) 457-3764

## PROFIL

Product Data:

**Fiberglass Reinforced Impact Modified, Chemically Coupled, Polypropylene**

| Properties | Unit | ASTM Test | J-62/ 10/E4 | J-62/ 20/E4 | J-62/ 30/E4 | J-62/ 40/E4 |
|---|---|---|---|---|---|---|
| **Mechanical** | | | | | | |
| Tensile Strength @ 73°F | psi | D638 | 5300 | 7000 | 8500 | 9200 |
| Elongation @ 73°F | % | D638 | 6 | 4.8 | 4.6 | 3.8 |
| Flexural Strength @ 73°F | psi | D790 | 6500 | 9200 | 11,000 | 12,300 |
| Flexural Modulus @ 73°F | psi × $10^5$ | D790 | 2.7 | 4.4 | 6.3 | 7.8 |
| Izod Impact Strength @ 73°F (1/4″) | ft-lb/in | D256 | 3.0 | 3.3 | 3.6 | 3.8 |
| Izod Impact Strength @ 32°F (1/4″ bars) | ft-lb/in | D256 | | 2.3 | 2.6 | 2.0 |
| Unnotched Impact Strength @ 73°F (1/4″) | ft-lb/in | D256 | 10.0 | 12.2 | 13.4 | 14.7 |
| Gardner Impact, Method B @ 32°F | in-lbs | D2444 | | 10.0 | 9.0 | 6.0 |
| Gardner Impact, Method B @ 73°F | in-lbs | D2444 | | 17.0 | 12.0 | 7.0 |
| Gardner Impact, Method B @212°F | in-lbs | D2444 | | 21.0 | 14.5 | 9.0 |

*(continued)*

NOTE—Information provided in this book has been condensed. The publisher recommends that the product manufacturer be contacted, at the address or phone number provided, for more complete information on product safety, processing parameters, and applications not provided in this information sheet.

**Fiberglass Reinforced Impact Modified, Chemically Coupled, Polypropylene *(continued)***

| Properties | Unit | ASTM Test | J-62/ 10/E4 | J-62/ 20/E4 | J-62/ 30/E4 | J-62/ 40/E4 |
|---|---|---|---|---|---|---|
| **Physical** | | | | | | |
| Water Absorption, 24 hours | % | D570 | .04 | .04 | .04 | .04 |
| Linear Mold Shrinkage | | | | | | |
| 1/8″ Average Section | in/in | | .0047 | .0028 | .0016 | .0014 |
| 1/4″ Average Section | in/in | | .0050 | .0024 | .0015 | .0013 |
| Specific Gravity | | D792 | 0.96 | 1.04 | 1.13 | 1.20 |
| **Thermal** | | | | | | |
| Deflection Temp. Under Load | °F | D648 | | | | |
| @ 264 psi | | | 233 | 267 | 275 | 279 |
| @ 66 psi | | | 300 | 309 | 315 | 315 |

# Wilson-Fiberfil International

## Contact Address:

Wilson-Fiberfil International
P.O. Box 3333
2267 West Mill Road
Evansville, IN 47732

Phone: 1 (800) 457-3764

## PROFIL®/FR

### Product Data:

**Polypropylene (Homopolymer) With Flame Retardant Additive**

| Properties | Unit | ASTM Test | J-60/10/FR | J-60/20/FR |
|---|---|---|---|---|
| **Mechanical** | | | | |
| Tensile Strength @ 73°F | psi | D638 | 7000 | 9000 |
| Elongation @ 73°F | % | D638 | 3.0 | 3.0 |
| Modulus of Elasticity @ 73°F | psi × $10^5$ | D638 | 5.0 | 8.0 |
| Flexural Strength @ 73°F | psi | D790 | 10,000 | 12,000 |
| Flexural Modulus @ 73° | psi × $10^5$ | D790 | 5.0 | 8.0 |
| Izod Impact Strength @ 73° (1/4″ bars) | ft-lb/in | D256 | 1.1 | 1.2 |
| **Physical** | | | | |
| Water Absorption, 24 hours | % | D570 | 0.03 | 0.02 |
| Linear Mold Shrinkage | | | | |
| 1/8″ Average Section | in/in | | .004 | .002 |
| 1/4″ Average Section | in/in | | .005 | .002 |
| 1/2″ Average Section | in/in | | .006 | .003 |
| Specific Gravity | | D792 | 1.32 | 1.40 |
| UL Subject 94 Vertical Test (1/16″) | | | V-0 | V-0 |
| UL Subject 94 Vertical Test (1/8″) | | | V-0 | V-0 |
| **Thermal** | | | | |
| Deflection Temp. Under Load @ 264 psi | °F | D648 | 285 | 295 |

*(continued)*

NOTE—Information provided in this book has been condensed. The publisher recommends that the product manufacturer be contacted, at the address or phone number provided, for more complete information on product safety, processing parameters, and applications not provided in this information sheet.

# Wilson-Fiberfil International

## Contact Address:

Wilson-Fiberfil International
P.O. Box 3333
2267 West Mill Road
Evansville, IN 47732

Phone: 1 (800) 457-3764

## PROFIL®/FR

### Product Data:

**Polypropylene (Homopolymer) With Flame Retardant Additive**

| Properties | Unit | ASTM Test | J-60/30/FR | J-60/40/FR | PP-60/FR |
|---|---|---|---|---|---|
| **Mechanical** | | | | | |
| Tensile Strength @ 73°F | psi | D638 | 10,000 | 12,000 | 3600 |
| Elongation @ 73°F | % | D638 | 2.0 | 2.0 | 5.0 |
| Modulus of Elasticity @ 73°F | psi × $10^5$ | D638 | 10.0 | 12.0 | 4.0 |
| Flexural Strength @ 73°F | psi | D790 | 13,000 | 15,000 | 5800 |
| Flexural Modulus @ 73° | psi × $10^5$ | D790 | 11.0 | 14.0 | 3.8 |
| Izod Impact Strength @ 73° (1/4″ bars) | ft-lb/in | D256 | 1.3 | 1.4 | 0.5 |
| **Physical** | | | | | |
| Water Absorption, 24 hours | % | D570 | 0.02 | 0.02 | 0.03 |
| Linear Mold Shrinkage | | | | | |
| 1/8″ Average Section | in/in | | .001 | .001 | .011 |
| 1/4″ Average Section | in/in | | .001 | .001 | .015 |
| 1/2″ Average Section | in/in | | .002 | .002 | .018 |
| Specific Gravity | | D792 | 1.47 | 1.57 | 1.42 |
| UL Subject 94 Vertical Test (1/16″) | | | V-0 | V-0 | V-0 |
| UL Subject 94 Vertical Test (1/8″) | | | V-0 | V-0 | |
| **Thermal** | | | | | |
| Deflection Temp. Under Load @ 264 psi | °F | D648 | 300 | 300 | 192 |

*(continued)*

NOTE – Information provided in this book has been condensed. The publisher recommends that the product manufacturer be contacted, at the address or phone number provided, for more complete information on product safety, processing parameters, and applications not provided in this information sheet.

# Appendix A

## FORMULATION, COMPOUNDING, AND MARKETS

Blends and alloys are the most active research and development area in polymers. The following tables give some examples of blends and alloys formulations, methods of compounding, including process aids and impact modifiers, properties of some alloys, and 1986 blends production.

### Ultra High Impact Nylons

| Formulation | |
|---|---|
| Nylon 66 | 45 |
| Nylon 6 | 36 |
| EPDM—Fumaric Acid | 10 |
| EPDM | 9 |
| **Properties** | |
| Notched Izod (ft-lbs/in) | |
| at 23°C | 23.7 |
| at 0°C | 21.7 |
| at −20°C | 21.0 |
| at −40°C | 12.8 |

Other high impact nylons based on MBS's, EPDM's and ionomers. Roura US 4,346,194

### High Impact Nylon/Polyethylene Blends

Maleic anhydride radical grafted onto low density PE (0.4% by wt MA graft).

Compounded with Nylon 66 on a twin screw extruder to achieve good dispersion. 30–40% PE for high impact.

Hobbs, Bopp, and Watkins, *Polym. Eng. and Sci.*, 23:380 (1983).

### ABS/Polycarbonate and ABS/Polysulfone

- ABS cuts cost
- Improves processability
- Provides good impact

### PVC/Acrylic Blends

- Completely miscible
- Acrylic raises PVC HDT
- Good appearance and low smoke

**Process Aids and Impact Modifiers**

Low levels of acrylics used as process aids for PVC

MBS and all-acrylic impact modifiers both have PMMA shells which are miscible with the PVC

### ASA (Acrylic-Styrene-Acrylonitrile) Materials

Can be blended with a number of materials including PVC as weatherable impact modifiers.

Main use is most likely as a capstock

US 3,944,631 and 4,168,285 and Canadian 1,186,838

### Approximate 1986 US Blends Production

| Blend | Pounds (MM) | $ (MM) |
|---|---|---|
| Impact PVC | 900 | 600 |
| PPO(PPE)/PS | 170 | 260 |
| PP/EPDM | 65 | 65 |
| ABS/PVC | 57 | 65 |
| Impact Nylon | 35 | 72 |
| PC/PE | 18 | 36 |
| PBT/PET | 13 | 20 |
| PVC/Acrylic | 13 | 15 |
| PC/ABS | 11 | 14 |
| Impact PBT | 6 | 10 |
| PC/PBT | 5 | 8 |
| PE/Nylon | 5 | 8 |
| ABS/PSF | .6 | 1.4 |
| Modified Ionomer | .6 | 1 |

### Polyethylene/Nylon Barrier Materials

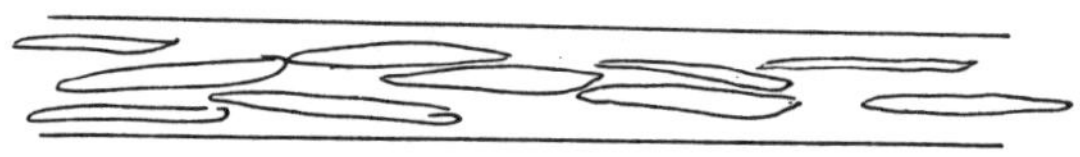

- Low shear mixing important.

| Typical Composition | Parts |
|---|---|
| Linear Polyethylene | 84 |
| Nylon 6/66 copolymer (22.5/77.5) | 15 |
| Fumaric Acid grafted Polyethylene (.9%) | 1 |

US 4,410,482

### Polycarbonate/Polyester Alloys

| Impact Modified Blends | |
|---|---|
| **Formulation** | |
| PBT | 63 |
| Polycarbonate | 15 |
| MBS Impact Modifier | 20 |
| Mold Release/Stabilizers | 2 |
| **Properties** | |
| HDT (°F) | 129 |
| Notched Izod (ft-lbs/in) | |
| 0.125″ thickness | 21 |
| 0.250″ thickness | 16 |

Cohen and Dieck, US 4,257,937

### Polycarbonate/Polyester Alloys

| Polycarbonate | Polybutylene terephthalate |
|---|---|
| Tough | Solvent resistant<br>Rapid crystallization |
| Poor stress crack resistance | Warps badly especially when reinforced |
| **PC/PBT Blend** | |
| Most of the advantages of both resins | |

Largest application of PC/PBT to date is in bumpers

### Polyethylene/Nylon Barrier Materials

**Effect of Morphology on Bottle Wall Strength and Permeation**

| Material/Morphology | Wall Strength (kg-cm/mil) | Gasoline loss (gm) |
|---|---|---|
| Laminar 77/20/3 | 2.4 | .25 |
| Well dispersed 77/20/3 | 9.8 | 8 |
| 100% PE Control | 2.1 | 19 |

US 4,410,482

### ABS/PVC Blends

- Blends are compatible but not miscible
- Most contain from 40 to 80% ABS
- ABS contributes impact resistance, HDT and processability
- PVC contributes flame retardancy, UV stability and excellent economics

### Polycarbonate/Polyolefin Blends

Advantages: Improved impact in thick sections and lower temperatures

Applications: Automotive instrument panels, spoilers, bumper flares, and safety helmets.

Examples:

| | |
|---|---|
| Lexan 4291 and 191 | Approx $2.00/lb |
| Merlon T70, T85 and T95 | ($.40 over GP PC) |

### PC/PO Blends

Compounding Technology: O'Connell
US 3,801,673
Melt viscosities of the polycarbonate and PE should be equal at processing temps and shear rates (560°F, 100/sec).

Effect of PC endcap and compatibility
Quinn US 4,358,563

**4% HDPE/Lexan—140 PC**

| Chainstopper | 0.25″ Notched Izod |
|---|---|
| Phenol | 10.1 |
| Oleic Acid | 11.2 |
| Stearic Acid | 13.1 |

# Appendix B

## REACTIVE POLYMERS

Related to blends and alloys is a group of products called Reactive Polymers. Dow Chemical U.S.A. is one company which makes such a product. Dow grafts an oxazoline group to polystyrene, making reactive polystyrene (RPS). This polymer can then be alloyed with such polymers as polyethylene, polypropylene, elastomer/rubber, polyester, epoxy, and acrylics.

Keys to RPS alloying are:

- Adequate functionality level
- Good mixing
- High melt flow
- Moderate shear
- Comparable viscosity
- 20–50% RPS (discontinuous phase)

Possible reactive substrates are:

- Film/sheet
- Oxazoline reactive surface
- Composites
- Coatings/paint

**Co-Reactive Polymers**

| Company | Product | Approx. Polymer | Modifier | Cost/lb |
|---|---|---|---|---|
| Dow | PRIMACOR | LDPE<br>epoxies | Acid | $0.89 |
| DuPont | VAMAC<br>NUCREL<br>CXA<br>ELVAX<br>ZYTEL | Acrylic rubber<br>EMMA<br>EPDM<br>EVA<br>Nylon/Surlyn | Acid<br>Acid<br>Anhydride<br>Acid<br>Acid | |
| BFGoodrich | HYCAR | B/AN | Acid/Amine | $1.60 |
| Allied | | Nylon | Acid | |
| Hercules | PROFAX | PP | Anhydride | |
| Himont | HERCOPRIME | PP | Anhydride | $2.00 |
| Reichhold | POLYBOND | LDPE, HDPE, PP, EPDM | Acid | $0.80 |
| Norchem | PLEXAR | HDPE, PP, LDPE, EVA | Anhydride | $1.00 |
| Mitsui | ADMER | HDPE, PP | Anhydride | $1.75 |
| Mitsubishi | MODIC | PP | Anhydride | |
| Monsanto | CADON | PS | Anhydride | $0.70 |
| ARCO | DYLARK | PS | Anhydride | $0.70 |

*(continued)*

### Co-Reactive Polymers *(continued)*

| Company | Product | Approx. Polymer | Modifier | Cost/lb |
|---|---|---|---|---|
| Shell | KRATON | Rubber SBS<br>PB | Anhydride | |
| Union Carbide | BAKELITE | PS | Acid | |
| Exxon | VISTALON | E/P Rubber | Anhydride | |
| Goodyear | VITEL | PET | Acid | |
| Nippon Zeon | NIPOL N-34 | AN/B Rubber | Acid | |
| Morton | MORPRIME | PE | Acid | |

### HDPE/RPS Alloy with ABS Properties

| | RPS | HDPE* | Alloy 60% HDPE* 40% RPS | ABS 9010 | |
|---|---|---|---|---|---|
| | | | | Injection** | Compression |
| Tensile Yield (psi) | 5300 | 2700 | 4100 | 6000 | 5400 |
| Tensile Modulus (psi) | 420,000 | 160,000 | 290,000 | 320,000 | 284,000 |
| % Elongation (break) | 2 | 900 | 60 | 35 | 2 |
| Notched Izod Impact (ft-lb/in) | .2 | 2 | 4 | 4.0 | 3.6 |
| Vicat Heat Distortion (°F) | 222 | 260 | 243 | 226 | 226 |
| DTUL (264 psi) (°F) | 183 | 107 | 185 | 222 | 205 |
| Melt Flow (g/10 min) (200°C/5 kg) | 8 | 6 | 3 | 2.7 | 2.7 |
| Specific Gravity | 1.04 | .96 | .98 | 1.05 | 1.05 |
| ESCR (2500 psi) (freon) (min) | | | 700 | | 300 |

*Anhydride functionalized.
**Literature value (others are compression molded and lab tested).

### Example Alloys

| | | | | | | |
|---|---|---|---|---|---|---|
| | 50% RPS | 50% RPS | 35% RPS | 35% RPS | 25% RPS | 25% RPS |
| | 50% PRIMACOR 1430 | 50% SURLYN 9450 | 55% HDPE | 40% PRIMACOR XU-61515.01 | 55% POLYBOND 1001 (12 ml) | 3% HERCOPRIME G |
| | | | 10% VISTALON RUBBER 24-3 | 25% HDPE | 20% PLEXAR 2284 | 72% PROFAX 7623 |
| Continuous polymer | LDPE | Ionomer | HDPE | HDPE | PP | PP |
| Tensile yield (psi) | 2900 | 2900 | 2800 | 2700 | 4000 | 3400 |

**Example Alloys *(continued)***

| | 50% RPS | 50% RPS | 35% RPS | 35% RPS | 25% RPS | 25% RPS |
|---|---|---|---|---|---|---|
| | 50% PRIMACOR 1430 | 50% SURLYN 9450 | 55% HDPE | 40% PRIMACOR XU-61515.01 | 55% POLYBOND 1001 (12 ml) | 3% HERCOPRIME G |
| | | | 10% VISTALON RUBBER 24-3 | 25% HDPE | 20% PLEXAR 2284 | 72% PROFAX 7623 |
| Tensile modulus (psi) | 110,000 | 120,000 | 160,000 | 140,000 | 350,000 | 180,000 |
| % Elongation | >50 | >50 | >50 | >50 | 13 | 13 |
| Izod Impact (ft-lb/in) | 13 | 10 | 5 | 10 | .5 | .7 |
| Melt flow (200°C/5 kg) | 3.7 | 3.5 | 3.7 | 8.9 | 9.6 | 10 |
| Vicat Heat Distortion (°F) | 191 | 189 | 237 | 228 | 271 | 263 |
| Approximate cost/lb | .90 | 1.00 | .70 | 1.05 | .90 | .60 |

All data on compression molded samples.

# Index

## APPLICATIONS


Abrasion Resistance
 Triax 1000 series, pp. 395–406
Acoustical Barrier
 AB 6000, 7000, pp. 407–416
Adhesives
 Kraton, pp. 463–479
Antistatic
 Terblend, p. 59
Appliance, Molded Components
 Kraton Rubber
  G-2700, G-7000, pp. 470, 472
  G-8000, p. 480
 Geloy XP 2001, p. 240
Automotive Bodies, pp. 118, 128
Automotive Components
 Zytel ST 801 HS, p. 211
 ETA, RTA, pp. 407–448
 Thermofil, pp. 485–526
 Pulse, pp. 191–196
Automotive Trim and Utility
 Bayblend DP-2-1452, 1483, 1484, pp. 357–361
 Makroblend DP-4-1350, 1351, 1354, pp. 378–382
  DP-4-1356, p. 385
 Ferroflex FF-400, 600, 1900, pp. 226, 227, 232
 Pulse 700, 710, pp. 187, 189
 Geloy, pp. 236–246

Blow Mold Parts
 Geloy, pp. 233–247
 Noryl GTX 625, p. 248
 Xenloy, 289–304
 Santoprene, p. 392
 Triax 1000 series, pp. 395–406
 ETA, pp. 420, 426, 432, 441–445
 T Rubber, p. 447
Business Machines
 Bayblend DP-2-1439, 1440, 1441, pp. 351–355

Chemical Resistance
 Xenoy, pp. 305–348
 Triax 1000 series, pp. 395–406
 Bayblend MJ-2500, p. 375
 Makroblend DP-4-1350, 1351, 1354, pp. 378–382
 Noryl GTX, pp. 248–257
 Noryl Plus CRN, pp. 258–260
 Valox, pp. 280–288
 Xenloy, pp. 289–304
Coloring Custom
 Zytel 408, p. 206
 Ferroflex FF series, pp. 223–232
 Valox, pp. 280–288
 Bur-A-Loy, p. 349
 ETA, RTA, pp. 407–448
 Kraton, pp. 463–479
Compression Molding
 Azdel, p. 57
Creep Under Stress – Low
 Noryl GTX, pp. 248–257
 Noryl Plus, p. 258

Decorating
 Bayblend MH-6570 p. 369
Dimensional Stability
 Bayblend MH-6570, p. 369
 MJ-2500, p. 375
 RPI-136, p. 449
 Noryl GTX, pp. 248–257
 Xenloy, pp. 289–304
 Xenoy, pp. 305–348
 Minlon, p. 204
 Terblend, p. 59

Electrical Hardware
 Geloy XP 2001, p. 240
Electrical Insulation
 Kraton Rubber (G-8000), p. 550
 Royaltherm, p. 527–529
 Delrin, pp. 198, 201
 Bur-A-Loy, p. 349
Extrusion Applications
 Profil, pp. 541–548
 Delrin, pp. 198, 201
 Geloy, pp. 233–247
 Valox, pp. 280–288
 Xenloy, pp. 289–304
 Santoprene, p. 392
 Triax 1000 series, pp. 395–406
 ETA, pp. 420, 426, 432, 441–445
 T Rubber, p. 447
 RPI-503EP, 603EP, pp. 451, 459
 Arloy 1000, 1100, 1200, pp. 51–55
 Terblend S KR 2861/1, p. 59
 Grilon BT-40, p. 221
 Ferroflex FF series, pp. 223–232
 Prevex series, pp. 61–116
 TPR series, pp. 117–168

Film and Sheet
 TPR series, p. 118
Flame Retardant
 Noryl Plus RFN430, p. 264
 Valox, pp. 280–288
 Nylafil/FR, Nylafil/VO, pp. 536–539
 Profil/FR, p. 549
Food Contact
 Bayblend MJ-2500, p. 375
 Zytel 408 L NC-10FC, p. 206
 Arloy, p. 51

Gaskets, Seals, Weatherstripping
 TPR series, pp. 118, 132, 135
 Royaltherm, pp. 527–529

Heat Resistance
 Royaltherm, p. 527
 Bayblend MH-6570, p. 369
 Noryl GTX, pp. 248–257
 Noryl Plus, pp. 258–277
 Triax 1000 series, pp. 295–406
 Royaltherm, pp. 527–529
Heat Stabilized
 Zytel 408 HS NC-10, p. 206
 ST 801 HS, p. 211
Hose and Tubing
 Bur-A-Loy, pp. 135, 349
 Royaltherm, pp. 527–529
 Hydrocarbon Resistant
 Makroblend UT-1018, p. 390
 DP-4-1350, 1351 pp. 378, 380

Impact Strength
 Bayblend MH-6570, p. 369
 MJ-2500, p. 375
 Xenoy, pp. 305–348
 Nylafil, pp. 530–535
Industrial/Mechanical Components
 Makroblend DP-4-1350, 1351, pp. 378, 380
 DP-4-1356, p. 385
 Delrin, pp. 198–203
 Torlon, pp. 7–47
Injection Molded Parts
 ETA 3104, 3121, 3125, 3131, pp. 435–441
 3163, 3183, pp. 445–447
 RTA 3184, 3223, 3263, pp. 449–453
 T Rubber, p. 463
 Geloy, pp. 233–247
 Valox, pp. 280–288
 Xenloy, pp. 289–304
 Santoprene, p. 392
 Triax 1000 series, pp. 395–406
 AB-6000, 7000 series, pp. 407–416
 ETA 3011/3010, 3031, 3041C, pp. 417–419
 ETA 3051, 3055, 3081, pp. 421–425
 RPI-504EP, p. 453
 RPI-507EP, p. 455
 RPI-136EP, p. 449
 RPI-602EP, p. 457
 Arloy 1000, 1100, 1200, pp. 51–55
 Terblend SKR2861/1, p. 59
 Bayblend MC 2500, 3200, pp. 363, 366
 DP-2-1439, 1440, 1441, pp. 351–355
 Makroblend UT-1018, p. 390

DP-4-1350, 1351, 1354, pp. 378–382
Grilon BT-40, p. 221
Ferroflex FF series, pp. 223–232
Prevex series, pp. 61–116
Pulse series, pp. 185–196
Delrin, pp. 198, 201
Zytel ST 901 L, p. 215

Large Molded Parts
Makroblend UT-1018, p. 390
Lawn and Garden
Makroblend UT-1018, p. 390
DP-4-1350, 1351, pp. 378, 380
DP-4-1357, p. 388
ETA, RTA, pp. 407–448
Low Temperature Impact Resistance
Noryl GTX, pp. 248–257

Marine Parts
Geloy XP 2003, p. 242
Military Equipment
Makroblend DP-4-1350, 1351, pp. 378, 380
DP-4-1356, p. 385

Molding Resin, High Temperature
Torlon, pp. 7–47

Ozone Resistant
Kraton G2700, p. 470–472
G7000, p. 472
G8000 (Elexar), p. 480

Plumbing, Pumps and Meters
Prevex P2A, P3A, pp. 61, 65

Recreational Vehicle Accessories
Geloy 1020, p. 234
XP 2003, p. 242
RTA, ETA, pp. 407–448

Stress Crack Resistance
ETA, RTA, pp. 407–448
Structural Foam Molding
Prevex series, pp. 61–116
Surgical Applications
Arloy, p. 51

Thermoform
Arloy 1000, 1100, 1200, pp. 51–55
Xenloy, pp. 289–304
Triax 1000 series, pp. 395–406
ETA, RTA, pp. 407–448
Thermoset Rubber
Bur-A-Loy, p. 349
Tool Housings
Nylafil, pp. 530–535
Arloy, p. 51
Zytel, pp. 211, 217

Ultra-Violet Resistant
Azdel, p. 57
Xenloy, pp. 289–304
Xenoy, pp. 305–348

Water Absorption – Low
Noryl GTX, pp. 248–257
Xenloy, pp. 289–304
Xenoy, pp. 305–348

# Index

## COMPANIES


Amoco Chemicals Corp., pp. 7–50
ARCO Chemical Company, pp. 51–56
Azdel, Inc., pp. 57–58

BASF Corporation, pp. 59–60
Borg-Warner Chemicals Inc., pp. 61–116
BP Performance Polymers Inc., pp. 117–168

Celanese Engineering Resins, Inc., pp. 169–184

Dow Chemical Company, pp. 185–197

E. I. du Pont de Nemours & Co. Inc., pp. 198–220
EMS-Emser Industries, pp. 221, 222

Ferro Corporation, pp. 223–232

General Electric Company, pp. 233–348

MACH-1 Compounding, pp. 349, 350

Mobay Chemical Corporation, pp. 351–391
Monsanto Chemical Company, pp. 392–406

Republic Plastics Company, pp. 407–448
Research Polymers International, pp. 449–462

Shell Chemical Company, pp. 463–484

Thermofil, Inc., pp. 485–526

Uniroyal Chemical Company Inc., pp. 527–529

Wilson-Fiberfil International, pp. 530–550

# Index

## COMPOSITION


Acetyl Resin, pp. 198, 201
Acrylic-Styrene-Acrylonitrile (ASA)
 Geloy, pp. 233–247

Engineering Polymers
 Bayblend MJ-2500, p. 375
 Pulse series, pp. 185–196
EPDM Elastomer (silicone modified)
 Royaltherm, p. 527

Nitrile Rubber/PVC Blend
 Bur-A-Loy, p. 349
Nylon/ABS Alloys
 Triax 1000 series, pp. 395–406
Nylon (glass filled), Flame Retardant
 Nylafil/FR, Nylafil/VO, pp. 536–539
Nylon, non-plasticized
 RPI-609EP, p. 459
Nylon Polymer, pp. 206–220
Nylon 6/6
 RPI-602EP, p. 459
 Nylafil (toughened), pp. 530–535
Nylon 6/6 (Al flake filled), p. 497
Nylon 6/6 Alloy (glass filled), p. 204
 Thermofil N7-series, pp. 499–511
Nylon 6/6 (graphite reinforced), pp. 31–35
PBT/PET Alloy (glass filled), pp. 485–496
PET/PBT Thermoplastic Blend
 Valox, pp. 280–288
Polyamide Thermoplastic
 Grilon BT40, p. 221
Polyamide-Imide/Fluorocarbon Alloy
 Torlon, pp. 7–47
Polycarbonate Alloy
 RPI-503EP, 504EP, 507EP, pp. 451–456
 Xenloy, pp. 289–304
Polycarbonate/ABS Alloys
 Bayblend MC-2500, 3200, pp. 363, 366
 MH-6570, 6580, 6595, pp. 369–373
 MJ-2500, p. 375
 DP-2-1439, 1440, 1441, pp. 351–355
 DP-2-1482, 1483, 1484, pp. 357–361
Polycarbonate Alloy (glass filled),
 pp. 519–526
Polycarbonate/LURAN Blend
 Terblend S KR 2861/1, p. 59
Polycarbonate/PET Blends
 Makroblend UT-1018, p. 390
Polycarbonate/PET Blends (reinforced)
 Makroblend DP-4-1350, 1351, 1354,
 pp. 378–382
 DP-4-1356, 1357, pp. 385, 388
Polyester Alloy Composite (reinforced)
 Azdel, p. 57
Polyolefin Copolymer Alloy (reinforced)
 RPI-136EP, p. 449
Polyolefin, Modified
 Ferroflex FF series, pp. 223–232
Polyolefin Thermoplastic, Elastomer
 TPR series, pp. 118–162
 AB, ETA, pp. 407–435, 441–445
Polyolefin Thermoplastic, Rigid
 RTA, pp. 436–440
Polyphenylene Oxide/Polyamide Alloys
 Noryl GTX, pp. 248–257
 Noryl Plus, pp. 258–277
Polypropylene
 Profil (reinforced), pp. 541–548
 Profil/FR (homopolymer), pp. 549–550
Polypropylene/EPDM Thermoplastic
 Santoprene, p. 392
Polystyrene, Oxazoline Reactive, p. 197
Polysulfone Alloy (glass filled)
 Plastalloy, p. 540

Thermoplastic Resin
 Bur-A-Loy, p. 349
 Prevex, pp. 61–116

# Index

## TRADE NAMES


Arloy pp. 51–55
Azdel, p. 57

Bayblend 351–375
Bur-A-Loy, p. 349

Delrin pp. 198, 201

Ferroflex FF series, pp. 223–232

Geloy, pp. 233–247
Grilon BT40, p. 221

Kraton Rubber series, p. 463–479

Makroblend, pp. 378–390

Minlon 12T, p. 204

Noryl GTX, pp. 248–257
Noryl Plus, pp. 258–277
Nylafil, pp. 530–539

Plastalloy, p. 540
Prevex, pp. 61–116
Profil, pp. 541–550
Pulse, pp. 185–196

Reactive Polystyrene, p. 197
Royaltherm, p. 527
RPI, pp. 449–462

Santoprene, p. 392

Terblend S KR 2861/1, p. 59
Thermofil, pp. 485–526
Thermoplastic Alloy
  Elastomeric (ETA), pp. 407–448
  Rigid (RTA), pp. 407–448
Torlon, pp. 7–47
TPR, pp. 117–168
Triax 1000 series, pp. 395–406

Valox, pp. 278–288
Vandar, pp. 169–184

Xenloy, pp. 289–304
Xenoy, pp. 305–348

Zytel, pp. 206–220

EniChem
Americas